T0074777

The Courant–Friedrichs–Lewy (CFL) Condition

Carlos A. de Moura · Carlos S. Kubrusly
Editors

The Courant–Friedrichs–Lewy (CFL) Condition

80 Years After Its Discovery

Editors
Carlos A. de Moura
Mathematics Institute
Rio de Janeiro State University (UERJ)
Rio de Janeiro, Brazil

Carlos S. Kubrusly
Department of Electrical Engineering
Catholic University of Rio de Janeiro
Rio de Janeiro, Brazil

Additional material to this book can be downloaded from http://extras.springer.com

ISBN 978-0-8176-8393-1 ISBN 978-0-8176-8394-8 (eBook)
DOI 10.1007/978-0-8176-8394-8
Springer New York Heidelberg Dordrecht London

Library of Congress Control Number: 2012952407

Mathematics Subject Classification: 35-XX, 65-XX, 68-XX

Printed on acid-free paper

Springer is part of Springer Science+Business Media (www.birkhauser-science.com)

Poster of *CFL condition—80 years gone by* meeting

Having failed to obtain any **CFL** picture, we have switched to a **CLL** one: Richard Courant, Hans Lewy, and Jean Leray at the Arden Conference Center around 1950. It may be thought as a first order approximation... (From Peter Lax files)

Richard Courant, Kurt Friedrichs, and Hans Lewy

Foreword

Despite being largely disseminated nowadays, "impact factors" do not need to be quoted to assure the depth and importance—in so many areas of science and technology—of the article submitted in 1927 by Richard Courant, Kurt Friedrichs, and Hans Lewy to *Mathematische Annalen* and published therein the following year.[1]

The authors' keen view of finite difference methods applied to approximate solutions of partial differential equations has provided the right hand hold to deal with numerical algorithms within this environment. The idea is to first look for how the studied schemes mimic the main properties of the operators they are intended to approximate—signal propagation speed being the first point to look at—and then to estimate the distance between the continuous model, which lives within the real line, and the discrete one, immersed in real life and, consequently, being tied to treating only numbers we are bound to operate with. They realized how this question is related to the answer to a puzzle posed for a long time to numerical analysts of PDEs: mesh refinements do not always improve the approximations, they can even make approximations worse. They discovered that everything amounts to a desperate need for stability—small changes in input data must never throw output far away from its true habitat. The constraint the discrete schemes must satisfy to guarantee stability became known as the CFL-condition, honoring the three authors.

In March 1967, to celebrate the article's 40th anniversary, *IBM Journal*[2] published a special issue, Vol. **11**(2), which featured the paper's translation into English,[3] as well as three articles that report the outcome of numerical methods for PDEs after that historical publication. Each of them has roughly chosen as its focus

[1]Über die partiellen Differenzengleichungen der mathematischen Physik, Vol. **100**, pp. 32–74. See Appendix B for a reprint of the paper original version.

[2]Now *IBM Journal of Research and Development.*

[3]See Appendix C for a reprint of this translation.

one of the three types of partial differential equations: elliptic,[4] hyperbolic,[5] and parabolic.[6]

Around 80 years had gone by since the CFL paper was printed when a meeting was held in Rio de Janeiro, in May 2010, to once again celebrate its outcome. Hosted by Rio de Janeiro State University (UERJ), it was organized with the participation of Rio's main institutions that deal with computational sciences (see the report in Appendix D). The meeting atmosphere was quite cozy, and it is a pleasure for the organizing committee to thank around 100 attendees that have made *CFL-condition, 80 years gone by* a scientifically rewarding encounter. Our thanks go also to the publishers of these proceedings. We further acknowledge the contributions by Jacqueline Telles (secretarial chores), Jhoab P. de Negreiros (LaTeX expertise), Sandra Moura (website design), and Tania Rodrigues (graphic designer). Additional information about the meeting—in particular some texts from the refereed contributed papers, as well as many pictures taken at the meeting—may be retrieved from its site at

http://www.ime.uerj.br/cfl80

Before summarizing the scientific papers contained in this volume, let us mention one of its special features: the musical piece, recorded especially for these proceedings, authored by Hans Lewy—who was also a composer before turning to mathematics—and played by Leonore (Lori) Lax, one of Richard Courant's daughters. She has also written a text with some recollection of Lewy's visits to Courant's home (see Appendix A, which contains some photos). The recording may be accessed through SpringerExtras at extras.springer.com/978-0-8176-8393-1.

The book opens with an article by Peter D. Lax, the meeting's main speaker. He dwells a little on the CFL paper and after some quick, sharp remarks—quite his writing style—shows some results to corroborate his main assertion: "The theory of difference schemes is much more sophisticated than the theory of differential equations."

Reuben Hersh's contribution deals with a "mysterious" question: Numerical analysts spend their lifetime to reach convergence results that become valid only when the parameters involved turn out to be extremely large. But in everyday life, why are they quite happy getting results drawn from real-life computers, therefore with not so overwhelming numbers?

The article by Rolf Jeltsch and Harish Kumar discusses a model for the different phenomena that occur at current interruption in a circuit breaker. They propose the equations of resistive magneto-hydrodynamics (RMHD), and it turns out that this is the first time a model based on RMHD has been used to simulate plasma arc in three dimensions.

---

[4] Seymour V. Parter, *Elliptic equations*, pp. 244–247.

[5] Peter D. Lax, *Hyberbolic Difference Equations: A Review of the Courant–Friedrichs–Lewy Paper in the Light of Recent Developments*, pp. 235–238.

[6] Olof B. Widlund, *On Difference Methods for Parabolic Equations and Alternating Direction Implicit Methods for Elliptic Equations*, pp. 239–243.

Sander Rhebergen and Bernardo Cockburn apply a novel space-time extension of the hybridizable discontinuous Galerkin (HDG) finite element method to the advection–diffusion equation. The resulting method combines the advantages of a space-time DG method with sensible improvement in efficiency and accuracy for the HDG methods.

The paper by J. Teixeira, Cal Neto, and Carlos Tomei indicates how a global Lyapunov–Schmidt decomposition, introduced within a *bona fide* theoretical context, gives rise to a quite effective numerical algorithm (which makes use of the finite element method) for the nonlinear equation $-\Delta u - f(u) = g$ with Dirichlet conditions on a bounded n-dimensional domain.

A filtering technique for the one-dimensional wave equation is proposed and tested in the article by Aurora Marica and Enrique Zuazua. Their concern is the failure of observability from the boundary for the quadratic classical finite element approximation.

Margarete O. Domingues, Sônia M. Gomes, Olivier Roussel, and Kai Schneider have authored an article which studies a wavelet-based multiresolution method. It deals with space-time grid adaptive techniques for a finite volume being the time discretization explicit. Their purpose, both to reduce the memory requirement and to speed-up computing, is reached through an efficient self-adaptive grid refinement and a controlled time-stepping.

Philippe G. LeFloch obtains a parabolic-type system for late-time asymptotics of solutions to nonlinear hyperbolic systems of balance laws with stiff relaxation. For these stiff problems, an approximation based on a finite volume is then introduced which preserves the late-time asymptotic regime. This method carries an important feature; namely, it requires the CFL condition associated with the hyperbolic system under study, rather than the more restrictive parabolic-type stability condition.

Kai Schneider, Dmitry Kolomenskiy, and Erwan Deriaz pose the question: "Is the CFL condition sufficient?". Their numerical results, using a spectral discretization in space, illustrate that the CFL condition is not sufficient for stability and that the time step is limited by non-integer powers (larger than one) of the spatial grid size.

The collection is closed with a paper by Uri Ascher and Kees van den Doel: "Fast Chaotic Artificial Time Integration". The authors claim that some faster gradient-descent methods generate chaotic dynamical systems for the normalized residual vectors. The fastest practical methods of this family in general appear to be the chaotic, two-step ones, but, despite their erratic behavior, these methods may also be used as smoothers, or regularization operators. Besides, their results also highlight the need for a better theory for these methods.

The meeting has also held a special session honoring Peter Lax.

Rio de Janeiro, Brazil

Carlos A. de Moura
Carlos S. Kubrusly

Contents

Stability of Difference Schemes

Peter D. Lax

Abstract The most powerful and most general method for constructing approximate solutions of hyperbolic partial differential equations with prescribed initial values is to discretize the space and time variables and solve the resulting finite system of equations. How to discretize is a subtle matter, as we shall demonstrate.

In this report, some of the proofs are only sketched; details can be found in Chap. 8 of my monograph "Hyperbolic Partial Differential Equations", 2006, AMS.

Keywords Hyperbolic PDE's · Finite difference schemes · Convergence · Stability

One of the seminal observations of the Courant–Friedrichs–Lewy paper of 1928 was that in order for solutions of a difference equation to converge to the solution of the partial differential equation the difference scheme must use all the information contained in the initial data that influence the solution. To satisfy this condition, the ratio of the spatial discretization to the time discretization must be at least as large as the largest velocity with which signals propagate in solutions of the partial differential equation. This inequality is called the CFL condition.

I well remember from the early days of computing, when physicists and engineers first undertook to solve numerically initial value problems, their utter astonishment to see the numerical solution blow up because they have unwittingly violated the CFL condition.

The CFL condition is only a necessary condition for the convergence of difference schemes. Here is an example: discretize the scalar equation

$$u_t + u_x = 0$$

by replacing the time derivative with a forward difference, and the space derivative with a symmetric difference. This scheme diverges, no matter how small the time discretization is compared to the space discretization.

P.D. Lax (✉)
Courant Institute of Mathematical Sciences, New York University, New York, USA
e-mail: lax@cims.nyu.edu

C.A. de Moura, C.S. Kubrusly (eds.), *The Courant–Friedrichs–Lewy (CFL) Condition*,
DOI 10.1007/978-0-8176-8394-8_1,

In this talk, I will report on sufficient conditions for the convergence of various difference schemes. I shall discuss a class of equations studied by K.O. Friedrichs, first order symmetric hyperbolic systems of the form

$$u_t = Au_x + Bu_y, \tag{1}$$

where $u(x, y, t)$ is a vector-valued function, and A and B are real symmetric matrices that may be smooth functions of x and y. The theory of these equations is fairly straightforward: let $u(x, y, t)$ be a solution in the whole (x, y)-space that dies down fast as x and y tend to infinity. Take the scalar product of the equation with u and integrate it over all x and y; we get

$$\int (u \cdot u_t)\, dx\, dy = \int (u \cdot Au_x + u \cdot Bu_y)\, dx\, dy. \tag{2}$$

If A and B are constant matrices, the integrand on the right is

$$\frac{d}{dx}(u, Au)/2 + \frac{d}{dy}(u, Bu)/2, \tag{3}$$

a sum of perfect x and y derivatives; therefore the integral is zero. The integrand on the left side is the t derivative $(1/2)d(u, u)/dt$ and can be regarded as the t derivative of

$$E(t) = \int \frac{1}{2}(u, u)\, dx\, dy.$$

Since the derivative of $E(t)$ is zero, it follows that $E(t)$ is independent of time, the law of conservation of energy.

When A and B are functions of x and y, the integrand on the right equals the sum (3) minus

$$(u, A_x u)/2 + (u, B_y u)/2,$$

a quantity bounded by $cE(t)$, c some constant. Therefore, we conclude that $dE(t)/dt \leq cE(t)$, which implies that

$$E(t) \leq E(0)e^{ct}. \tag{4}$$

Such an inequality for solutions of hyperbolic equation is called an energy inequality.

We turn now to two-level explicit difference schemes. We look first at the case of one space dimension, that is,

$$u_t = Au_x. \tag{5}$$

The difference schemes are of the form

$$u^{n+1} = C_h u^n, \tag{6}$$

where u^n denotes the values of u at time t and u^{n+1} the values of u at the next time level $t+h$, n an integer multiple of h. We take the space discretization to be h, the same as the time discretization. The operator C_h is a finite sum of the form

$$C_h = \sum C_j T^j, \tag{7}$$

where T is translation by h: $(Tu)(x) = u(x+h)$, and T^j is the j-th power of T. Since we have chosen the space discretization and time discretization to be equal, the CFL condition requires that signals for solutions of the differential equation (5) propagate with speed not greater than 1. Since the signal speeds for solutions of (5) are the eigenvalues of A, this requires that the eigenvalues of A lie between -1 and 1.

In order for the difference scheme (6)–(7) to be consistent with the differential equation (5), the coefficients C_j have to satisfy the conditions

$$\sum C_j = I, \qquad \sum jC_j = A. \tag{8}$$

These conditions are satisfied by

$$C_{-1} = (I-A)/2, \qquad C_1 = (I+A)/2, \quad \text{all other } C_j = 0. \tag{8$'$}$$

Since the CFL condition requires that the eigenvalues of A be not greater than 1, it follows that the coefficients C_j in (8) are non-negative. We appeal now to:

Lemma 1 *Suppose that the coefficients C_j in the operator C_h in* (7) *are non-negative real symmetric matrices that depend smoothly on $x = jh$. Then the L^2-norm of the operator C_h is less than $1 + \mathbf{O}(h)$.*

We sketch the proof: abbreviate u_n as u, $u_{(n+1)}$ as v; then the difference scheme reads

$$v = C_h u = \sum C_j T_j u.$$

Take the scalar product with v:

$$(v, v) = \sum \left(v, C_j T^j u\right).$$

Since the C_j are non-negative, we can apply the Cauchy–Schwarz inequality:

$$\left(v, C_j T^j u\right) \le \sqrt{(v, C_j v)}\sqrt{\left(T^j u, C_j T^j u\right)}.$$

Using the arithmetic–geometric mean inequality on the product on the right, we get

$$(v, v) \le (1/2)\sum (v, C_j v) + (1/2)\sum \left(T^j u, C_j T^j u\right).$$

According to the first consistency condition in (8), $\sum C_j = I$; therefore, the first sum on the right is $(v, v)/2$. If the coefficients C_j are independent of x, the second sum is (u,u), so we get the desired inequality. If the C_j depend on x, it gives rise to a term $\mathbf{O}(h)(u,u)$.

The proof of this lemma is somewhat analogous, though a little trickier than the energy estimate for solutions of the partial differential equation presented above.

It follows from the lemma that the n-th power of C_h is bounded by $(1+\mathbf{O}(h))^n$; these quantities are uniformly bounded if nh is less than some specified number T. This proves the stability of the difference scheme. According to the standard theory of difference approximations, this guarantees the convergence of the difference scheme.

Condition (8) is first order consistency, and it only guarantees that the solution of the difference scheme differs from the solution of the differential equation by $\mathbf{O}(h)$. For higher order accuracy, we need higher order consistency:

$$\sum C_j = I, \qquad \sum jC_j = A, \quad \text{and} \quad \sum j^2 C_j = A^2. \tag{9}$$

These conditions are satisfied by the Lax–Wendroff scheme:

$$C_{-1} = \left(A^2 - A\right)/2, \qquad C_0 = I - A^2, \qquad C_1 = \left(A^2 + A\right)/2, \tag{9$'$}$$

plus terms of order h. According to the CFL condition, the eigenvalues of A are less than 1; it follows that the eigenvalues of A^2 are less than those of A. If A has a positive eigenvalue, the corresponding eigenvalue of C_{-1} is negative; if A has a negative eigenvalue, the corresponding eigenvalue of C_{-1} is negative.

More generally it is not hard to show that, except for trivial cases, it is not possible to satisfy condition (9) by non-negative matrices C_j. Therefore, for second and higher order schemes, we need a proof of stability different from the proof for schemes with positive coefficients.

Our starting point is a stability criterion due to von Neumann. We associate with the difference scheme (6)–(7) its symbol defined as the matrix-valued trigonometric polynomial

$$C(s) = \sum C_j e^{ijs}. \tag{10}$$

When the coefficients depend on $x = hj$, so does the symbol $C(s,x)$. The stability criterion of von Neumann says that if the scheme (6)–(7) is stable, the absolute value of the eigenvalues of its symbol $C(s,x)$ are not greater than 1.

Sketch of proof In the constant coefficient case, $C(s)$ is independent of x. If we take the Fourier transform of the difference equation (6), we get

$$U^{n+1}(s) = C(s)U^n(s), \tag{11}$$

where $U^n(s) = \sum e^{ijs}$ and similarly for U^{n+1}. We can iterate (11) and obtain the operator linking the initial value U^0 to U^n:

$$U^n(s) = C(s)^n U_0(s).$$

Clearly, if for some value of s, $C(s)$ has an eigenvalue greater than 1, the L^2-norm of $U^n(s)$ blows up exponentially. This shows that if the von Neumann condition is violated, the difference scheme is unstable. This argument can be modified for schemes with variable coefficients.

Von Neumann raised the question whether the stability criterion, possibly somewhat sharpened, is sufficient for stability. For schemes in one space variable, the answer is yes; here is a sketch of the argument.

The consistency conditions for schemes (8) and (9) in one space variable relate combinations of the coefficients C_j to powers of the matrix A in the differential equation (5). It follows that the natural choice for the coefficients C_j will be polynomials in A and its powers. Since A is a symmetric matrix, the symbol $C(s)$ defined in (10) is for each s and x a normal matrix. A normal matrix whose eigenvalues are not greater than 1 in absolute value has norm ≤ 1; therefore, $C^*(s)C(s) \leq I$. From this one can deduce the stability of the scheme.

We turn now to difference approximations of hyperbolic equations (1) in two space variables. We write the difference equation as

$$u_k^{n+1} = \sum c_j u_{k+j}^n, \tag{12}$$

where $k = (k_1, k_2)$ and $j = (j_1, j_2)$ are multi-indices, and u_k^n is the value of the approximate solution at time $t = nh$ and at the lattice point $(k_1, k_2)h$. The analogs of the first order consistency conditions (8) are

$$\sum C_j' = I, \qquad \sum j_1 C_j = A, \qquad \sum j_2 C_j = B. \tag{13}$$

These relations can be satisfied analogously to (9) by setting all C_j equal to zero except at the four neighboring lattice points. We denote these as C_W, C_E, C_S, C_N. We choose, in analogy with (8′),

$$\begin{aligned} C_W &= 1/2(I/2 - A), \qquad & C_E &= 1/2(I/2 + A), \\ C_S &= 1/2(I/2 - B), \qquad & C_N &= 1/2(I/2 + B). \end{aligned} \tag{13′}$$

If the norms of A and B are less than $1/2$, all four matrices above are positive. The argument given in the one dimensional case can be used to prove that this scheme with positive coefficient is stable.

The domain of dependence of the point $(0, 0, 1)$ for the difference scheme (13) is the rectangle $|x + y| \leq 1$ and $|x - u| \leq 1$. The domain of dependence of the point $(0, 0, 1)$ for the differential equation is contained in that rectangle if the norm of $A + B$ and $A - B$ does not exceed 1. This shows that the sufficient conditions $|A| < 1/2$, $|B| < 1/2$ are more stringent than the CFL condition. □

One can modify the scheme (13) so that the new scheme is positive under the CFL conditions

$$|A + B| \leq 1, \qquad |A - B| \leq 1.$$

We turn now to second order schemes for Eq. (1). A straightforward way of constructing such a scheme is to use the Taylor approximation:

$$u(t+h) = u(t) + u_t(t)h + u_{tt}(t)h^2 + \mathbf{O}(h^3),$$

and then use the differential equation (1) to express u_t and u_{tt} in terms of space derivatives. If we then approximate the first and second space derivatives with symmetric difference quotients, we get the nine point Lax–Wendroff difference scheme. The symbol of this scheme,

$$C(s,r) = \sum C_j \, e^{i(j_1 s + j_2 r)},$$

is

$$\begin{aligned} C(s,r) = I &- A^2(1-\cos s) - B^2(1-\cos r) \\ &- 1/2(AB+BA)\sin s \sin r + i(A\sin s + B\sin r). \end{aligned} \tag{14}$$

The von Neumann stability condition is that the eigenvalues of $C(s,r)$ should be less than 1. But since the matrices A and B do not commute, except in trivial cases, $C(s,r)$ is no longer a normal matrix, so we cannot conclude that the norm of $C(s,r)$ is less than 1; in fact, it is not less than 1.

Instead of looking at the norm of $C(s,r)$, Burt Wendroff and the author have shown that if the matrices A and B have norm less than $1/8$, the numerical range of $C(s,r)$ is contained in the unit circle. We recall that the numerical range of a matrix M is the set of all complex numbers of the form $w \cdot Mw$, w any unit vector with complex entries. Since the spectrum belongs to the numerical range, it follows that the scheme with symbol $C(s,r)$ satisfies the von Neumann criterion. The condition on the numerical range is more than the von Neumann condition and has as a consequence the

Stability Theorem *Suppose a difference scheme of form* (12) *has the following properties*:

(i) *The coefficients $C_j(x)$ are twice differentiable functions of x.*
(ii) *For every s and x, the numerical range of the symbol $C(s,x)$ lies in the unit circle.*

Then the numerical range of the difference operator C_h is less than $1+\mathbf{O}(h)$.

Such an operator is stable; for, according to the Halmos–Pearcy theorem, the numerical range of C_j^n is less than $(1+\mathbf{O}(h))^n$, and so the norm of C_j^n is less than twice that.

The key tool used in the proof is this result of Louis Nirenberg and the author: Let P_h be a difference operator of the form

$$P_h = \sum P_j(x) T^j,$$

where the coefficients $P_j(x)$ are twice differentiable functions of x. Suppose that the symbol of P_h,

$$P(s,x) = \sum P_j(x)\, e^{i(j_1 s + j_2 r)},$$

is Hermitian and non-negative for all s and x:

$$P(s,x) \geq 0.$$

Then the Hermitian part of P_h, Re $P_h = (P_h + P_h^*)/2$, satisfies

$$\text{Re}\, P_h \geq -\text{constant}\, h.$$

The proof of this estimate is tricky, as indicated by the requirement that the coefficients of the scheme be not once but twice differentiable functions.

The results described above hold for any number of space variables, not just two.

I hope that the discussion presented proves the claim that the theory of difference schemes is much more sophisticated than the theory of differential equations.

Mathematical Intuition: Poincaré, Pólya, Dewey

Reuben Hersh

Abstract Practical calculation of the limit of a sequence often violates the definition of convergence to a limit as taught in calculus. Together with examples from Euler, Pólya and Poincaré, this fact shows that in mathematics, as in science and in everyday life, we are often obligated to use knowledge that is derived, not rigorously or deductively, but simply by making the best use of available information–plausible reasoning. The "philosophy of mathematical practice" fits into the general framework of "warranted assertibility", the pragmatist view of the logic of inquiry developed by John Dewey.

Keywords Intuition · Induction · Pragmatism · Approximation · Convergence · Limits · Knowledge

1 Introduction

In Rio de Janeiro in May 2010, I spoke at a meeting of numerical analysts honoring the 80th anniversary of the famous paper by Courant, Friedrichs, and Lewy. In order to give a philosophical talk appropriate for hard-core computer-oriented mathematicians, I focused on a certain striking paradox that is situated right at the heart of analysis, both pure and applied. (That paradox was presented, with considerable mathematical elaboration, in Phil Davis's excellent article, *The Paradox of the Irrelevant Beginning*, cf. [5].) In order to make this paradox cut as sharply as possible, I performed a little dialogue, with help from Carlos Motta. With the help of Jody Azzouni, I used that dialogue again, to introduce this talk in Rome.

To set the stage, recall the notion of a convergent sequence, which is at the heart of both pure analysis and applied mathematics. In every calculus course, the student learns that whether a sequence converges to a limit and what that limit is depend only on the "end" of the sequence—that is, the part that is "very far out"—in the tail, so to speak, or in the infinite part. Yet, in a specific instance when the limit is

R. Hersh (✉)
Department of Mathematics and Statistics, University of New Mexico, Albuquerque, USA
e-mail: rhersh@gmail.com

C.A. de Moura, C.S. Kubrusly (eds.), *The Courant–Friedrichs–Lewy (CFL) Condition*,
DOI 10.1007/978-0-8176-8394-8_2,

actually needed, usually all that is considered is the beginning of the sequence—the first few terms—the finite part, so to speak. (Even if the calculation is carried out to a hundred or a thousand iterations, this is still only the first few, compared to the remaining, neglected, infinite tail.)

In this little drama of mine, the hero is a sincere, well-meaning student, who has not yet learned to accept life as it really is. A second character is the *Successful Mathematician*—the Ideal Mathematician's son-in-law. His mathematics is ecumenical: a little pure, a little applied, and a little in-between. He has grants from federal agencies, a corporation here and there, and a private foundation or two. His conversation with the *Stubborn Student* is somewhat reminiscent of a famous conversation between his Dad, the Ideal Mathematician, and a philosophy grad student, who long ago asked, "What is a mathematical proof, really?"

2 A Dialogue

The Successful Mathematician (*SM*) is accosted by the Stubborn Student (*SS*) from his Applied Analysis course.

SS: Sir, do you mind if I ask a stupid question?

SM: Of course not. There is no such thing as a stupid question.

SS: Right. I remember, you said that. So here's my question. What is the real definition of "convergence"? Like, convergence of an infinite sequence, for instance?

SM: Well, I'm sure you already know the answer. The sequence converges to a limit, L, if it gets within a distance epsilon of L, and stays there, for any positive epsilon, no matter how small.

SS: Sure, that's in the book, I know that. But then, what do people mean when they say, keep iterating till the iteration converges? How does that work?

SM: Well, it's obvious, isn't it? If after a hundred terms your sequence stays at 3, correct to four decimal places, then the limit is 3.

SS: Right. But how long is it supposed to stay there? For a hundred terms, for two hundred, for a hundred million terms?

SM: Of course, you wouldn't go on for a hundred million. That really would be stupid. Why would you waste time and money like that?

SS: Yes, I see what you mean. But what then? A hundred and ten? Two hundred? A thousand?

SM: It all depends on how much you care. And how much it is costing, and how much time it is taking.

SS: All right, that's what I would do. But when does it converge?

SM: I told you. It converges if it gets within epsilon ...

SS: Never mind about that. I am supposed to go on computing "until it converges", so how am I supposed to recognize that "it has converged"?

SM: When it gets within four decimal points of some particular number and stays there.

SS: Stays there how long? Till when?

SM: Whatever is reasonable. Use your judgment! It's just plain common sense, for Pete's sake!

SS: But what if it keeps bouncing around within four decimal points and never gets any closer? You said any epsilon, no matter how small, not just 0.0001. Or if I keep on long enough, it might finally get bigger than 3, even bigger than 4, way, way out, past the thousandth term.

SM: Maybe this, maybe that. We haven't got time for all these *maybes*. Somebody else is waiting to get on that machine. And your bill from the computing center is getting pretty big.

SS: (mournfully) I guess you're not going to tell me the answer.

SM: You just don't get it, do you? Why don't you go bother that Reuben Hersh over there, he looks like he has nothing better to do.

SS: Excuse me, Professor Hersh. My name is–

RH: That's OK. I overheard your conversation with Professor Successful over there. Have a seat.

SS: Thank you. So, you already know what my question is.

RH: Yes, I do.

SS: So, what is the answer?

RH: He told you the truth. The real definition of convergence is exactly what he said, with the epsilon in it, the epsilon that is arbitrarily small but positive.

SS: So then, what does it mean, "go on until the sequence converges, then stop"?

RH: It's meaningless. It's not a precise mathematical statement. As a precise mathematical statement, it's meaningless.

SS: So, if it's meaningless, what does it mean?

RH: He told you what it means. Quit when you can see, when you can be pretty sure, what the limit must be. That's what it means.

SS: But that has nothing to do with convergence!

RH: Right.

SS: Convergence only depends on the last part, the end, the infinite part of the sequence. It has nothing to do with the front part. You can change the first hundred million terms of the sequence, and that won't affect whether it converges, or what the limit is.

RH: Right! Right! Right! You really are an **A** student.

SS: I know... So it all just doesn't make any sense. You teach us some fancy definition of convergence, but when you want to compute a number, you just forget about it and say it converges when common sense, or whatever you call it, says something must be the answer. Even though it might not be the answer at all!

RH: Excellent. I am impressed.

SS: Stop patronizing me. I'm not a child.
RH: Right. I will stop patronizing you because you are not a child.
SS: You're still doing it.
RH: It's a habit. I can't help it.
SS: Time to break a bad habit.
RH: OK. But seriously, you are absolutely right. I agree with every word you say.
SS: Yes, and you also agree with every word Professor Successful says.
RH: He was telling the truth, but he couldn't make you understand.
SS: All right. You make me understand.
RH: It's like theory and practice. Or the ideal and the actual. Or Heaven and Earth.
SS: How is that?
RH: The definition of convergence lives in a theoretical world. An ideal world. Where things can happen as long as we can clearly imagine them. As long as we can understand and agree on them. Like really being positive and arbitrarily small. No number we can write down is positive and arbitrarily small. It has to have some definite size if it is actually a number. But we can imagine it getting smaller and smaller and smaller while staying positive, and we can even express that idea in a formal sentence, so we accept it and work with it. It seems to convey what we want to mean by converging to a limit. But it's only an ideal, something we can imagine, not something we can ever really do.
SS: So you're saying mathematics is all a big fairy tale, a fiction, it doesn't actually exist?
RH: NO! I never said fairy tale or fiction. I said imaginary. Maybe I should have said consensual. Something we can all agree on and work with because we all understand it the same way.
SS: That's cool. We all. All of you. Does that include me?
RH: Sure. Stay in school a few more years. Learn some more. You'll get into the club. You've got what it takes.
SS: I'm not so sure. I have trouble believing two opposite things at once.
RH: Then how do you get along in daily life? How do you even get out of bed in the morning?
SS: What are you talking about?
RH: How do you know someone hasn't left a bear trap by your bedside that will chop off your foot as soon as you step down?
SS: That's ridiculous.
RH: It is. But how do you know it is?
SS: Never mind how I know. I just know it's ridiculous. And so do you.
RH: Exactly. We know stuff, but we don't always know how we know it. Still, we do know it.
SS: So you're saying, we know that what looks like a limit really is a limit, even though we can't prove it, or explain it, still we know it.
RH: We know it the same way you know nobody has left a bear trap by your bedside. You just know it.

SS: Right.

RH: But it's still possible that you're wrong. It is possible that something ridiculous actually happens. Not likely, not worth worrying about. But not impossible.

SS: Then math is really just like everything else. What a bummer! I like math because it's not like everything else. In math, we know for sure. We prove things. One and one is two. Pi is irrational. A circle is round, not square. For sure.

RH: Then why are you upset? Everything is just fine, isn't it?

SS: Why don't you admit it? If you don't have a proof, you just don't know if L is the limit or not.

RH: That's a fair question. So what is the answer?

SS: Because you really want to think you know L is the limit, even if it's not true.

RH: Not that it's not true, just that it might not be true.

Thanks for your kind attention. What is supposed to be the meaning of this performance? What am I getting at? In this talk, I am NOT attempting to make a contribution to the "problem of induction". Therefore, I may be allowed to omit a review of its 2,500-year literature. I am reporting and discussing what people really do, in practical convergence calculations, and in the process of mathematical discovery. I am going into a discussion of practical knowledge in mathematics, as a kind of real knowledge, even though it is not demonstrative or deductive knowledge. I try to explain why people must do what they do, in order to accomplish what they are trying to accomplish. I will conclude by arguing that the right broader context for the philosophy of mathematical practice is actually the philosophy of pragmatism, as expounded by John Dewey.

But first of all, just this remarkable fact. What we do when we want actual numbers may be totally unjustified, according to our theory and our definition. And even more remarkable-nobody seems to notice, or to worry about it!

Why is that? Well, the definition of convergence taught in calculus classes, as developed by those great men Augustin Cauchy and Karl Weierstrass, seems to actually convey what we want to mean by limit and convergence. It is a great success. Just look at the glorious edifice of mathematical analysis! On the other hand, in specific cases, it often is beyond our powers to give a rigorous error estimate, even when we have an approximation scheme that seems perfectly sound. As in the major problems of three-dimensional continuum mechanics with realistic nonlinearities, such as oceanography, weather prediction, stability of large complex structures like big bridges and airplanes.... And even when we could possibly give a rigorous error estimate, it would often require great expenditure of time and labor. Surely it's OK to just use the result of a calculation when it makes itself evident and there's no particular reason to expect any hidden difficulty.

In brief, we are virtually compelled by the practicalities to accept the number that computation seems to give us, even though, by the standards of rigorous logic, there is still an admitted possibility that we may be mistaken. This computational result is a kind of mathematical knowledge! It is practical knowledge, knowledge sound enough to be the basis of practical decisions about things like designing bridges and airplanes-matters of life and death.

In short, I am proclaiming that in mathematics, apart from and distinct from the so-called deductive or demonstrative knowledge, there is also ordinary, fallible knowledge, of the same sort as our daily knowledge of our physical environment and our own bodies. "Anything new that we learn about the world involves plausible reasoning, which is the only kind of reasoning for which we care in everyday affairs", cf. [19]. This sentence of Pólya's makes an implicit separation between mathematics and everyday affairs. But nowadays, in many different ways, for many different kinds of people, mathematics blends into every-day affairs. In these situations, the dominance of plausible over demonstrative reasoning applies even to mathematics itself, as in the daily labors of numerical analysts, applied mathematicians, design engineers.... Controlling a rocket trip to the moon is not an exercise in mathematical rigor. It relies on a lack of malice on the part of that Being referred to by Albert Einstein as *der lieber Gott*.

(For fear of misunderstanding, I explain—this is not a confession of belief in a Supreme Being. It's just Einstein's poetic or metaphoric way of saying, Nature is not an opponent consciously trying to trick us.)

But it's not only that we have no choice in the matter. It's also that, to tell the truth, it seems perfectly reasonable! Believing what the computation tells us is just what people have been doing all along, and (nearly always) it does seem to be OK. What's wrong with that?

This kind of reasoning is sometimes called "plausible", and sometimes called "intuitive". I will say a little more about those two words pretty soon. But I want to draw your attention very clearly to two glaring facts about this kind of plausible or intuitive reasoning. First of all, it is pretty much the kind of reasoning that we are accustomed to in ordinary empirical science, and in technology, and in fact in everyday thinking, dealing with any kind of practical or realistic problem of human life. Secondly, it makes no claim to be demonstrative, or deductive, or conclusive, as is often said to be the essential characteristic of mathematical thinking. We are face to face with mathematical knowledge that is not different in kind from ordinary everyday commonplace human knowledge. Fallible! But knowledge, nonetheless!

Never mind the pretend doubt of philosophical skepticism. We are adults, not infants. Human adults know a lot! How to find their way from bed to breakfast-and people's names and faces—and so forth and so on. This is real knowledge. It is not infallible, not eternal, not heavenly, not Platonic, it is just what daily life depends on, that's all. That's what I mean by ordinary, practical, everyday knowledge. Based not mainly on rigorous demonstration or deduction, but mainly on experience properly interpreted. And here we see mathematical knowledge that is of the same ordinary, everyday kind, based not on infallible deduction, but on fallible, plausible, intuitive thinking.

Then what justifies it in a logical sense? That is, what fundamental presupposition about the world, about reality, lies behind our willingness to commit this logical offense, of believing what isn't proved?

I have already quoted the famous saying of Albert Einstein that supplies the key to unlocking this paradox. My friend Peter Lax supplied the original German, I only remembered the English translation.

Raffinniert ist der lieber Gott, aber boshaft ist Er nicht.

The Lord God is subtle, but He is not malicious.

Of course, Einstein was speaking as a physicist struggling to unravel the secrets of Nature. The laws of Nature are not always obvious or simple, they are often subtle. But we can believe, we must believe, that Nature is not set up to trick us, by a malicious opponent. God, or Nature, must be playing fair. How do we know that? We really don't know it, as a matter of certainty! But we must believe it, if we seek to understand Nature with any hope of success. And since we do have some success in that search, our belief that Nature is subtle but not malicious is justified.

This problem of inferring generalizations from specific instances is known in logic as "the problem of induction". My purpose is to point out that such generalizations in fact are made, and must be made, not only in daily life and in empirical science, but also in mathematics.

That is, in the practice of mathematics also we must believe that we are not dealing with a malicious opponent who is seeking to trick us. We experiment, we calculate, we draw diagrams. And eventually, using caution and the experience of the ages, we see the light. Gauss famously said, "I have my theorems. Now I have to find my proofs".

But is it not naïve, for people who have lived through the hideous twentieth century, to still hope that God is not malicious? Consider, for example, people who for thousands of years have lived safely on some atoll in the South Pacific. Today an unforeseen tsunami drowns them all. Might they not curse God in their last breath?

Here is an extensive quote from Leonhard Euler, by way of George Pólya. Euler is speaking of a certain beautiful and surprising regularity in the sum of the divisors of the integers.

> This law, which I shall explain in a moment, is in my opinion, so much more remarkable as it is of such a nature that we can be assured of its truth without giving it a perfect demonstration. Nevertheless, I shall present such evidence for it as might be regarded as almost equivalent to a rigorous demonstration... Anybody can satisfy himself of its truth by as many examples as he may wish to develop. And since I must admit that I am not in a position to give it a rigorous demonstration, I will justify it by a sufficiently large number of examples... I think these examples are sufficient to discourage anyone from imagining that it is by mere chance that my rule is in agreement with the truth... The examples that I have just developed will undoubtedly dispel any qualms which we might have had about the truth of my formula... It seems impossible that the law which has been discovered to hold for 20 terms, for example, would not be observed in the terms that follow.
> (Taken from [19].)

Observe two things about this quote from Euler. First of all, for him the plausible reasoning in this example is so irresistible that it leaves no room for doubt. He is certain that anyone who looks at his examples is bound to agree. Yet secondly, he strongly regrets his inability to provide a demonstration of the fact, and still hopes to find one.

But since he is already certain of the truth of his finding, why ask for a demonstrative proof? The answer is easy, for anyone familiar with mathematical work. The demonstration would not just affirm the truth of the formula, it would show why the formula MUST be true. That is the main importance of proof in mathematics! A plausible argument, relying on examples, analogy and induction, can be very strong, can carry total conviction. But if it is not demonstrative, it fails to show why the result MUST be true. That is to say, it fails to show that it is rigidly connected to established mathematics.

At the head of Chap. V of [19], Pólya placed the following apocryphal quotation, attributed to "the traditional mathematics professor":

"When you have satisfied yourself that the theorem is true, you start proving it".

This faith—that experience is not a trap laid to mislead us—is the unstated axiom. It lets us believe the numbers that come out of our calculations, including the canned programs that engineers use every day as black boxes. We know that it can sometimes be false. But even as we keep possible tsunamis in mind, we have no alternative but to act as if the world makes sense. We must continue to act on the basis of our experience. (Including, of course, experiences of unexpected disasters.)

Consider this recollection of infantile mathematical research by the famous physicist Freeman Dyson, who wrote in 2004:

> One episode I remember vividly, I don't know how old I was; I only know that I was young enough to be put down for an afternoon nap in my crib... I didn't feel like sleeping, so I spent the time calculating. I added one plus a half plus a quarter plus an eighth plus a sixteenth and so on, and I discovered that if you go on adding like this forever you end up with two. Then I tried adding one plus a third plus a ninth and so on, and discovered that if you go on adding like this forever you end up with one and a half. Then I tried one plus a quarter and so on, and ended up with one and a third. So I had discovered infinite series. I don't remember talking about this to anybody at the time. It was just a game I enjoyed playing.

Yes, he knew the limit! How did he know it? Not the way we teach it in high school (by getting an exact formula for the sum of n terms of a geometric sequence, and then proving that as n goes to infinity, the difference from the proposed limit becomes and remains arbitrarily small.) No, just as when we first show this to tenth-graders, he saw that the sums follow a simple pattern that clearly is "converging" to 2. The formal, rigorous proof gives insight into the reason for a fact we have already seen plainly.

Can we go wrong this way? Certainly we can. Another quote from Euler.

> There are even many properties of the numbers with which we are well acquainted, but which we are not yet able to prove; only observations have led us to their knowledge..., the kind of knowledge which is supported only by observations and is not yet proved must be carefully distinguished from the truth; it is gained by induction, as we usually say. Yet we have seen cases in which mere induction led to error. Therefore, we should take great care not to accept as true such properties of the numbers which we have discovered by observation and which are supported by induction alone. Indeed, we should use such a discovery as an opportunity to investigate more exactly the properties discovered and to prove or disprove them. (In [19], p. 3.)

Notice how Euler distinguishes between "knowledge" and "truth"! He does say "knowledge", not mere "conjecture".

There is a famous theorem of Littlewood concerning a pair of number-theoretic functions $\pi(x)$ and $Li(x)$. All calculation shows that $Li(x)$ is greater than $\pi(x)$, for x as large as we can calculate. Yet Littlewood proved that eventually $\pi(x)$ becomes greater than $Li(x)$, and not just once, but infinitely often! Yes, mathematical truth can be very subtle. While trusting it not to be malicious, we must not underestimate its subtlety. ($\pi(x)$ is the prime counting function and $Li(x)$ is the logarithmic integral function.)

3 Mathematical Intuition

We are concerned with "the philosophy of mathematical practice". Mathematical practice includes studying, teaching and applying mathematics. But I suppose we have in mind first of all the discovery and creation of mathematics-mathematical research. We start with Jacques Hadamard, go on to Henri Poincaré, move on to George Pólya, and then to John Dewey.

Hadamard had a very long life and a very productive career. His most noted achievement (shared independently by de la Vallée Poussin) was proving the logarithmic distribution of the prime numbers. I want to recall a famous remark of Hadamard's. "The object of mathematical rigor is to sanction and legitimize the conquests of intuition, and there never was any other object for it." (See [20].)

From the viewpoint of standard "philosophy of mathematics", this is a very surprising, strange remark. Isn't mathematical rigor-that is, strict deductive reasoning-the most essential feature of mathematics? And indeed, what can Hadamard even mean by this word, "intuition"? A word that means one thing to Descartes, another thing to Kant. I think the philosophers of mathematics have pretty unanimously chosen to ignore this remark of Hadamard. Yet Hadamard did know a lot of mathematics, both rigorous and intuitive. And this remark was quoted approvingly by both Borel and Pólya. It seems to me that this bewildering remark deserves to be taken seriously.

Let's pursue the question a step further, by recalling the famous essay *Mathematical Discovery*, written by Hadamard's teacher, Henri Poincaré, cf. [16]. Poincaré was one of the supreme mathematicians of the turn of the nineteenth and twentieth century. We've been hearing his name recently, in connection with his conjecture on the 3-sphere, just recently proved by Grisha Perelman of St. Petersburg. Poincaré was not only a great mathematician, he was a brilliant essayist. And in the essay *Mathematical Discovery*, Poincaré makes a serious effort to explain mathematical intuition. He tells the famous story of how he discovered the Fuchsian and Theta-Fuchsian functions. He had been struggling with the problem unsuccessfully when he was distracted by being called up for military service:

> At this moment I left Caën, where I was then living, to take part in a geological conference arranged by the School of Mines. The incidents of the journey made me forget my mathematical work. When we arrived at Coutances, we got into a bus to go for a drive, and, just as I put my foot on the step the idea came to me, though nothing in my former thoughts seemed to have prepared me for it, that the transformations I had used to define Fuchsian functions were identical with those of non-Euclidean geometry. I made no verification, and had no time to do so, since I took up the conversation again as soon as I had sat down in the bus, but I felt absolute certainty at once. When I got back to Caën, I verified the result at my leisure to satisfy my conscience. (Cf. [16].)

What a perfect example of rigor "merely legitimizing the conquests of intuition"! How does Poincaré explain it? First of all, he points out that some sort of subconscious thinking must be going on. But if it is subconscious, he presumes it must be running on somehow at random. How unlikely, then, for it to find one of the very few good combinations, among the huge number of useless ones! To explain further, he writes:

> If I may be permitted a crude comparison, let us represent the future elements of our combinations as something resembling Epicurus' hooked atoms. When the mind is in complete repose these atoms are immovable; they are, so to speak, attached to the wall... On the other hand, during a period of apparent repose, but of unconscious work, some of them are detached from the wall and set in motion. They plough through space in all directions, like a swarm of gnats, for instance, or, if we prefer a more learned comparison, like the gaseous molecules in the kinetic theory of gases. Their mutual impacts may then produce new combinations. (Cf. [16].)

The preliminary conscious work "detached them from the wall". The mobilized atoms, he speculated, would therefore be "those from which we might reasonably expect the desired solution... My comparison is very crude, but I cannot well see how I could explain my thought in any other way". (Cf. [16].)

What can we make of this picture of "Epicurean hooked atoms", flying about somewhere-in the mind? A striking, suggestive image, but one not subject even in principle to either verification or disproof. Our traditional philosopher remains little interested. This is fantasy or poetry, not science or philosophy. But this is Poincaré! He knows what he's talking about. He has something important to tell us. It's not easy to understand, but let's take him seriously, too.

To be fair, Poincaré proposed his image of gnats or gas molecules only after mentioning the possibility that the subconscious is actually more intelligent than the conscious mind. But this, he said, he was not willing to contemplate. However, other writers have proposed that the subconscious is less inhibited, more imaginative, more creative than the conscious. (Poincaré's essay title is sometimes translated as *Mathematical Creation* rather than *Mathematical Discovery*.) David Hilbert supposedly once said of a student who had given up mathematics for poetry, "Good! He didn't have enough imagination for mathematics". Hadamard [12] in 1949 carefully analyzed the role of the subconscious in mathematical discovery and its connection with intuition. It is time for contemporary cognitive psychology to pay attention to Hadamard's insights. See the reference quoted in the Appendix about current scientific work on the creative power of the subconscious.

Before going on, I want to mention the work of Carlo Cellucci, Emily Grosholz, and Andrei Rodin. Cellucci strongly favors plausible reasoning, but he rejects intuition. However, the intuition he rejects isn't what I'm talking about. He's rejecting the old myth, of an infallible insight straight into the Transcendental. Of course, I'm not advocating that outdated myth. Emily Grosholz, on the other hand, takes intuition very seriously. Her impressive historical study of what she calls "internal intuition" is in the same direction as my own thinking being presented here. Andrei Rodin has recently written a remarkable historical study of intuition, see [24]. He shows that intuition played a central role in Lobachevski's non-Euclidean geometry, in Zermelo's axiomatic set theory, and even in up-to-date category theory. (By the way, in category theory he could also have cited the standard practice of proof by "diagram chasing" as a blatant example of intuitive, visual proof.) His exposition makes the indispensable role of intuition clear and convincing. But his use of the term "intuition" remains, one might say, "intuitive", for he offers no definition of the term, nor even a general description, beyond his specific examples.

4 Pólya

My most helpful authority is George Pólya. I actually induced Pólya to come give talks in New Mexico, for previously, as a young instructor, I had met him at Stanford where he was an honored and famous professor. Pólya was not of the stature of Poincaré or Hilbert, but he was still one of the most original, creative, versatile, and influential mathematicians of his generation. His book with Gábor Szegö [21] made them both famous. It expounds large areas of advanced analytic function theory by means of a carefully arranged, graded sequence of problems with hints and solutions. Not only does it teach advanced function theory, it also teaches problem-solving. And by example, it shows how to teach mathematics by teaching problem-solving. Moreover, it implies a certain view of the nature of mathematics, so it is a philosophical work in disguise.

Later, when Pólya wrote his very well-known, influential books on mathematical heuristic, he admitted that what he was doing could be regarded as having philosophical content. He writes, "I do not know whether the contents of these four chapters deserve to be called philosophy. If this is philosophy, it is certainly a pretty low-brow kind of philosophy, more concerned with understanding concrete examples and the concrete behavior of people than with expounding generalities". (Cf. [19], page viii.) Unpretentious as Pólya was, he was still aware of his true stature in mathematics. I suspect he was also aware of the philosophical depth of his heuristic. He played it down because, like most mathematicians (I can only think of one or two exceptions), he disliked controversy and arguing, or competing for the goal of becoming top dog in some cubbyhole of academia. The Prince of Mathematicians, Carl Friedrich Gauss, kept his monumental discovery of non-Euclidean geometry hidden in a desk drawer to avoid stirring up the Bœotians, as he called them, meaning the post-Kantian German philosophy professors of his day. (In ancient Athens, "Bœotian" was slang for "ignorant country hick".) Raymond Wilder was a leading topologist who wrote extensively on mathematics as a culture. He admitted to me that his writings implicitly challenged both formalism and Platonism. "Why not say so?" I asked. Because he didn't relish getting involved in philosophical argument.

Well, how does Pólya's work on heuristic clarify mathematical intuition? Pólya's heuristic is presented as pedagogy. Pólya is showing the novice how to solve problems. But what is "solving a problem"? In the very first sentence of the preface to [20], he writes: "Solving a problem means finding a way out of a difficulty, a way around an obstacle, attaining an aim which was not immediately attainable. Solving problems is the specific achievement of intelligence, and intelligence is the specific gift of mankind: solving problems can be regarded as the most characteristically human activity". "Problem" is simply another word for any project or enterprise which one cannot immediately take care of with the tools at hand. In mathematics, something more than a mere calculation. Showing how to solve problems amounts to showing how to do research!

Pólya's exposition is never general and abstract, he always uses a specific mathematical problem for the heuristic he wants to teach. His mathematical examples are always fresh and attractive. And his heuristic methods? First of all, there is what he calls "induction". That is, looking at examples, as many as necessary, and using them to guess a pattern, a generalization. But be careful! Never just believe your guess! He insists that you must "Guess and test, guess and test". Along with induction, there is analogy, and there is making diagrams, graphs and every other kind of picture, and then reasoning or guessing from the picture. And finally, there's the "default hypothesis of chance"—that an observed pattern is mere coincidence.

(Mark Steiner has the distinction among philosophers of paying serious attention to Pólya. After quoting at length from Pólya's presentation of Euler's heuristic derivation of the sum of a certain infinite series, Steiner comes to an important conclusion: In mathematics we can have knowledge without proof! Based on the testimony of mathematicians, he even urges philosophers to pay attention to the question of mathematical intuition.)

I have two comments about Pólya's heuristic that I think he would have accepted. First of all, the methods he is presenting, by means of elementary examples, are methods he used himself in research. "In fact, my main source was my own research, and my treatment of many an elementary problem mirrors my experience with advanced problems". (Cf. [20], page xi.) In teaching us how to solve problems, he's teaching us about mathematical practice: How it works. What is done. To find out "What is mathematics?" we must simply reinterpret Pólya's examples as descriptive rather than pedagogical.

Secondly, with hardly any stretching or adjustment, the heuristic devices that he's teaching can be applied for any other kind of problem-solving, far beyond mathematics. He actually says that he is bringing to mathematics the kind of thinking ordinarily associated with empirical science. But we can go further. These ways of thinking are associated with every kind of problem-solving, in every area of human life! Someone needed to get across a river or lake and had the brilliant idea of "a boat"—whether it was a dugout log or a birch bark canoe. Someone else, needing shelter from the burning sun in the California Mojave, thought of digging a hole in the ground. And someone else, under the piercing wind of northern Canada, thought of making a shelter from blocks of ice.

How does anyone think of such a thing, solve such a concrete problem? By some kind of analogy with something else he has seen, or perhaps been told about. By plausible thinking. And often by a sudden insight that arises "from below". *Intuitively*, you might say.

5 Mental Models

It often happens that a concrete problem, whether in science or in ordinary daily life, is pressing on the mind, even when the particular materials or objects in question are not physically present. You keep on thinking about it, while you're walking, and when you're waking from sleep. Productive thought commonly takes place, in the absence of the concrete objects or materials being thought about. This thinking about something not present to sight or touch can be called "abstract thinking". Abstract thinking about a concrete object. How does that work? How can our mind/brain think productively about something that's not there in front of the eyes? Evidently, it operates on something mental, what we may call a mental image or representation. In the current literature of cognitive psychology, one talks about "a mental model". In this article, I use the term "mental model" to mean a mental structure built from recollected facts (some expressed in words), along with an ensemble of sensory memories, perhaps connected, as if by walking around the object in question, or by imagining the object from underneath or above, even if never actually seen in these views. A rich complex of connected knowledge and conjecture based on verbal, visual, kinesthetic, even auditory or olfactory information, but simplified, to exclude irrelevant details. Everything that's helpful for thinking about the object of interest when the object isn't here. Under the pressure of a strong desire or need to solve a specific problem, we assemble a mental model which the mind-brain can manipulate or analyze.

Subconscious thinking is not a special peculiarity of mathematical thinking, but a common, taken-for-granted, part of every-day problem-solving. When we consider this commonplace fact, we aren't tempted to compare it to a swarm of gnats hooking together at random. No, we assume, as a matter of course, that this subconscious thinking follows rules, methods, habits or pathways, that somehow, to some extent, correspond to the familiar plausible thinking we do when we're wide awake. Such as thinking by analogy or by induction. After all, if it is to be productive, what else can it do? If it had any better methods, then those better methods would also be what we would follow in conscious thinking! And subconscious thinking in mathematics must be much like subconscious thinking in any other domain, carrying on plausible reasoning as enunciated by various writers, above all by George Pólya. This description of subconscious thinking is not far from Michael Polanyi's "tacit dimension".

When applied to everyday problem solving, all this is rather obvious, perhaps even banal. My goal is to clarify mathematical intuition, in the sense of Hadamard and Poincaré. "Intuition" in the sense of Hadamard and Poincaré is a fallible psychological experience that has to be accounted for in any realistic philosophy of mathematics. It simply means guesses or insights attained by plausible reasoning, either fully conscious or partly subconscious. In this sense, it is a specific phenomenon of common experience. It has nothing to do with the ancient mystical myth of an intuition that surpasses logic by making a direct connection to the Transcendental.

The term "abstract thinking" is commonplace in talk about mathematics. The triangle, the main subject of Euclidean geometry, is an abstraction, even though it's idealized from visible triangles on the blackboard. Thinking of a physical object in its absence, like a stream to be crossed or a boat to be imagined and then built, is already "abstract" thinking, and the word "abstract" connects us to the abstract objects of mathematics.

Let me be as clear and simple as I can be about the connection. After we have some practice drawing triangles, we can think about triangles, we discover properties of triangles. We do this by reasoning about mental images, as well as images on paper. This is already abstract thinking. When we go on to regular polygons of arbitrarily many sides, we have made another departure. Eventually, we think of the triangle as a 2-simplex, and abstract from the triangle to the n-simplex. For $n = 3$ this is just the tetrahedron, but for $n = 4$, 5, or 6, it is something never yet seen by human eye. Yet these higher simplexes also can become familiar and, as it were, concrete-seeming. If we devote our waking lives to thinking about them, then we have some kind of "mental model" of them. Having this mental model, we can access it, and thereby we can reason intuitively-have intuitive insights-by which I mean simply insights not based on consciously known reasoning. An "intuition" is then simply a belief (possibly mistaken!) arising from internal inspection of a mental image or representation—a "model". It may be assisted by subconscious plausible reasoning, based on the availability of that mental image. We do this in practical life. We do it in empirical science, and in mathematics. In empirical science and ordinary life, the image may stand for either an actual object, a physical entity, or a potential one that could be realized physically. In mathematics, our mental model is sometimes idealized from a physical object-for example, from a collection of identical coins or buttons when we're thinking about arithmetic. But in mathematics we also may possess a mental model with no physical counterpart. For example, it is generally believed that Bill Thurston's famous conjectures on the classification of four-manifolds were achieved by an exceptional ability, on the part of Thurston, to think intuitively in the fourth dimension. Perhaps Grisha Perelman was also guided by some four-dimensional intuition, in his arduous arguments and calculations to prove the Thurston program.

To summarize, mathematical intuition is an application of conscious or subconscious heuristic thinking of the same kind that is used every day in ordinary life by ordinary people, as well as in empirical science by scientists. This has been said before, by both Hadamard and Pólya. In fact, this position is similar to Kurt Gödel's, who famously wrote, "I don't see any reason why we should have less confidence in this kind of perception, i.e., in mathematical intuition, than in sense perception". Why, indeed? After all, both are fallible, but both are plausible, and must be based on plausible reasoning.

For Gödel, however, as for every writer in the dominant philosophy of mathematics, intuition is called in only to justify the axioms. Once the axioms are written down, the role of mathematical intuition is strictly limited to "heuristic"—to formulating conjectures. These await legitimation by deductive proof, for only deductive proof can establish "certainty". Indeed, this was stated as firmly by Pólya as by any analytic philosopher. But what is meant by "mathematical certainty"? If it simply means deductive proof, this statement is a mere circular truism. However, as I meant to suggest by the little dialog at the beginning of this paper, there is also practical certainty, even within mathematics! We are certain of many things in ordinary daily life, without deductive proof, and this is also the case in mathematics itself. Practical certainty is a belief strong enough to lead to serious practical decisions and actions. For example, we stake our lives on the numerical values that went into the engineering design of an Airbus or the Golden Gate Bridge. Mainstream philosophy of mathematics does not recognize such practical certainty. Nevertheless, it is an undeniable fact of life.

It is a fact of life not only in applied mathematics but also in pure mathematics. For example, the familiar picture of the Mandelbrot set, a very famous bit of recent pure mathematics, is generated by a machine computation. By definition, any particular point in the complex plane is inside the Mandelbrot set if a certain associated iteration stays bounded. If that iteration at some stage produces a number with absolute value greater than two, then, from a known theorem, we can conclude that the iteration goes to infinity, and the parameter point in question is outside the Mandelbrot set. What if the point is inside the Mandelbrot set? No finite number of iterations in itself can guarantee that the iteration will never go beyond absolute value 2. If we do eventually decide that it looks like it will stay bounded, we may be right, but we are still cheating. This decision is opportunistic and unavoidable, just as in an ordinary calculation about turbulent flow.

Computation (numerics) is accepted by purists only as a source of conjectures awaiting rigorous proof. However, from the pragmatic, non-purist viewpoint, if numerics is our guide to action, then it is in effect a source of knowledge. Dewey called it "warranted assertibility". (Possibly even a "truth". A "truth" that remains open to possible reconsideration.)

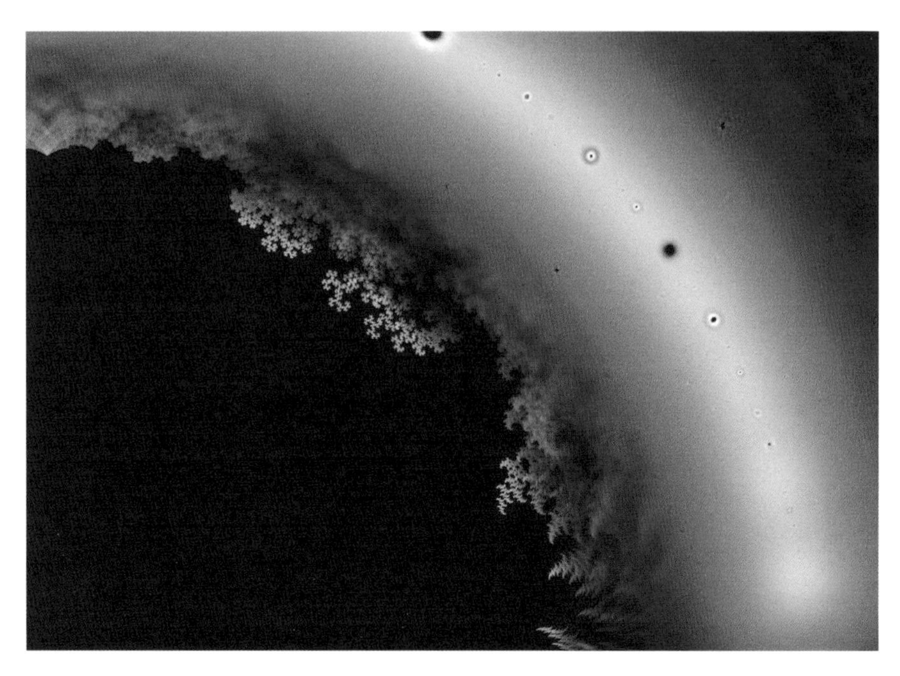

Fig. 1 400 million polynomial roots

Another example from pure mathematics appeared on John Baez's blog ([1]) where it is credited to Sam Derbyshire. His pictures plot the location in the complex plane of the roots of all polynomials of degree 24 with coefficients plus one or minus one. The qualitative features of these pictures are absolutely convincing—i.e., impossible to disbelieve. Baez wrote, "That's 2^{24} polynomials, and about 24×2^{24} roots—or about 400 million roots! It took *Mathematica* 4 days to generate the coordinates of the roots, producing about 5 gigabytes of data". (Figure 1 shows the part of the plot in the first quadrant, for complex roots with non-negative real and imaginary parts.)

There is more information in this picture than can even be formulated as conjectures, let alone seriously attacked with rigor. Since indeed we cannot help believing them (perhaps only believing with 99.999 % credence) then (pragmatically) we give them "warranted assertibility", just like my belief that I can walk out my door without encountering sudden death in one form or another. The distinction between rigorous math and plausible math, pure math and applied math, etc., becomes blurred. It is still visible, certainly, but not so sharp. It's a little fuzzy. Purely computational results in pure mathematics, when backed up by sophisticated checking against a relevant theory, have a factual status similar to that of accepted facts from empirical science. The distinction between what is taken to be "known", and what is set aside as merely guessed or "conjectured," is not so cut and dried as the usual discussions claim to believe.

6 Mental Models Subject to Social Control

"Plausible" or "heuristic" thinking is applied, either consciously or subconsciously, to mental models. These mental models may correspond to tangible or visible physical objects in ordinary life and empirical science. Or they may not correspond to any such things, but may be pure mental representations, as in much of contemporary analysis, algebra, and even geometry. By pure mental models I mean models not obtained directly by idealization of visual or other sensual experience.

But what controls these mental models? If they have no physical counterpart, what keeps them from being wildly idiosyncratic and incommunicable? What we have omitted up to this point, and what is the crux of the matter: mathematical images are not private, individual entities. From the origin of mathematics in bartering, buying and selling, or in building the Parthenon and the Pyramids, this subject has always been a social, an "inter-subjective" activity. Its advances and conquests have always been validated, corrected and absorbed in a social context-first of all, in the classroom. Mathematicians can and must talk to each other about their ideas. One way or the other, they do communicate, share and compare their conceptions of mathematical entities, which means precisely these models, these images and representations I have been describing. Discrepancies are recognized and worked out, either by correcting errors, reconciling differences, or splitting apart into different, independent pathways. Appropriate terminology and symbols are created as needed.

Mathematics depends on a mutually acknowledging group of competent practitioners, whose consensus decides at any time what is regarded as correct or incorrect, complete or incomplete. That is how it always worked, and that is how it works today. This was made very clear by the elaborate process in which Perelman's proposed proof of the Thurston program (including the Poincaré conjecture) was vetted, examined, discussed, criticized, and finally accepted by the "Ricci flow community", and then by its friends in the wider communities of differential geometry and low-dimensional topology, and then by the prize committees of the Fields Medal and the Clay Foundation.

Thus, when we speak of a mathematical concept, we speak not of a single isolated mental image, but rather of a family of mutually correcting mental images. They are privately owned, but publicly checked, examined, corrected, and accepted or rejected. This is the role of the mathematical research community, how it indoctrinates and certifies new members, how it reviews, accepts or rejects proposed publication, how it chooses directions of research to follow and develop, or to ignore and allow to die. All these social activities are based on a necessary condition: that the individual members have mental models that fit together, that yield the same answers to test questions. A new branch of mathematics is established when consensus is reached about the possible test questions and their answers. That collection of possible questions and answers (not necessarily explicit) becomes the means of accepting or rejecting proposed new members.

If two or three mathematicians do more than merely communicate about some mathematical topic, but actually collaborate to dig up new information and understanding about it, then the matching of their mental models must be even closer.

They may need to establish a congruence between their subconscious thinking about it as well as their conscious thinking. This can be manifested when they are working together, and one speaks the very thought that the partner was about to speak.

And to the question

"What is mathematics?"

the answer is

"It is socially validated reasoning about these mutually congruent mental models".

What makes mathematics possible? It is our ability to create mental models which are "precise", meaning simply that they are part of a shared family of mutually congruent models. In particular, such an image as a line segment, or two intersecting line segments, and so on. Or the image of a collection of mutually interchangeable identical objects (ideal coins or buttons). And so on. To understand better how that ability exists, both psychologically and neurophysiologically, is a worthy goal for empirical science. The current interactive flowering of developmental psychology, language acquisition, and cognitive neuroscience shows that this hope is not without substance. (See, e.g., [2, 6, 13, 14, 26, 27].)

The existence of mathematics shows that the human mind is capable of creating, refining, and sharing such precise concepts, which admit of reasoning that can be shared, mutually checked, and confirmed or rejected. There are great variations in the vividness, completeness, and connectivity of different mental images of the "same" mathematical entity as held by different mathematicians. And also great variations in their ability to concentrate on that image and squeeze out all of its hidden information. Recall that well-known mathematician, Sir Isaac Newton. When asked how he made his discoveries in mathematics and physics, he answered simply, "By keeping the problem constantly before my mind, until the light gradually dawns". Indeed, neither meals nor sleep were allowed to interrupt Newton's concentration on the problem. Mathematicians are notoriously absent-minded. Their concentration, which outsiders call "absent-mindedness", is just the open secret of mathematical success.

Their reasoning is qualitatively the same as the reasoning carried out by a hunter tracking a deer in the Appalachian woodland a thousand years ago. "If the deer went to the right, I would see a hoof print here. But I don't see it. There's only one other way he could have gone. So he must have gone to the left". Concrete deductive reasoning, which is the basis for abstract deductive reasoning.

To sum up! I have drawn a picture of mathematical reasoning which claims to make sense of intuition according to Hadamard and Poincaré, and which interprets Pólya's heuristic as a description of ordinary practical reason, applied to the abstract situations and problems of mathematics, working on mental models in the same way that ordinary practical reasoning in absentia works on a mental model. (We may assist our mental images by creating images on paper-drawing pictures-that to some extent capture crucial features of the mental images.)

7 Dewey and Pragmatism

Before bringing in John Dewey, the third name promised at the beginning, I must first mention Dewey's precursor in American pragmatism, Charles Sanders Peirce, for Peirce was also a precursor to Pólya. To deduction and induction, Peirce added a third logical operation, "abduction", something rather close to Pólya's "intelligent guessing".

The philosophy of mathematics as practiced in many articles and books is a thing unto itself, hardly connected either to living mathematics or to general philosophy. But how can it be claimed that the nature of mathematics is unrelated to the general question of human knowledge? There has to be a fit between your beliefs about mathematics and your beliefs about science and about the mind. I claim that Dewey's pragmatism offers the right philosophical context for the philosophy of mathematical practice to fit into. I am thinking especially of *Logic—the Theory of Inquiry*. For Dewey, "inquiry" is conceived very broadly and inclusively. It is "the controlled or directed transformation of an indeterminate situation into one that is so determinate in its constituent distinctions and relations as to convert the elements of the original situation into a unified whole". So broadly understood, inquiry is one of the primary attributes of our species. Only because of that trait have we survived, after we climbed down from the trees. I cannot help comparing Dewey's definition of inquiry with Pólya's definition of problem solving. It seems to me they are very much pointing in the same direction, taking us down the same track. With the conspicuous difference that, unlike Dewey, Pólya is concise and memorable.

Dewey makes a radical departure from standard traditional philosophy (following on from his predecessors Peirce and William James, and his contemporary George Herbert Mead). He does not throw away the concept of truth, but he gives up the criterion of truthfulness, as the judge of useful or productive thinking. Immanuel Kant made clear once and for all that while we may know the truth, we cannot know for certain that we do know it. We must perforce make the best of both demonstrative and plausible reasoning. This seems rather close to "warranted assertibility", as Dewey chooses to call it. But Pólya or Poincaré are merely talking about mathematical thinking, Dewey is talking about human life itself.

What about deductive thinking? From Dewey's perspective of "warranted assertibility", deductive proof is not a unique, isolated mode of knowledge. A hunter tracking a deer in the North American woodland a thousand years ago concluded, "So it must have gone to the left". Concrete deductive reasoning, the necessary basis of theoretical deductive reasoning. And it never brings certainty, simply because any particular deductive proof is a proof in practice, not in principle. Proof in practice is a human artifact, and so it can't help leaving some room for possible question, even possible error. (And that remains true of machine proof, whether by analog, digital, or quantum computer. What changes is the magnitude of the remaining possible error and doubt, which can never vanish finally.) In this way, we take our leave, once and for all, of the Platonic ideal of knowledge-indubitable and unchanging-in favor, one might say, of an Aristotelian view, a scientific and empirical one. And while deductive proof becomes human and not divine or infallible, non-deductive plausible reasoning and intuition receive their due as a source of knowledge in mathematics, just as in every other part of human life. Dewey's breadth of vision—seeing philosophy always in the context of experience, that is to say, of humanity at large—brings a pleasant breath of fresh air into this stuffy room.

Nicholas Rescher writes in [23]:

> The need for understanding, for "knowing one's way about", is one of the most fundamental demands of the human condition... Once the ball is set rolling, it keeps on going under its own momentum-far beyond the limits of strictly practical necessity... The discomfort of unknowing is a natural aspect of human sensibility. To be ignorant of what goes on about us is almost physically painful for us... The requirement for information, for cognitive orientation within our environment, is as pressing a human need as for food itself.

The need for understanding is often met by a story of some kind. In our scientific age, we expect a story built on a sophisticated experimental-theoretical methodology. In earlier times, no such methodology was available, and a story might be invented in terms of gods or spirits or ancestors. In inventing such explanations, whether in what we now call mythology or what we now call science, people have always been guided by a second fundamental drive or need. Rescher does not mention it, but Dewey does not leave it out. That is the need to impart form, beauty, appealing shape or symmetry to our creations, whether they be straw baskets, clay pots, wooden spears and shields, or geometrical figures and algebraic calculations. In *Art as Experience*, Dewey shows that the esthetic, the concern for pleasing form, for symmetry and balance, is also an inherent universal aspect of humanity. In mathematics, this is no less a universal factor than the problem-solving drive. In *Mathematical Discovery*, Poincaré takes great pains to emphasize the key role of esthetic preference in the development of mathematics. We prefer the attractive looking problems to work on, we strive for diagrams and graphs that are graceful and pleasing. Every mathematician who has talked about the nature of mathematics has portrayed it as above all an art form. So this is a second aspect of pragmatism that sheds light on mathematical practice.

Rescher's careful development omits mathematical knowledge and activity. And Dewey himself doesn't seem to have been deeply interested in the philosophy of mathematics, although there are interesting pages about mathematics in *Logic*, as well as in his earlier books *The Quest for Certainty* and *The Psychology of Number*. He may have been somewhat influenced by the prevalent view of philosophy of mathematics as an enclave of specialists, fenced off both from the rest of philosophy and from mathematics itself.

But if we take these pragmatist remarks of Rescher's seriously and compare them to what mathematicians do, we find a remarkably good fit. Just as people living in the woodland just naturally want to know and find out about all the stuff they see growing—what makes it grow, what makes it die, what you can do with it to make a canoe or a tent—so people who get into the world of numbers, or the world of triangles and circles, just naturally want to know how it all fits together, and how it can be stretched and pulled this way or that. "Guess and test", is the way George Pólya put it. "Proofs and refutations" was the phrase used by another mathematically trained Hungarian philosopher, following up an investigation started by Pólya. Whichever way you want to put it, it is nothing more or less than the exploration of the mathematical environment, which we create and expand as we explore it. We are manifesting in the conceptual realm one of the characteristic behaviors of *homo sapiens*.

Even though we lack claws or teeth to match beasts of prey, or fleetness to overtake the deer, or heavy fur or a thick shell, we long ago adapted to virtually every environment on Earth. We invented swimming, paddling and sailing, cooking and brewing and baking and preserving, and we expanded our social groups from families to clans to tribes to kingdoms to empires. All this by "inquiry", or by problem-solving. Dewey shows that this inquiry is an innate specific drive or need of our species. It was manifested when, motivated by practical concerns, we invented counting and the drawing of triangles. That same drive, to find projects, puzzles, and directions for growth, to make distinctions and connections, and then again make new distinctions and new connections, has resulted in the Empire of mathematics we inhabit today.

Acknowledgements In this work, I benefited from suggestions and criticisms by Carlo Cellucci, Richard Epstein, Russell Goodman, Cleve Moler, Peter Lax, Ulf Persson, Vera John-Steiner, and members of the study group on mathematical thinking in Santa Fe, New Mexico.

Appendix

Ap Dijksterhuis and Teun Meurs, Where creativity resides: the generative power of unconscious thought, *Social Psychology Program*, University of Amsterdam, Roetersstraat 15, 1018 WB Amsterdam, The Netherlands, 2004, 2005.

Abstract In three experiments, the relation between different modes of thought and the generation of "creative" and original ideas was investigated. Participants were asked to generate items according to a specific instruction (e.g., generate place names starting with an "A"). They either did so immediately after receiving the instruction, or after a few minutes of conscious thought, or after a few minutes of distraction during which "unconscious thought" was hypothesized to take place. Throughout the experiments, the items participants listed under "unconscious thought" conditions were more original. It was concluded that whereas conscious thought may be focused and convergent, unconscious thought may be more associative and divergent.

References

1. Baez, J.: http://math.ucr.edu/home/baez/week285.html (2010)

2. Carey, S.: The Origin of Concepts. Oxford University Press, New York (2009)
3. Cellucci, C.: Introduction to *Filosofia e matematica*. In: Hersh, R. (ed.) 18 Unconventional Essays on the Nature of Mathematics. Springer, New York (2006)
4. Davies, E.B.: Whither Mathematics? Not. Am. Math. Soc. **52**, 1350–1356 (2004)
5. Davis, P.: The paradox of the irrelevant beginning. Unpublished manuscript (2007)
6. Dehaene, S.: The Number Sense. Oxford University Press, New York (1997)
7. Dewey, J.: The Quest for Certainty. Minton, Balch and Company, New York (1929)
8. Dewey, J.: Art as Experience. G. P. Putnam, New York (1934)
9. Dewey, J.: Logic—The Theory of Inquiry. Henry Holt, New York (1938)
10. Gödel, K.: What is Cantor's continuum problem? Am. Math. Mon. **54**, 515–525 (1947)
11. Grosholz, E.R.: Representation and Productive Ambiguity in Mathematics and the Sciences. Oxford University Press, New York (2007)
12. Hadamard, J.: An Essay on the Psychology of Invention in the Mathematical Field. Princeton University Press, Princeton (1949)
13. Johnson-Laird, P.N.: Mental Models: Towards a Cognitive Science of Language, Inference, and Consciousness. Harvard University Press, Cambridge (1983)
14. Lakoff, G., Nunez, R.: Where Mathematics Comes From: How the Mind Brings Mathematics Into Being. Basic Books, New York (2000)
15. McLellan, J.A., Dewey, J.: The Psychology of Number and Its Applications to Methods of Teaching Arithmetic. D. Appleton and Company, New York (1897)
16. Poincaré, H.: Science and Method. Dover Publications, New York (1952)
17. Polanyi, M.: The Tacit Dimension. Doubleday and Company, Gloucester (1983)
18. Pólya, G.: How To Solve It. Princeton University Press, Princeton (1945)
19. Pólya, G.: Mathematics and Plausible Reasoning. Princeton University Press, Princeton (1954)
20. Pólya, G.: Mathematical Discovery. Wiley, New York (1980)
21. Pólya, G., Szegö, G.: Aufgaben und Lehrsatze aus der Analysis. Springer, Berlin (1970)
22. Ratner, S.: John Dewey, empiricism and experimentalism in the recent philosophy of mathematics. J. Hist. Idcas **53**(3), 467–479 (1992)
23. Rescher, N.: Cognitive Pragmatism. University of Pittsburgh Press, Pittsburgh (2001)
24. Rodin, A.: How mathematical concepts get their bodies. Topoi **29**, 53–60 (2010)
25. Steiner, M.: Mathematical Knowledge. Cornell University Press, Ithaca (1975)
26. Zwaan, R.A.: Aspects of Literary Comprehension: A Cognitive Approach. J. Benjamins Publishing Company, Amsterdam (1993)
27. Zwaan, R.A.: Grounding Cognition: The Role of Perception and Action in Memory, Language and Thinking. Cambridge University Press, New York (2005)

Three-Dimensional Plasma Arc Simulation Using Resistive MHD

Rolf Jeltsch and Harish Kumar

Abstract We propose a model for simulating the real gas, high current plasma arc in three dimensions based on the equations of resistive MHD. These model equations are discretized using Runge–Kutta Discontinuous Galerkin (RKDG) methods. The Nektar code is used for the simulation which is extended to include Runge–Kutta time stepping, accurate Riemann solvers, and real gas data. The model is then shown to be suitable for simulating a plasma arc by using it to generate a high current plasma arc. Furthermore, the model is used to investigate the effects of the external magnetic field on the arc. In particular, it is shown that the external magnetic field forces the plasma arc to rotate.

Keywords Plasma arc · Resistive Magnetohydrodynamics (MHD) · Runge–Kutta Discontinuous Galerkin (RKDG) methods

1 Introduction

A circuit breaker is an electrical switch designed to protect electrical circuits from the damage that can be caused by high fault current or voltage fluctuations. Once a circuit breaker detects a fault, contacts within the circuit breaker open to interrupt the circuit. When the fault current is interrupted, a plasma arc is generated. This arc must be cooled, and extinguished in a controlled way, to protect connected circuits and the device itself. Hence, plasma arcs provide a safe way of diffusing the energy of fault current. Consequently, the study of the arc behavior is of great importance to the power industry.

Many physical phenomena occur during interruption of fault current in the circuit breaker, e.g., movement of contacts, pressure build up, radiative transfer, convection, heat conduction, melting of contact material, magnetic and electric effects.

R. Jeltsch (✉)
ETH Zürich, Zurich, Switzerland
e-mail: rolf.jeltsch@math.ethz.ch

H. Kumar
INRIA, Bordeaux, France
e-mail: harish.kumar@inria.fr

C.A. de Moura, C.S. Kubrusly (eds.), *The Courant–Friedrichs–Lewy (CFL) Condition*,
DOI 10.1007/978-0-8176-8394-8_3, © Springer Science+Business Media New York 2013

Due to the presence of these wide ranging phenomena, simulation of plasma arc is a difficult task. To overcome these difficulties, extensive approximations related to the geometry, description of arc movements, and the influence of magnetic fields on the plasma arc are made. Several authors propose models for the simulation of the plasma arc. In [1], authors present a three-dimensional model for arc simulations at 100 A current. In [2], effects of the external magnetic fields and the gas materials on a three-dimensional high current arc is simulated. However, the position of the arc root stays the same during temporal evolution and an external magnetic field is imposed, not calculated. In [3] and [4], the external magnetic field is calculated using Biot–Savart law, and the arc root is not fixed.

The mathematical models proposed in [1–4] are based on Navier–Stokes equations for fluid flow and Maxwell's equations for the electromagnetism which are solved simultaneously. They are coupled by adding the source terms in momentum balance due to Lorentz force and Joule heating in energy balance equation. These models, although suitable for small magnetic Reynolds number simulations, are highly unstable for large magnetic Reynolds number simulations.

In this work, we are interested in developing a mathematical model for a plasma arc with very high currents (100–200 kA). At these high currents, very high temperatures are expected. This gives rise to a high magnetic Reynolds number (in particular, close to the contacts). Consequently, we consider a model based on the equation of resistive magnetohydrodynamics (MHD). We believe that this is the first time a model based on the resistive MHD has been used to simulate a plasma arc in three dimensions (see [5–7]).

The equations of resistive magnetohydrodynamics (MHD) govern the evolution of a quasi-neutral, conductive fluid and the magnetic field within it, neglecting the magnetization of individual particles, the Hall current, and the time rate of change of the electric field in Maxwell's equations. The complete details about these equations can be found in [8]. Numerical discretization of these equations is a complicated task due to the presence of nonlinearities in the convection flux. In addition to these difficulties, for the plasma arc simulations we need to consider a complicated geometry, real gas data for physical parameters, and mixed boundary conditions.

We use Runge–Kutta Discontinuous Galerkin (RKDG) methods for the discretization of MHD equations. Discontinuous Galerkin (DG) methods were first introduced by Hill and Reed in [9] for the neutron transport equations. These methods were then generalized for systems of hyperbolic conservation laws by Cockburn, Shu and co-workers (see [10]). In DG methods, the solution in space is approximated using piecewise polynomials on each element. Exact or approximate Riemann solvers from finite volume methods are used to compute the numerical fluxes between elements. Due to the assumed discontinuity of the solution at element interfaces, DG methods can easily handle adaptive strategies and can be easily parallelized.

To simulate a plasma arc in a circuit breaker, we proceed as follows:

1. First, we assume that the domain is filled with hot gas. An arc is imposed between the contacts by specifying appropriate initial and boundary conditions.

2. This initial arc is then evolved till a steady state is reached. The principle idea is that with time, gas will radiate, which will result in temperature reduction everywhere except where gas is heated by the current in the arc. The resulting solution is now considered as an actual arc.
3. We then apply the external magnetic field by suitably modifying the magnetic field and the boundary conditions.
4. We show that using an appropriate external magnetic field it is possible to manipulate the arc. In particular, we show that the external magnetic field can be used to force the arc to rotate.

The article is organized as follows: In Section 2, we present the model equations of resistive MHD in non-dimensional variables. In Section 3, RKDG methods for resistive MHD equations in three dimensions are described. We present the variational formulation using a model equation. We then describe the three-dimensional basis functions for different types of elements. In Section 4, we first present initial and boundary conditions for arc generations and discuss the simulation results. We then investigate the effect of external magnetic field on the arc.

2 Equations of Resistive MHD

For non-dimensional conservative variables, the resistive MHD equations are

$$\frac{\partial \rho}{\partial t} + \nabla \cdot (\rho \mathbf{v}) = 0, \tag{1a}$$

$$\frac{\partial (\rho \mathbf{v})}{\partial t} + \nabla \cdot \left(\rho \mathbf{v}\mathbf{v} - \mathbf{B}\mathbf{B} + \left(p + \frac{1}{2}|\mathbf{B}|^2 \right)\mathbf{I} - \frac{1}{\mathrm{Re}}\Pi \right) = 0, \tag{1b}$$

$$\frac{\partial \mathbf{B}}{\partial t} + \nabla \times \left(\mathbf{v} \times \mathbf{B} + \frac{1}{S_r}(\nabla \times \mathbf{B}) \right) = 0, \tag{1c}$$

$$\frac{\partial E}{\partial t} + \nabla \cdot \left((E + p)\mathbf{v} + \left(\frac{1}{2}|\mathbf{B}|^2\mathbf{I} - \mathbf{B}\mathbf{B} \right) \cdot \mathbf{v} \right.$$
$$\left. - \frac{1}{\mathrm{Re}}\Pi \cdot \mathbf{v} + \frac{1}{S_r}\left(\mathbf{B} \cdot \nabla \mathbf{B} - \nabla\left(\frac{1}{2}|\mathbf{B}|^2 \right) \right) - \frac{1}{G_r}\nabla T \right) = ST^4 \tag{1d}$$

$$\nabla \cdot \mathbf{B} = 0, \tag{1e}$$

with the equation of state for energy

$$E = \frac{p}{\gamma - 1} + \frac{1}{2}\rho|\mathbf{v}|^2 + \frac{1}{2}|\mathbf{B}|^2, \tag{2}$$

and the stress tensor

$$\Pi = \nu\left(\nabla \mathbf{v} + (\nabla \mathbf{v})^\top \right) - \nu\frac{2}{3}(\nabla \cdot \mathbf{v})\mathbf{I}. \tag{3}$$

Here ρ is the density, $\mathbf{v}$ is the velocity, p is the pressure, $\mathbf{B}$ is the magnetic field, E is the total energy, and T is the temperature of the plasma. Equation (1a) is the equation for the mass conservation. Equations (1b)–(1d) are equations of balance law for the momentum, the magnetic field, and the total energy, respectively. Equation (1e) is the divergence free condition for the magnetic field, representing non-existence of magnetic monopoles.

The non-dimensionalization was carried out using the reference length L_0, the reference pressure P_0, and the reference temperature T_0. Using these parameters, we use gas data to calculate the reference density ρ_0 at temperature T_0 and pressure P_0. Furthermore, the reference velocity is calculated using $V_0 = \sqrt{P_0/\rho_0}$ and the reference magnetic field is calculated using $B_0 = \sqrt{P_0 \mu_0}$, where μ_0 is magnetic permeability. The non-dimensional parameters appearing in the above equations are Reynolds number $\mathrm{Re} = \frac{\rho_0 V_0 L_0}{\nu}$, Lundquist number $S_r = \frac{\mu_0 V_0 L_0}{\eta}$, Prandlt number $G_r = \frac{\rho_0 V_0 L_0 R_0}{\kappa}$, and scaled Stefan's radiation constant $S = \frac{s L_0 T_0^4}{V_0 P_0}$. Here ν is viscosity of the fluid, η $(= 1/\sigma)$ is the resistivity of fluid (σ is the conductivity of fluid), κ is the heat diffusion constant, R_0 is the gas constant at temperature T_0, and γ is the ratio of specific heats. In general, all these quantities depend on the pressure and temperature. However, we ignore their dependence on pressure. This is due to the negligible variation in these values due to the pressure change when compared to the variation due to the temperature change. Also, s is Stefan's radiation constant.

3 RKDG Methods for Resistive MHD

In this section, we present spatial and temporal discretization of the MHD equations (1a)–(1e). The spatial discretization is based on DG methods. Note that it is enough to consider DG methods for the scalar advection–diffusion equation,

$$\frac{\partial u}{\partial t} + \sum_{1 \le i \le n} \frac{\partial}{\partial x_i} \left(f_i(u) - \sum_{i \le j \le n} a_{ij} \frac{\partial}{\partial x_j} u \right) = 0, \tag{4}$$

as we can apply similar spatial discretization to each component of (1a)–(1e). In (4), f_i is convection flux, a_{ij} are diffusion coefficients with the condition that the matrix $(a_{ij})_{ij}$ is symmetric and semi-positive definite, so there exists a symmetric matrix (b_{ij}) such that

$$a_{ij} = \sum_{1 \le l \le d} b_{il} b_{lj}.$$

3.1 Variational Formulation

Following [10], we introduce an auxiliary variable $q_l = \sum_{1\le j\le n} b_{lj} \frac{\partial u}{\partial x_j}$ and rewrite (4) as

$$\frac{\partial u}{\partial t} + \sum_{1\le i\le n} \frac{\partial}{\partial x_i}\left(f_i(u) - \sum_{i\le l\le n} b_{il} q_l\right) = 0, \tag{5a}$$

$$q_l - \sum_{1\le j\le n} \frac{\partial g_{lj}}{\partial x_j} = 0, \quad \text{for } l = 1, \ldots, n, \tag{5b}$$

where $g_{lj} = \int_0^u b_{lj}\, ds$. We set $w = (u, q_1, q_2, \ldots, q_n^\top)$, and introduce the flux,

$$\mathbf{h}_i(w) = \left(f_i(u) - \textstyle\sum_{1\le l\le n} a_{il} q_l, -g_{1i}, \ldots, -g_{ni}\right)^\top. \tag{6}$$

Multiplying with a test function and integrating by parts results in

$$\int_K \frac{\partial u}{\partial t} v_u\, dx - \sum_{1\le i\le n} \int_K h_{iu} \frac{\partial}{\partial x_i} v_u\, dx + \int_{\partial K} \hat{h}_u(w, \mathbf{n}) v_h\, dx = 0, \tag{7a}$$

$$\int_K q_l v_{q_l}\, dx - \sum_{1\le j\le n} \int_K h_{jq_l} \frac{\partial}{\partial x_j} v_{q_l}\, dx + \int_{\partial K} \hat{h}_{q_l}(w, \mathbf{n}) v_h\, dx = 0. \tag{7b}$$

This is the variational formulation which we need to approximate. The flux $\hat{\mathbf{h}}(w, \mathbf{n})$ is divided into two parts,

$$\hat{\mathbf{h}} = \hat{\mathbf{h}}_{\text{conv}} + \hat{\mathbf{h}}_{\text{diff}},$$

where convective flux is given by

$$\hat{\mathbf{h}}_{\text{conv}}\left(w^-, w^+, \mathbf{n}\right) = \left(\hat{f}\left(u^+, u^-, \mathbf{n}\right), 0\right)^\top.$$

Here $\hat{f}$ is calculated using exact or approximated Riemann solvers. In these simulations, we use local Lax–Friedrichs numerical flux given by

$$\mathbf{f}_{LF}\left(u^-, u^+\right) = \frac{1}{2}\left(f\left(u^-\right) + f\left(u^+\right)\right) - \frac{\max_i(\max(|\lambda_i(u^-)|, |\lambda_i(u^+)|))}{2}\left(u^+ - u^-\right), \tag{8}$$

where λ_i are the eigenvalues of the Jacobian of the MHD convection flux f. We use Bassi–Rebay flux (see [11]) to approximate the diffusion flux $\hat{\mathbf{h}}_{\text{diff}}$, i.e., the averages of diffusion fluxes across the interface.

3.2 Three-Dimensional Basis Functions

The RKDG method we use is implemented in the Nektar code, developed by Karniadakis et al. (see [12–14]). The original code has been extended to include Runge–

Table 1 Local collapsed coordinates for three-dimensional elements

Element type	Upper limits	Local collapsed coordinates		
Hexahedron	$-1 \leq \xi_1, \xi_2, \xi_3 \leq 1$	ξ_1	ξ_2	ξ_3
Prism	$\xi_1 < 1, \xi_2 + \xi_3 \leq 0$ with $-1 \leq \xi_1, \xi_2, \xi_3 \leq 1$	$\bar{\eta}_1 = \frac{2(1+\xi_1)}{1-\xi_2} - 1$	ξ_2	ξ_3
Pyramid	$\xi_1 + \xi_3, \xi_2 + \xi_3 \leq 0$ with $-1 \leq \xi_1, \xi_2, \xi_3 \leq 1$	$\bar{\eta}_1 = \frac{2(1+\xi_1)}{1-\xi_2} - 1$	$\eta_2 = \frac{2(1+\xi_2)}{1-\xi_2} - 1$	$\eta_3 = \xi_3$
Tetrahedron	$\xi_1 + \xi_2 + \xi_3 \leq -1$ with $-1 \leq \xi_1, \xi_2, \xi_3 \leq 1$	$\eta_1 = \frac{2(1+\xi_1)}{-\xi_2-\xi_3} - 1$	$\eta_2 = \frac{2(1+\xi_2)}{1-\xi_2} - 1$	$\eta_3 = \xi_3$

Kutta time stepping, slope limiters, and accurate Riemann solvers, among other features (see [15]). In the DG discretization, functions are approximated by using basis functions

$$f = \sum_i a_i \phi_i, \tag{9}$$

where the basis functions ϕ_i's are simple functions, e.g., polynomials. These functions are chosen in such a way that the whole algorithm is computationally efficient. The set of polynomial basis functions used in **Nektar** were proposed by Dubiner in [16] for two dimensions and were extended to three dimensions in [12]. They are based on the tensor product of one-dimensional basis functions which are derived using Jacobi polynomials. Here we describe three-dimensional basis functions.

The one-dimensional basis functions are defined on bounded intervals; therefore, an implicit assumption on the tensor product basis functions for higher dimension is that coordinates in two and three-dimensional regions are bounded by constant limits. But in two or three dimensions, that is not true in general, e.g., for a triangle. To overcome this difficulty, we define a collapsed coordinate system for three dimensions which maps elements without this property (Tetrahedral) to the element (Hexahedral) bounded by constant limits. These coordinates for various types of elements are given in Table 1.

Under these transformed coordinates, three-dimensional elements are bounded by constant limits. For example, a tetrahedron $\hat{T}_3$ which, in Cartesian coordinates, is given by

$$\hat{T}_3 = \{-1 \leq \xi_1, \xi_2, \xi_3 \leq 1 \text{ such that } \xi_1 + \xi_2 + \xi_3 \leq -1\},$$

is transformed to

$$\hat{T}_3 = \{-1 \leq \eta_1, \eta_2, \eta_3 \leq 1\}$$

Table 2 Basis functions for three-dimensional elements

Hexahedron Basis	$\phi_{pqr}(\xi_1,\xi_2,\xi_3)=\psi_p^a(\xi_1)\psi_q^a(\xi_2)\psi_r^a(\xi_3)$
Prism Basis	$\phi_{pqr}(\xi_1,\xi_2,\xi_3)=\psi_p^a(\bar{\eta}_1)\psi_q^a(\xi_2)\psi_{pr}^b(\xi_3)$
Pyramid Basis	$\phi_{pqr}(\xi_1,\xi_2,\xi_3)=\psi_p^a(\bar{\eta}_1)\psi_q^a(\eta_2)\psi_{pqr}^c(\eta_3)$
Tetrahedron Basis	$\phi_{pqr}(\xi_1,\xi_2,\xi_3)=\psi_p^a(\eta_1)\psi_{pq}^b(\eta_2)\psi_{pqr}^c(\eta_3)$

in local collapsed coordinates. To define three-dimensional basis functions, we first define functions

$$\psi_p^a(z)=P_p^{0,0}(z),\qquad \psi_{pq}^b(z)=\left(\frac{1-z}{2}\right)^p P_q^{2p+1,0}(z), \tag{10}$$

$$\psi_{pqr}^c(z)=\left(\frac{1-z}{2}\right)^{p+q} P_r^{2p+2q+2,0}(z), \tag{11}$$

where $P_n^{\alpha,\beta}$ is the nth-order Jacobi polynomial with weights α and β. Then using local collapsed coordinates, the three-dimensional basis functions for various elements are given in Table 2.

These basis functions are orthogonal under Legendre inner product over each element, resulting in a diagonal mass matrix. The functions are polynomials in both the Cartesian and non-Cartesian coordinates. It was proved in [17] that the coefficients of the basis functions for a solution decay exponentially with polynomial order, thus the numerical solution converges exponentially as the maximum polynomial order of the approximation is increased.

3.3 Time Stepping

To advance solutions in time, the RKDG method uses a Runge–Kutta (RK) time marching scheme. Here we present the second-, third-, and fourth-order accurate RKDG schemes. For second- and third-order simulations, we present the TVD RK schemes of Shu (see [18]). For the fourth-order simulations we use the classic scheme. Consider the semi-discrete ODE

$$\frac{du_h}{dt}=L_h(u_h).$$

Let u_h^n be the discrete solution at time t^n, and let $\Delta t^n=t^{n+1}-t^n$. In order to advance a numerical solution from time t^n to t^{n+1}, the RK algorithm runs as follows:

1. Set $u_h^{(0)}=u_h^n$.
2. For $i=1,\dots,k+1$, compute,

$$u_h^{(i)}=\sum_{l=0}^{i-1}\alpha_{il}u_h^{(l)}+\beta_{il}\Delta t^n L_h\big(u_h^{(l)}\big).$$

3. Set $u_h^{n+1}=u_h^{(k+1)}$.

Table 3 Parameters for Runge–Kutta time marching schemes

order	α_{il}	β_{il}
2	1	1
	1/2 1/2	0 1/2
3	1	1
	3/4 1/4	0 1/4
	1/3 0 2/3	0 0 2/3

The values of the coefficients used are shown in Table 3. For the linear advection equation, it was proved by Cockburn et al. in [19] that the RKDG method is L^∞-stable for piecewise linear ($k = 1$) approximate solutions if a second-order RK scheme is used with a time-step satisfying

$$c\frac{\Delta t}{\Delta x} \leq \frac{1}{3},$$

where c is the constant advection speed. The numerical experiments in [10] show that when approximate solutions of polynomial degree k are used, an order $k + 1$ RK scheme must be used, which simply corresponds to matching the temporal and spatial accuracy of the RKDG scheme. In this case, the L^∞-stability condition is

$$c\frac{\Delta t}{\Delta x} \leq \frac{1}{2k+1}.$$

For the nonlinear case, the same stability conditions are used but with c replaced by the maximum eigenvalue of the system.

4 Three-Dimensional Arc Simulations

To simulate the plasma arc, the Nektar code has been modified to implement real gas data for following physical parameters: electric conductivity, fluid viscosity, specific heats, gas constant, and thermal conductivity. The gas used in circuit breakers is SF_6. The real gas data is implemented by approximating it at a pressure of 10^6 Pa with piecewise smooth functions (see [6, 7]). An example of this is given in Fig. 1, where we have plotted the approximated electrical conductivity w.r.t. temperature. Note that the dependence of the gas data on temperature introduces further stiffness in the equations. All the results presented here are of the first order accuracy.

The domain for simulation is illustrated in Fig. 2. In Fig. 2(left), we have the three-dimensional domain for the computation which is the arc chamber of the circuit breaker. Figure 2(right) shows the XY plane cut of the three-dimensional geometry. The domain is axial symmetric along the y-axis. The radius of domain is 70 mm and length (y-axis) is 200 mm long. We assume that we have an arc attached to both electrodes which are 10 mm wide.

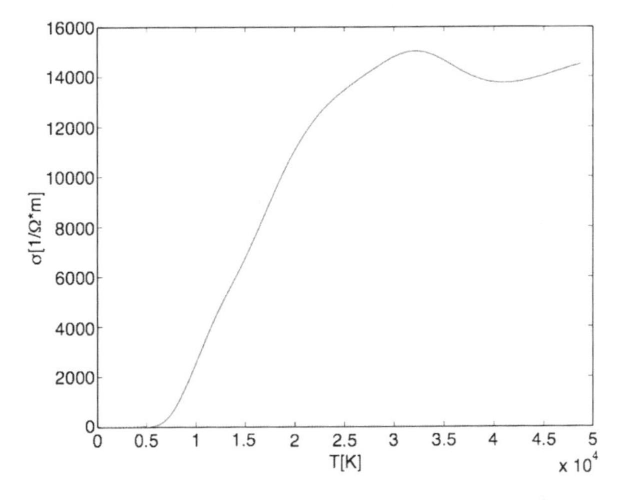

Fig. 1 Conductivity of the SF_6 gas at pressure $P = 10^6$ Pa

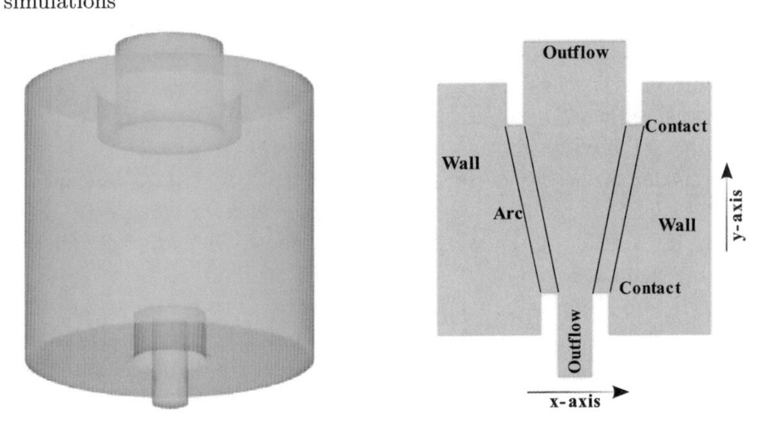

Fig. 2 Geometry of the Arc chamber

In a circuit breaker with a rotating arc, the current that flows inside the arc also goes through a coil located around the arc chamber. This process induces an external magnetic field in the y-direction. This external magnetic field interacts with the arc through the Lorentz force term in the momentum conservation equation. Observe that, in the design of arc chamber, the contacts at the arc root have different radii, which guarantees that the current in the resulting arc will not be parallel to the y-axis. Consequently, the Lorentz force term $\mathbf{J} \times \mathbf{B}$ will be nonzero.

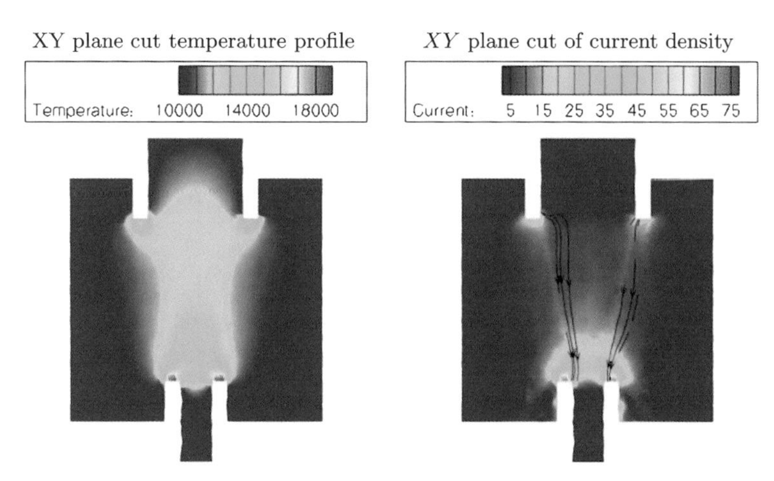

Fig. 3 Temperature and current density at time $t = 0.569$ ms

4.1 Arc Generation

Initially, we assume that the domain (see Fig. 2) is filled with SF_6 gas at the temperature of 20000 K and the pressure of 10^6 Pa. At these values of pressure and temperature, the density of the SF_6 gas is $0.0829\ \text{kg/m}^3$. The flow is considered to be steady initially, i.e., $\mathbf{v} = 0.0$ m/s. The magnetic field components B_x and B_z are computed using *Biot–Savart Law* and correspond to the total current of $I = 100$ kA in the initial arc of width 10 mm (see [6, 7]) joining both contacts.

We consider the reference length of $L_0 = 10^{-3}$ m. The reference pressure is $P_0 = 10^6$ Pa, and the reference temperature is $T_0 = 5000$ K. Using the gas data, we have the reference density $\rho_0 = 0.506\ \text{kg/m}^3$. The wall boundary conditions for the wall are the same as in the previous chapter. The wall temperature is $T = 10000$ K except at the arc roots where we put $T = 20000$ K. The wall boundary conditions are implemented for velocity by inverting the normal component of the velocity at the wall. The magnetic field conditions for the wall are implemented by assuming the condition of no current.

Using these reference variables and assuming that the minimum conductivity is $\sigma_{\text{min}} = 6000$, we would have a *Lundquist* number $S_r = \mu_0 V_0 L_0 \sigma_{\text{min}} = 1.06 \times 10^{-2}$. This value would give rise to an extremely stiff system and this, in turn, would make the computational time unreasonably large. We scale this by a factor of 1000. Similarly, we scale G_r by a factor of 20. We do realize that this can effect results quantitatively, but we believe that qualitatively the results still hold. We use 101044 tetrahedron elements in our computations. Computational time is 24 hours with 64 processors. At time $t = 0.569$ ms we have the following results:

Figure 3(left) is the temperature profile of the arc in the XY plane. We observe that most of the heating takes place at the center of the domain. Figure 3(right) is the current density profile of the arc in the XY plane with the current lines and the current moving downward. The current lines have moved towards the center of the

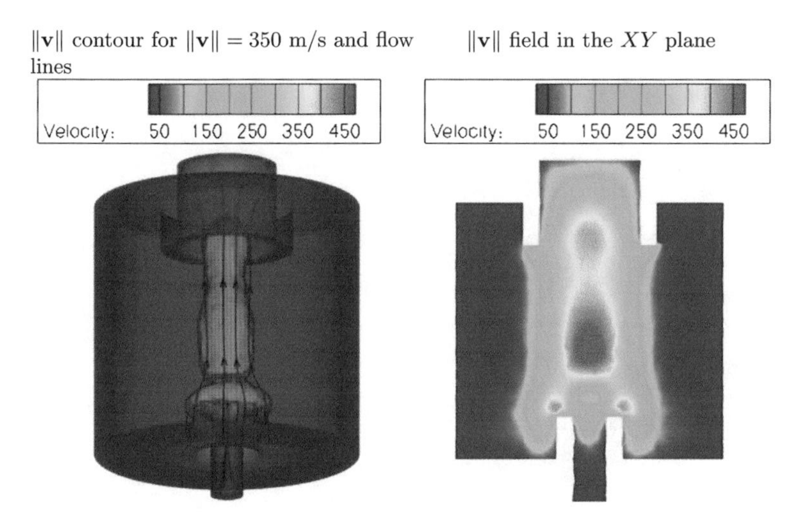

Fig. 4 $\|\mathbf{v}\|$ field of the arc at $t = 0.569$ ms

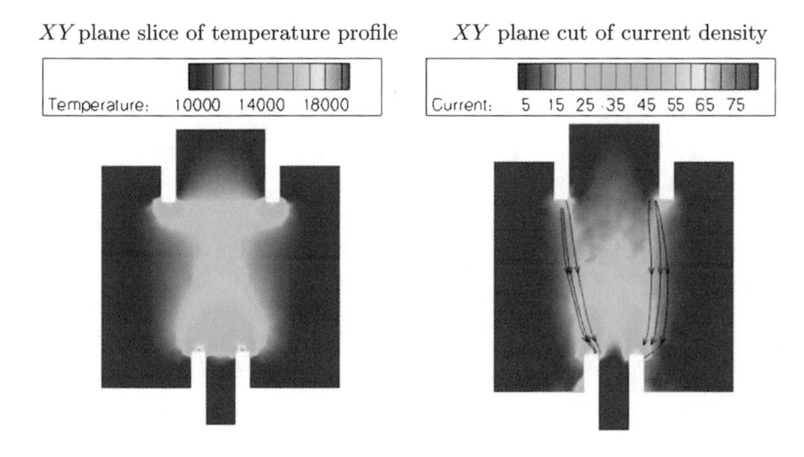

Fig. 5 Temperature with external magnetic field after time $t = 1.138$ ms

domain from its initial position due to higher temperature. Figure 4 is the profile of the velocity field. We also note that the gas is pushed away from the arc, through the outflow boundaries. There is also a bifurcation in velocity flow lines near the lower end of the domain.

4.2 Effects of External Magnetic Fields

The external magnetic field of $B_y = 0.5$ T is applied by adding it to the arc's magnetic field and then modifying the boundary conditions with the magnetic field

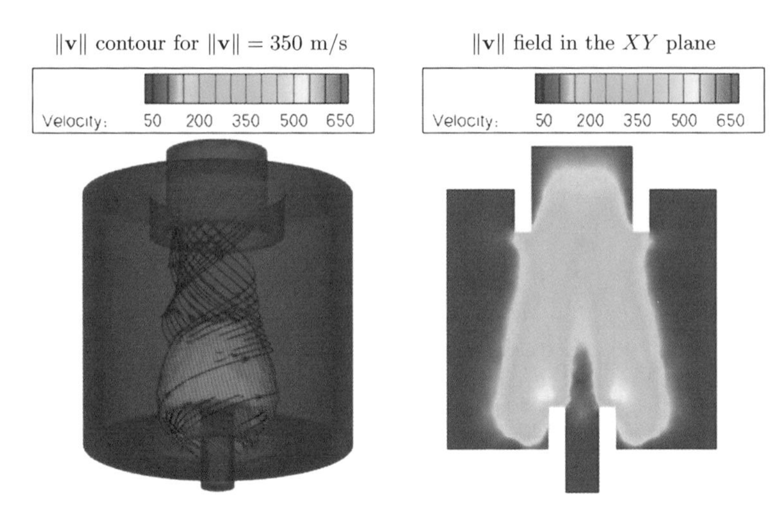

Fig. 6 $\|\mathbf{v}\|$ field with external magnetic field after time $t = 1.138$ ms

$B_y = 0.5$ T. Note that during the simulation the y-component B_y of the magnetic field is also simulated. The computational time was another 24 hours on 64 processors. After further $t = 0.569$ ms, we obtain the following results:

Figure 5(left) illustrates the temperature profile of the arc. We observe that the temperature is comparatively less than what it was before. Figure 5(right) represents the new current density profile. When compared with Fig. 3(right), we observe that there is a change in the shape of the current density close to the lower contact. The most important result is shown in Fig. 6. The streamlines of the velocity field show that the arc is rotating. In fact, the velocity profile is completely changed when compared with Fig. 4. Also, note the significant jump in the absolute value of the velocity. Without the external magnetic field, the maximum absolute velocity was 490 m/s, compared to 689 m/s with the external magnetic field. Furthermore, the maximum velocity is at the arc roots, instead of the center.

5 Conclusion

We show the suitability of the equations of resistive MHD for three-dimensional computations of the plasma arc in high current circuit breakers. These equations are used to generate the arc for a total current of 100 kA. We then apply the external magnetic field and use it to generate a rotation in the arc and observe a significant increase in the velocity. This can be used to minimize the operating energy of the circuit breaker. One of the major obstacles in simulating the real gas arc is the stiffness due to the low values of the conductivity. A possible solution for this can be the use of implicit time stepping for the simulations.

Acknowledgements The authors would like to acknowledge C. Schwab, V. Wheatley, M. Torrilhon, and R. Hiptmair for their support and constructive discussions on this work. G.E. Karniadakis provided the authors with the original version of Nektar and ABB Baden provided real gas data for SF_6 gas, which is gratefully acknowledged.

References

1. Schlitz, L.Z., Garimella, S.V., Chan, S.H.: Gas dynamics and electromagnetic processes in high-current arc plasmas. Part I: model formulation and steady state solutions. J. Appl. Phys. **85**(5), 2540–2546 (1999)
2. Schlitz, L.Z., Garimella, S.V., Chan, S.H.: Gas dynamics and electromagnetic processes in high-current arc plasmas. Part II: effects of external magnetic fields and gassing materials. J. Appl. Phys. **85**(5), 2547–2555 (1999)
3. Lindmayer, M.: Simulation of switching devices based on general transport equation. In: Int. Conference on Electrical Contacts, Zürich (2002)
4. Barcikowski, F., Lindmayer, M.: Simulations of the heat balance in low-voltage switchgear. In: Int. Conference on Electrical Contacts, Stockholm (2000)
5. Huguenot, P., Kumar, H., Wheatley, V., Jeltsch, R., Schwab, C.: Numerical simulations of high current arc in circuit breakers. In: 24th International Conference on Electrical Contacts (ICEC), Saint-Malo, France (2008)
6. Huguenot, P.: Axisymmetric high current arc simulations in generator circuit breakers based on real gas magnetohydrodynamics models. Diss., Eidgenössische Technische Hochschule, ETH, Zürich, No. 17625 (2008)
7. Kumar, H.: Three dimensional high current arc simulations for circuit breakers using real gas resistive magnetohydrodynamics. Diss., Eidgenössische Technische Hochschule, ETH, Zürich, No. 18460 (2009)
8. Goedbloed, H., Poedts, S.: Principles of Magnetohydrodynamics. Cambridge University Press, Cambridge (2004)
9. Hill, T.R., Reed, W.H.: Triangular mesh methods for neutron transport equation. Tech. Rep. LA-UR-73-479, Los Alamos Scientific Laboratory (1973)
10. Cockburn, B.: Advanced numerical approximation of nonlinear hyperbolic equations. In: An Introduction to the Discontinuous Galerkin Method for Convection-Dominated Problems. Lecture Notes in Mathematics, pp. 151–268. Springer, Berlin (1998)
11. Bassi, F., Rebay, S.: A high-order accurate discontinuous finite element method for the numerical solution of the compressible Navier–Stokes equations. J. Comput. Phys. **131**, 267–279 (1997)
12. Sherwin, S.J., Karniadakis, G.E.: A new triangular and tetrahedral basis for high-order(hp) finite element methods. Int. J. Numer. Methods Eng. **123**, 3775–3802 (1995)
13. Karniadakis, G.E., Sherwin, S.J.: Spectral/hp Element Methods for Computational Fluid Dynamics. Oxford University Press, Oxford (2005)
14. Lin, G., Karniadakis, G.E.: A discontinuous Galerkin method for two-temperature plasmas. Comput. Methods Appl. Mech. Eng. **195**, 3504–3527 (2006)
15. Wheatley, V., Kumar, H., Huguenot, P.: On the role of Riemann solvers in discontinuous Galerkin methods for magnetohydrodynamics. J. Comput. Phys. **229**, 660–680 (2010)
16. Dubiner, M.: Spectral methods on triangles and other domains. J. Sci. Comput. **6**, 345–390 (1991)
17. Warburton, T.C., Karniadakis, G.E.: A discontinuous Galerkin method for the viscous MHD equations. J. Comput. Phys. **152**, 608–641 (1999)
18. Shu, C.W.: TVD time discretizations. SIAM J. Math. Anal. **14**, 1073–1084 (1988)
19. Cockburn, B., Shu, C.W.: The Runge–Kutta local projection p^1-discontinuous Galerkin method for scalar conservation laws. Math. Model. Numer. Anal. **25**, 337–361 (1991)

Space-Time Hybridizable Discontinuous Galerkin Method for the Advection–Diffusion Equation on Moving and Deforming Meshes

Sander Rhebergen and Bernardo Cockburn

Abstract We present the first space-time hybridizable discontinuous Galerkin finite element method for the advection–diffusion equation. Space-time discontinuous Galerkin methods have been proven to be very well suited for moving and deforming meshes which automatically satisfy the so-called Geometric Conservation law, for being able to provide higher-order accurate approximations in both time and space by simply increasing the degree of the polynomials used for the space-time finite elements, and for easily handling space-time adaptivity strategies. The hybridizable discontinuous Galerkin methods we introduce here add to these advantages their distinctive feature, namely, that the only globally-coupled degrees of freedom are those of the approximate trace of the scalar unknown. This results in a significant reduction of the size of the matrices to be numerically inverted, a more efficient implementation, and even better accuracy. We introduce the method, discuss its implementation and numerically explore its convergence properties.

Keywords Discontinuous Galerkin methods · Advection–diffusion equations · Space-time methods

1 Introduction

Many applications in fluid dynamics require the solution of a set of partial differential equations in time-dependent flow domains. Examples include fluid-structure interaction, moving spatial configurations (e.g., helicopter rotors) and flows with free-surfaces (e.g., wave impacts on coastal and off-shore structures), see, e.g., [21] and [10]. The accurate solution of partial differential equations by a numerical method on moving and deforming meshes, however, is non-trivial. Many schemes fail to preserve the trivial solution of a uniform flow field on dynamic meshes. This condition, the so-called Geometric Conservation Law (GCL), was proved to be essential for the accuracy of the solution [13].

One class of numerical methods that automatically satisfies the GCL is the space-time Discontinuous Galerkin (DG) method. The main example is nothing but the first DG method [25], originally devised for the numerical simulation of neutron transport. Extensions have been obtained and successfully used in a wide variety of applications, e.g., the compressible Euler and Navier–Stokes equations [11, 24, 33],

C.A. de Moura, C.S. Kubrusly (eds.), *The Courant–Friedrichs–Lewy (CFL) Condition*, DOI 10.1007/978-0-8176-8394-8_4,

the shallow-water equations [1, 2, 31], two-phase flows [27, 29], hyperbolic non-conservative partial differential equations [26], and advection–diffusion and Oseen flows [30, 32]. In addition to their versatility, these methods can provide higher-order accurate approximations in both time and space and are ideally suited for hp-adaptivity. On the other hand, they are computationally expensive and so require the use of sophisticated solvers like Newton-GMRES solvers for the Navier–Stokes equations [22] and like optimized multigrid methods, see [12, 28, 34, 35] for the case of advection-dominated flows.

Recently, a new class of discontinuous Galerkin methods, namely, the hybridizable Discontinuous Galerkin method (HDG) [3, 6, 8], see also [5, 9], was introduced, in the framework of diffusion problems, with the *sole* purpose of reducing the computational complexity of these methods. In the HDG method, the approximate scalar variable and its corresponding flux are expressed in terms of an approximate trace of the scalar variable on the element faces. By enforcing the continuity of the normal component of the flux across the faces, a unique value for the approximate trace can be defined. A global system of equations for the approximate trace only is thus obtained, therefore significantly reducing the globally-coupled degrees of freedom of the discontinuous Galerkin method. The HDG method is computationally more efficient, can be more efficiently implemented, and is more accurate than all previously known discontinuous Galerkin methods. The method has been extended to time-dependent linear and nonlinear convection–diffusion in [4, 14, 15], to incompressible fluid flow [7, 17–19] and to the compressible Euler and Navier–Stokes equations [20]; see the recent review [16]. In all these papers, when dealing with time-dependent problems, implicit finite difference or Runge–Kutta time-marching methods were used. In this article, we extend the HDG method for the first time to a space-time setting. The resulting method thus combines the advantages of a space-time DG method with the efficiency and accuracy of the HDG methods.

The outline of this article is as follows. In Sect. 2, we introduce the advection–diffusion equation to which, in Sect. 3, we apply the space-time HDG method. A thorough numerical study of the convergence properties of the method is presented in Sect. 4. We end this article with some concluding remarks in Sect. 5.

2 The Advection–Diffusion Equation

We consider the following time-dependent advection–diffusion model problem:

$$\begin{aligned}
&u_{,0} + (a_k u - \kappa_{ks} u_{,s})_{,k} = f && \text{in } \mathcal{E}, \\
&u = u_0 && \text{on } \Omega(t_0), \\
&u = g_D && \text{on } \mathcal{Q}_D,
\end{aligned} \tag{1}$$

where a comma notation denotes differentiation with respect to the Cartesian coordinate x_k and the summation convention is used on repeated indices. Here $\mathcal{E} \in \mathbb{R}^{d+1}$ is the physical space-time domain (with d the spatial dimension), f is a source term,

$a(x) \in \mathbb{R}^d$ a given advective divergence-free velocity field, and $\kappa(x) \in \mathbb{R}^{d\times d}$ a positive definite diffusion tensor. The initial flow field is denoted by u_0 and the Dirichlet boundary data, g_D, is defined on the Dirichlet boundary $\mathcal{Q}_D$.

By introducing an auxiliary variable $\theta_k = -\kappa_{ks} u_{,s}$, we can rewrite (1) as a first-order system of equations:

$$u_{,0} + (a_k u + \theta_k)_{,k} = f \quad \text{in } \mathcal{E}, \tag{2a}$$

$$\theta_k + \kappa_{ks} u_{,s} = 0 \quad \text{in } \mathcal{E}, \tag{2b}$$

with the boundary conditions

$$\begin{aligned} u &= u_0 \quad \text{on } \Omega(t_0), \\ u &= g_D \quad \text{on } \mathcal{Q}_D. \end{aligned} \tag{2c}$$

3 The Space-Time HDG Method

In this section, we will present the space-time hybridizable discontinuous Galerkin method. We closely follow the notation of, e.g., [11, 26, 33] to highlight the similarities and differences between a space-time HDG and a space-time DG method.

3.1 *Space-Time Notation*

In a space-time method, space and time variables are not distinguished. A point at time $t = x_0$ with position vector $\bar{x} = (x_1, x_2, \ldots, x_d)$ has Cartesian coordinates $(x_0, \bar{x})$ in the open domain $\mathcal{E} \subset \mathbb{R}^{d+1}$. At time t the flow domain $\Omega(t)$ is defined as $\Omega(t) := \{\bar{x} \in \mathbb{R}^d : (t, \bar{x}) \in \mathcal{E}\}$. Let the initial and final time of the evolution of the space-time domain be denoted by t_0 and T, then the boundary of the space-time domain, $\partial\mathcal{E}$, consists of the hyper-surfaces

$$\begin{aligned} \Omega(t_0) &:= \{x \in \partial\mathcal{E} : x_0 = t_0\}, \\ \Omega(T) &:= \{x \in \partial\mathcal{E} : x_0 = T\}, \\ \mathcal{Q} &:= \{x \in \partial\mathcal{E} : t_0 < x_0 < T\}. \end{aligned}$$

The time interval $[t_0, T]$ is partitioned using the time levels $t_0 < t_1 < \cdots < T$, where the nth time interval is defined as $I_n = (t_n, t_{n+1})$ with length $\Delta t_n = t_{n+1} - t_n$. The space-time domain $\mathcal{E}$ is then divided into N_t space-time slabs $\mathcal{E}^n = \mathcal{E} \cap I_n$. Each space-time slab $\mathcal{E}^n$ is bounded by $\Omega(t_n)$, $\Omega(t_{n+1})$ and $\mathcal{Q}^n = \partial\mathcal{E}^n/(\Omega(t_n) \cup \Omega(t_{n+1}))$.

The flow domain $\Omega(t_n)$ is approximated by $\Omega_h(t_n)$, where $\Omega_h(t) \to \Omega(t)$ as $h \to 0$, with h the radius of the smallest sphere completely containing the largest space-time element. The domain $\Omega_h(t_n)$ is divided into N_n non-overlapping spatial elements $K_j(t_n)$. Similarly, $\Omega(t_{n+1})$ is approximated by $\Omega_h(t_{n+1})$. The space-time elements $\mathcal{K}_j^n$ are constructed by connecting K_j^n with K_j^{n+1} by using linear interpolation in time. In case of curved boundaries, a higher order accurate interpolation is used for elements connected to the domain boundary. The space-time elements $\mathcal{K}^n$ are connected to a master element $\widehat{\mathcal{K}}$ by an iso-parametric mapping $G_{\mathcal{K}}^n$. The tessellation $\mathcal{T}_h^n$ of the space-time slab $\mathcal{E}_h^n$ consists of all space-time elements $\mathcal{K}_j^n$; thus the tessellation $\mathcal{T}_h$ of the discrete flow domain $\mathcal{E}_h := \bigcup_{n=0}^{N_t-1} \mathcal{E}_h^n$ then is defined as $\mathcal{T}_h := \bigcup_{n=0}^{N_t-1} \mathcal{T}_h^n$.

The element boundary $\partial\mathcal{K}_j^n$, which is the union of open faces of $\mathcal{K}_j^n$, consists of three parts: $K_j(t_n^+) = \lim_{\epsilon\downarrow 0} K_j(t_n+\epsilon)$, $K_j(t_{n+1}^-) = \lim_{\epsilon\downarrow 0} K_j(t_{n+1}-\epsilon)$, and $\mathcal{Q}_j^n = \partial\mathcal{K}_j^n/(K_j(t_n^+)\cup K_j(t_{n+1}^-))$. We define $\mathcal{S}_h^n$ as the set of surfaces $\mathcal{S}$ of the form $\mathcal{Q}_j^n \cap \partial\mathcal{E}$ or of the form $\mathcal{Q}_j^n \cap \mathcal{Q}_{j'}^n$. We set $\mathcal{S}_h := \bigcup_{n=0}^{N_t-1} \mathcal{S}_h^n$.

To obtain the Arbitrary Lagrangian Eulerian (ALE) formulation, we have to introduce the grid velocity $v \in \mathbb{R}^d$. Let $\bar{x}(t^n)$ be a point on $\mathcal{Q}_j^n$ with $x_0 = t^n$. As the mesh moves, the point $\bar{x}(t^n)$ moves along $\mathcal{Q}_j^n$ to $\bar{x}(t^{n+1})$ according to some prescribed movement defined by $\bar{x}(t) = V(t;\bar{x}(t^n))$, $t \in I_n$, with V a given function. The grid velocity on $\mathcal{Q}_j^n$ is then defined by $v = \partial_t V$. The outward space-time normal vector at an element boundary point on $\partial\mathcal{K}_j^n$ can then be shown to be given by [33]:

$$n = \begin{bmatrix} (1,\bar{0}) & \text{at } K_j(t_{n+1}^-), \\ (-1,\bar{0}) & \text{at } K_j(t_n^+), \\ (-v_k\bar{n}_k,\bar{n}) & \text{at } \mathcal{Q}_j^n, \end{bmatrix} \tag{3}$$

where $\bar{0} \in \mathbb{R}^d$ and $\bar{n} \in \mathbb{R}^d$ the space-component of the space-time normal.

3.2 Approximation Spaces

Let $P^p(\mathcal{K})$ denote the space of polynomials of degree at most p on the reference element $\hat{\mathcal{K}}$ and consider $L^2(\Omega)$, that is, the space of square integrable functions on Ω. We introduce the discontinuous finite element spaces

$$W_h^p = \{\omega \in L^2(\mathcal{E}_h) : \omega|_{\mathcal{K}} \circ G_{\mathcal{K}} \in P^p(\hat{\mathcal{K}}), \forall\mathcal{K} \in \mathcal{T}_h\},$$

and

$$V_h^p = \{\nu \in (L^2(\mathcal{E}_h))^d : \nu|_{\mathcal{K}} \circ G_{\mathcal{K}} \in (P^p(\hat{\mathcal{K}}))^d, \forall\mathcal{K} \in \mathcal{T}_h\}.$$

We also introduce a traced finite element space:

$$M_h^p = \{\mu \in L^2(\mathcal{S}_h) : \mu|_{\mathcal{S}} \circ G_{\mathcal{S}} \in P^p(\hat{\mathcal{S}}), \forall\mathcal{S} \in \mathcal{S}_h\}.$$

We set $M_h^p(g_D) = \{\mu \in M_h^p : \mu = \mathsf{P}g_D \text{ on } \Gamma_D\}$, where P denotes the L^2-projection into the space $\{\mu|_{\partial\Omega}\ \forall\mu \in M_h^p\}$. Note that M_h^p consists of functions which are continuous inside the faces $\mathcal{S} \in \mathcal{S}_h$ and discontinuous at their borders.

3.3 Weak Formulation on Each Space-Time Element

Here, we are going to find the weak formulation on each of the space-time elements. Our objective is to be able to determine an approximation inside each space-time element only in terms of the data and on the numerical trace

$$\lambda := \hat{u}|_{\mathcal{S}_h} \in M_h^p. \tag{4}$$

We proceed as follows. Multiplying (2a) by a test function $\omega \in W_h^p$ and (2b) by a test function $\nu \in V_h^p$ and integrating by parts in space-time over an element $\mathcal{K} \in \mathcal{T}_h$, we obtain:

$$\begin{aligned} &-\int_{\mathcal{K}} \big(\omega_{,0} u + \omega_{,k}(a_k u + \theta_k)\big)\, dx \\ &\qquad + \int_{\partial\mathcal{K}} \omega^L \big(\hat{u}_{n_0} + (\widehat{a_k u} + \hat{\theta}_k)\bar{n}_k\big)\, ds = \int_{\mathcal{K}} f\omega\, dx, \end{aligned} \tag{5a}$$

$$\int_{\mathcal{K}} \nu_k \theta_k\, dx - \int_{\mathcal{K}} \nu_{k,s}\kappa_{ks} u\, dx + \int_{\mathcal{Q}} \nu_k \kappa_{ks} \hat{u}\bar{n}_s\, ds = 0. \tag{5b}$$

Here, the numerical traces $\widehat{a_k u} + \hat{\theta}_k$ and $\hat{u}$ are approximations to, respectively, $a_k u - \kappa_{ks} u_{,s}$ and u over $\partial\mathcal{K}$, and are introduced to couple local to global information as well as for stability purposes. These numerical traces will be defined later on.

To obtain the ALE-formulation to accommodate moving and deforming meshes, we follow [33] and use the definition of the space-time normal vector (3) to write the boundary integral in (5a) as:

$$\begin{aligned} &\int_{\partial\mathcal{K}} \omega\big(\hat{u}_{n_0} + (\widehat{a_k u} + \hat{\theta}_k)\bar{n}_k\big)\, ds \\ &\quad = \int_{K(t_{n+1}^-)} \omega\hat{u}\, d\bar{x} - \int_{K(t_n^+)} \omega\hat{u}\, d\bar{x} + \int_{\mathcal{Q}} \omega(\widehat{a_k u} - \hat{u}\nu_k + \hat{\theta}_k)\bar{n}_k\, ds. \end{aligned} \tag{6}$$

The numerical traces $\hat{u}$ on $K(t_{n+1}^-)$ and $K(t_n^+)$ are chosen inspired in a causality-in-time argument and are therefore defined as the upwind flux:

$$\hat{u} = \begin{cases} u_{n+1}^- & \text{at } K(t_{n+1}^-), \\ u_n^- & \text{at } K(t_n^+), \end{cases}$$

where u_n^- and u_{n+1}^- are the traces of u on $K(t_n)$ and $K(t_{n+1})$ from, respectively, the previous and the current space-time slab. The function u_0^- is nothing but the L^2-projection of the initial data u_0 into the space $\{\omega|_{\Omega_h(t_0)} : \omega \in W_h^p\}$.

To be able to solve (5a) and (5b) locally, the numerical trace must depend only on λ and on the traces obtained from the interior of the space-time element $\mathcal{K}$. To achieve this, we take the numerical traces $(\widehat{a_k u} + \hat{\theta}_k - \nu_k\hat{u})\bar{n}_k$ of the form

$$(\widehat{a_k u} + \hat{\theta}_k - \nu_k\hat{u})\bar{n}_k = (a_k - \nu_k)\bar{n}_k\lambda + \theta_k\bar{n}_k + \tau(u - \lambda) \quad \text{on } \mathcal{Q}, \tag{7}$$

for some positive function τ. The selection of τ shall be described later.

Finally, by using (6) in combination with the upwind flux on the time faces and the numerical trace (7), (5a) and (5b) become:

$$-\int_{\mathcal{K}}(\omega_{,0}u+\omega_{,k}(a_ku+\theta_k))\,dx+\Big(\int_{K(t_{n+1}^-)}\omega u_{n+1}^-\,d\bar{x}-\int_{K(t_n^+)}\omega u_n^-\,d\bar{x}\Big) \tag{8a}$$

$$+\int_{\mathcal{Q}}\omega^L\big((a_k-v_k)\bar{n}_k\lambda+\theta_k\bar{n}_k+\tau(u-\lambda)\big)\,ds=\int_{\mathcal{K}}f\omega\,dx, \tag{8b}$$

$$\int_{\mathcal{K}}v_k\theta_k\,dx-\int_{\mathcal{K}}v_{k,s}\kappa_{ks}u\,dx+\int_{\mathcal{Q}}v_k\kappa_{ks}\lambda\bar{n}_s\,ds=0, \tag{8c}$$

for all $(\omega,v)\in W_h^p\times V_h^p$.

3.4 The Global Weak Formulation for the Approximate Trace λ

We still need to determine λ. To do this, we require that the boundary conditions be weakly satisfied and that the normal component of the numerical trace of the flux $\widehat{a_ku}+\hat{\theta}_k-v_k\hat{u}$ given in (7) be single valued. In other words, we require that $\lambda\in M_h^p$ be the solution of

$$\lambda=\mathsf{P}(g_D)\quad\text{on }\mathcal{Q}_D^n, \tag{9a}$$

$$\sum_{\mathcal{K}\in\mathcal{T}_h^n}\int_{\mathcal{Q}}\mu(\widehat{a_ku}+\hat{\theta}_k-v_k\hat{u})\bar{n}_k\,ds=0, \tag{9b}$$

for all $\mu\in M_h^p(0)$; recall that this implies that $\mu=0$ on $\mathcal{Q}_D$.

3.5 The Geometric Conservation Law

We now prove that the Geometric Conservation Law (GCL) is automatically satisfied by the space-time HDG method. The GCL states that uniform flow must be preserved on a moving mesh. Let U denote a uniform flow field. In a uniform flow field, $\lambda=U$ and $\theta_k=0$. Substituting this into (8a) and considering the element $\mathcal{K}$ on the time interval $(t,t+\varepsilon)$, we obtain:

$$-\int_{\mathcal{K}}(\omega_{,0}U+\omega_{,k}a_kU)\,dx+\int_{K(t+\varepsilon)}\omega U\,d\bar{x}-\int_{K(t)}\omega U\,d\bar{x}$$
$$+\int_{\mathcal{Q}}\omega(a_k-v_k)U\bar{n}_k\,ds=0. \tag{10}$$

Note that this formulation is exactly the same as a standard space-time DG formulation in uniform flows. Since U is constant and arbitrary, U can be divided out of (10). Furthermore, we can rewrite (10) as:

$$\int_t^{t+\varepsilon}\left(-\int_{K(t)}(\omega_{,0}+\omega_{,k}a_k)\,d\bar{x}+\int_{\partial K(t)}\omega(a_k-v_k)\bar{n}_k\,d\bar{s}\right)dt$$
$$+\int_{K(t+\varepsilon)}\omega\,d\bar{x}-\int_{K(t)}\omega\,d\bar{x}=0. \tag{11}$$

With the following equality

$$\int_{K(t+\varepsilon)}\omega\,d\bar{x}-\int_{K(t)}\omega\,d\bar{x}=\int_t^{t+\varepsilon}\left(\frac{d}{dt}\int_{K(t)}\omega\,d\bar{x}\right)dt,$$

noting that t, $t+\varepsilon$ are arbitrary, and considering a constant polynomial approximation, we obtain the GCL:

$$\frac{d}{dt}\int_{K(t)}d\bar{x}-\int_{\partial K(t)}v_k\bar{n}_k\,d\bar{s}=0, \tag{12}$$

using the fact that integration over a closed surface $\partial K(t)$ of a constant is equal to zero. This law states that to preserve uniform flow on a moving mesh, the change in area/volume of each element must be equal to the area/volume swept by the element boundary [13].

3.6 Existence and Uniqueness of the Approximate Solution

Next, we present a result that shows that when the stabilization function τ is suitably defined, the approximation of our space-time HDG method is well defined.

Theorem 1 *Assume that the matrix-valued function κ is symmetric and positive definite and constant on each space-time element $\mathcal{K}\in\mathcal{T}_h$. Assume that the advective velocity a is divergence-free. Then, if we take the stabilization function τ such that*

$$\tau\geq\frac{1}{2}(a_k-v_k)\bar{n}_k+\tau_0\quad\text{on }\mathcal{Q}\;\forall\mathcal{K}\in\mathcal{T}_h,$$

where τ_0 is a strictly positive constant, the approximate solution of the HDG method under consideration is well defined.

Proof We only have to show that if the data is equal to zero, the only solution of the weak formulation (8a)–(8c) relating λ to (θ,u) and the equations determining λ (9a), (9b) is the trivial one. It is easy to see that we only need to work on any time slab $\mathcal{E}^n$ assuming that $u_n^-=0$.

Thus, taking $\omega:=u$ in (8a), we get

$$-\sum_{\mathcal{K}\in\mathcal{T}_h^n}\int_{\mathcal{K}}\big(u_{,0}u+u_{,k}(a_ku+\theta_k)\big)\,dx+\sum_{\mathcal{K}\in\mathcal{T}_h^n}\int_{K(t_{n+1}^-)}\left(u_{n+1}^-\right)^2d\bar{x}$$
$$+\sum_{\mathcal{K}\in\mathcal{T}_h^n}\int_{\mathcal{Q}}u(\widehat{a_ku}-\hat{u}v_k+\hat{\theta}_k)\bar{n}_k\,ds=0.$$

Integrating by parts and rearranging terms, we obtain

$$
\begin{aligned}
&-\sum_{\mathcal{K}\in\mathcal{T}_h^n}\int_{\mathcal{K}} u_{,k}\theta_k\,dx + \frac{1}{2}\sum_{\mathcal{K}\in\mathcal{T}_h^n}\int_{K(t_{n+1}^-)} \left(u_{n+1}^-\right)^2 d\bar{x}\\
&+\frac{1}{2}\sum_{\mathcal{K}\in\mathcal{T}_h^n}\int_{K(t_n^+)} \left(u_n^+\right)^2 d\bar{x} - \frac{1}{2}\sum_{\mathcal{K}\in\mathcal{T}_h^n}\int_{\mathcal{Q}} u^2(a_k-v_k)\bar{n}_k\,ds\\
&+\sum_{\mathcal{K}\in\mathcal{T}_h^n}\int_{\mathcal{Q}} u\,(\widehat{a_k u} - \hat{u}v_k + \hat{\theta}_k)\bar{n}_k\,ds = 0.
\end{aligned}
$$

Since the tensor-valued function κ is piecewise constant, we can take $v := \kappa^{-1}\theta$ in (8c) to get

$$
\sum_{\mathcal{K}\in\mathcal{T}_h^n}\int_{\mathcal{K}} \left(\kappa^{-1}\right)_{ks}\theta_k\theta_s\,dx - \sum_{\mathcal{K}\in\mathcal{T}_h^n}\int_{\mathcal{K}} \theta_{s,s}u\,dx + \sum_{\mathcal{K}\in\mathcal{T}_h^n}\int_{\mathcal{Q}} \theta_s\hat{u}\bar{n}_s\,ds = 0.
$$

Adding this equation to the previous one, we obtain

$$
\begin{aligned}
&\frac{1}{2}\sum_{\mathcal{K}\in\mathcal{T}_h^n}\int_{K(t_{n+1}^-)} \left(u_{n+1}^-\right)^2 d\bar{x} + \frac{1}{2}\sum_{\mathcal{K}\in\mathcal{T}_h^n}\int_{K(t_n^+)} \left(u_n^+\right)^2 d\bar{x}\\
&+\sum_{\mathcal{K}\in\mathcal{T}_h^n}\int_{\mathcal{K}} \left(\kappa^{-1}\right)_{ks}\theta_k\theta_s\,dx + \Theta_h = 0,
\end{aligned}
$$

where

$$
\Theta_h := \sum_{\mathcal{K}\in\mathcal{T}_h^n}\int_{\mathcal{Q}}\left(\theta_k(\hat{u}-u)\bar{n}_k - \frac{1}{2}u^2(a_k-v_k)\bar{n}_k + u(\widehat{a_k u} - \hat{u}v_k + \hat{\theta}_k)\bar{n}_k\right) ds.
$$

We claim that Θ_h is a dissipative term. To see this, note that, by (9b),

$$
\begin{aligned}
\Theta_h := \sum_{\mathcal{K}\in\mathcal{T}_h^n}\int_{\mathcal{Q}}\Big(&\theta_k(\hat{u}-u)\bar{n}_k - \frac{1}{2}u^2(a_k-v_k)\bar{n}_k\\
&+(u-\hat{u})(\widehat{a_k u} - \hat{u}v_k + \hat{\theta}_k)\bar{n}_k\Big)\,ds,
\end{aligned}
$$

and by the definition of the numerical trace (7),

$$
\begin{aligned}
\Theta_h &:= \sum_{\mathcal{K}\in\mathcal{T}_h^n}\int_{\mathcal{Q}}\left(-\frac{1}{2}u^2(a_k-v_k)\bar{n}_k + (u-\lambda)\left((a_k-v_k)\bar{n}_k\lambda + \tau(u-\lambda)\right)\right) ds\\
&= \sum_{\mathcal{K}\in\mathcal{T}_h^n}\int_{\mathcal{Q}}\left(-\frac{1}{2}\left((u-\lambda)^2+\lambda^2\right)(a_k-v_k)\bar{n}_k + \tau(u-\lambda)^2\right) ds\\
&= \sum_{\mathcal{K}\in\mathcal{T}_h^n}\int_{\mathcal{Q}}\left(\tau - \frac{1}{2}(a_k-v_k)\bar{n}_k\right)(u-\lambda)^2\,ds.
\end{aligned}
$$

We can now conclude that $u_{n+1}^- = 0$ on Ω_{n+1}, that $u_n^+ = 0$ on Ω_n, that $\theta_h = 0$ on $\mathcal{T}_h^n$, and that $u = \lambda$ on $\mathcal{S}_h^n$. Equation (8c) now gives that u is constant in space on the time-slab $\mathcal{T}_h^n$ and since $u = \lambda = 0$ on the Dirichlet boundary, we obtain that $u = 0$ on $\mathcal{T}_h^n$ and that $\lambda = 0$ on $\mathcal{S}_h^n$. This completes the proof. □

3.7 The Local Stabilization Parameter τ

In the rest of this article, we assume $\kappa_{11} = \kappa_{22} = \kappa$ and $\kappa_{ks} = 0$ otherwise. Then the local stabilization parameter τ is chosen similarly as done in [14]. We, however, slightly modify the local stabilization parameter to account for moving grids. Two options are discussed in [14], the centered scheme and the upwinded scheme. To account for the diffusion and advection effects, let $\tau = \tau_a + \tau_d$, where τ_a and τ_d are the local stabilization parameters related to the advection and diffusion, respectively. Consider an interior face $\mathcal{S} = \mathcal{Q}^L \cap \mathcal{Q}^R$ between the space-time elements $\mathcal{K}^L$ and $\mathcal{K}^R$ and denote by $(\cdot)^L$ the trace of $(\cdot)$ on $\mathcal{S}$ from $\mathcal{K}^L$, and similarly for $(\cdot)^R$. Furthermore, let $\bar{n}$ be the outward normal with respect to $\mathcal{K}^L$.

Centered Scheme To obtain a centered scheme, take on each face $\tau_a^L = \tau_a^R = \eta_a$ and $\tau_d^L = \tau_d^R = \eta_d$, where

$$\eta_a = \left|(a_k - v_k)\bar{n}_k\right|, \qquad \eta_d = \frac{\kappa}{\ell}, \tag{13}$$

and ℓ denotes a representative diffusive length scale.

Upwinded Scheme To obtain an upwinded scheme, choose $\tau_a^{L,R}$ and $\tau_d^{L,R}$ according to

$$\begin{aligned}
\left(\tau_a^L, \tau_d^L\right) &= (\eta_a, \eta_d)\frac{|(a_k - v_k)\bar{n}_k| + (a_k - v_k)\bar{n}_k}{2|(a_k - v_k)\bar{n}_k|},\\
\left(\tau_a^R, \tau_d^R\right) &= (\eta_a, \eta_d)\frac{|(a_k - v_k)\bar{n}_k| - (a_k - v_k)\bar{n}_k}{2|(a_k - v_k)\bar{n}_k|},
\end{aligned}$$

with η_a and η_d given in (13).

4 Numerical Results

In this section, we consider numerical results for the space-time HDG discretization of the advection–diffusion equation. For each test case, we show the convergence history of the flow field u, the auxiliary variables θ_1 and θ_2 and the mean of the flow field $\bar{u}$. Note that

$$\|\bar{u} - \bar{u}_h\|_{L^2(\Omega)} = \sqrt{\sum_K \frac{1}{|K|}\left(\int_K (u - u_h)\,d\bar{x}\right)^2}.$$

4.1 Steady-State Solution of the Advection and the Advection–Diffusion Equation on a Uniform Mesh

In this first test case, we consider both the advection and the advection–diffusion equations on a uniform mesh. For this we consider (2a), (2b) on the space-time domain $\mathcal{E} = (0, T) \times (0, 1)^2$ where the source term $f(x_1, x_2)$ and the Dirichlet boundary condition g are such that the exact solution is given by $u(x_1, x_2) = 4 + \sin(\pi x_1)\sin(\pi x_2) + \sin(2\pi x_1) + \sin(2\pi x_2)$. We take $a_1 = a_2 = 1$ and, in the case of the advection–diffusion equation, $\kappa = 0.01$. In the case of the advection equation, $\kappa = 0$. Therefore, for this test case, we modify the definition of θ such that $\theta_k = u_{,k}$.

We use a space-time HDG discretization using linear-, quadratic-, and cubic-polynomial approximations and obtain convergence orders. The local stabilization parameter τ is chosen such that we obtain a *central scheme*. For this steady-state problem, we take one physical time step of $T = \Delta t = 10^{15}$. In Tables 1 and 2, we show the convergence results obtained when $\kappa = 0.01$ and $\kappa = 0$, respectively.

For the advection–diffusion equation, from Table 1, we see the expected orders of convergence for the scalar variable u and the auxiliary variables θ_1 and θ_2, namely, for a P^p polynomial approximation we obtain the orders of convergence of $p + 1$. For the mean variable $\bar{u}$, for P^1 we obtain superconvergence with order $p + 2$. For P^2 and P^3 we seem to be achieving superconvergence with order $p + 3$!

For the advection equation, from Table 2, we obtain the expected order of convergence for the scalar variable u, namely, for a P^p polynomial approximation we obtain orders of convergence $p + 1$. For the "artificial" auxiliary variables θ_1 and θ_2, we only obtain orders of convergence p. For the mean variable $\bar{u}$, we find the strange behavior that for odd $p = 1, 3$ polynomial approximation we achieve superconvergence of orders $p + 2$ while for even $p = 2$, we only achieve a convergence order of $p + 1$.

4.2 Steady-State Boundary Layer Problem

Next, we consider a boundary layer problem. Consider (2a), (2b) on the space-time domain $\mathcal{E} = (0, T) \times (0, 1)^2$ where $f = 0$ and where $g(x_1, x_2)$ equals at the domain boundary the exact steady-state solution:

$$u(x_1, x_2) = \frac{1}{2}\left(\frac{\exp(a_1/\kappa) - \exp(a_1 x_1/\kappa)}{\exp(a_1/\kappa) - 1} + \frac{\exp(a_2/\kappa) - \exp(a_2 x_2/\kappa)}{\exp(a_2/\kappa) - 1}\right).$$

In the discretization, we use a Shishkin mesh in which the coordinates (x_1^u, x_2^u) of a uniform mesh are mapped onto a mesh suitable for dealing with boundary layers. The mapping is given by:

$$x_i = \begin{bmatrix} 2(1 - \sigma_i)x_i^u, & \text{for } x_i^u < 0.5, \\ 1 + 2\sigma_i(x_i^u - 1), & \text{for } x_i^u \geq 0.5, \end{bmatrix} \qquad i = 1, 2,$$

Table 1 History of convergence for the steady-state advection–diffusion equation on a uniform mesh with $\kappa = 0.01$

Degree	N_{cells}	$\|u - u_h\|_{L^2(\Omega)}$		$\|\theta_1 - \theta_1^h\|_{L^2(\Omega)}$		$\|\theta_2 - \theta_2^h\|_{L^2(\Omega)}$		$\|\bar{u} - \bar{u}_h\|_{L^2(\Omega)}$	
		Error	Order	Error	Order	Error	Order	Error	Order
1	8	3.53e–2	–	6.80e–1	–	6.80e–1	–	1.45e–2	–
	16	7.41e–3	2.3	2.62e–1	1.4	2.62e–1	1.4	1.64e–3	3.1
	32	1.66e–3	2.2	8.84e–2	1.6	8.84e–2	1.6	1.92e–4	3.1
	64	3.93e–4	2.1	2.67e–2	1.7	2.67e–2	1.7	2.38e–5	3.0
	128	9.59e–5	2.0	7.44e–3	1.8	7.44e–3	1.8	3.05e–6	3.0
2	8	2.52e–3	–	5.30e–2	–	5.30e–2	–	1.99e–3	–
	16	2.19e–4	3.5	9.46e–3	2.5	9.46e–3	2.5	1.50e–4	3.7
	32	1.59e–5	3.8	1.54e–3	2.6	1.54e–3	2.6	7.46e–6	4.3
	64	1.28e–6	3.6	2.25e–4	2.8	2.25e–4	2.8	2.90e–7	4.7
	128	1.29e–7	3.3	3.06e–5	2.9	3.06e–5	2.9	1.00e–8	4.9
3	8	9.71e–5	–	1.80e–3	–	1.80e–3	–	2.67e–5	–
	16	6.01e–6	4.0	1.29e–4	3.8	1.29e–4	3.8	3.66e–7	6.2
	32	3.64e–7	4.0	8.26e–6	4.0	8.26e–6	4.0	3.36e–9	6.8
	64	2.13e–8	4.1	5.34e–7	4.0	5.34e–7	4.0	2.74e–11	6.9
	128	1.27e–9	4.1	3.52e–8	3.9	3.52e–8	3.9	5.23e–13	5.7

Table 2 History of convergence for the steady-state advection equation on a uniform mesh in which $\kappa = 0$

Degree	N_{cells}	$\|u - u_h\|_{L^2(\Omega)}$		$\|\theta_1 - \theta_1^h\|_{L^2(\Omega)}$		$\|\theta_2 - \theta_2^h\|_{L^2(\Omega)}$		$\|\bar{u} - \bar{u}_h\|_{L^2(\Omega)}$	
		Error	Order	Error	Order	Error	Order	Error	Order
1	8	4.43e–2	–	9.71e–1	–	9.71e–1	–	2.22e–2	–
	16	1.01e–2	2.1	5.05e–1	0.9	5.05e–1	0.9	3.06e–3	2.9
	32	2.45e–3	2.0	2.55e–1	1.0	2.55e–1	1.0	3.97e–4	2.9
	64	6.06e–4	2.0	1.28e–1	1.0	1.28e–1	1.0	5.05e–5	3.0
	128	1.51e–4	2.0	6.39e–2	1.0	6.39e–2	1.0	6.35e–6	3.0
2	8	4.31e–3	–	8.57e–2	–	8.57e–2	–	3.89e–3	–
	16	5.76e–4	2.9	2.07e–2	2.0	2.07e–2	2.0	5.26e–4	2.9
	32	7.34e–5	3.0	5.13e–3	2.0	5.13e–3	2.0	6.73e–5	3.0
	64	9.24e–6	3.0	1.28e–3	2.0	1.28e–3	2.0	8.48e–6	3.0
	128	1.16e–6	3.0	3.20e–4	2.0	3.20e–4	2.0	1.06e–6	3.0
3	8	1.20e–4	–	4.77e–3	–	4.77e–3	–	8.77e–5	–
	16	5.82e–6	4.4	5.86e–4	3.0	5.86e–4	3.0	2.85e–6	4.9
	32	3.28e–7	4.1	7.30e–5	3.0	7.30e–5	3.0	9.08e–8	5.0
	64	1.99e–8	4.0	9.12e–6	3.0	9.12e–6	3.0	2.87e–9	5.0
	128	1.23e–9	4.0	1.14e–6	3.0	1.14e–6	3.0	9.16e–11	5.0

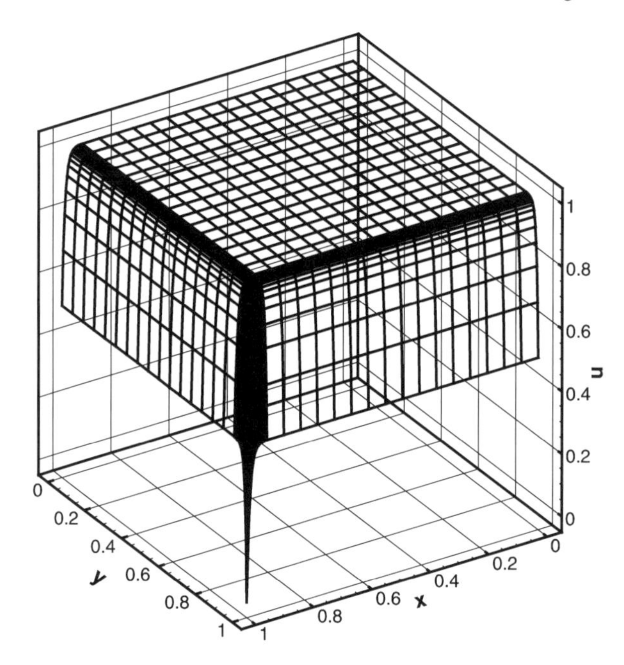

Fig. 1 The steady-state solution of the boundary layer problem using a cubic polynomial approximation on a grid with 32×32 elements (using a central flux)

where $\sigma_i = \min(0.5, 2\kappa/a_i \ln(N_i))$, and where N_i is the number of elements in the x_i direction (see, e.g., [28]). For this test case, we take $a_1 = a_2 = 1$ and $\kappa = 0.01$.

We solve the above problem using a space-time HDG discretization using linear-, quadratic-, and cubic- polynomial approximations in space. For steady-state test-cases, it is sufficient to take a constant polynomial approximation. The local stabilization parameter τ is chosen such that we obtain a *central scheme*. We will consider two cases for the diffusive length scale, ℓ, in (13), namely $\ell = 1$ and $\ell = \min(\sqrt{|K_L|}, \sqrt{|K_R|})$, in which $K_{L,R}$ are the areas of the two spatial elements adjacent to the face on which τ is evaluated. Furthermore, we set $\tau = 0$ on all boundaries. We remark that if $\tau \neq 0$ on the boundaries, for this test case we do not achieve expected convergence rates. For this steady-state problem, we take one physical time step of $T = \Delta t = 10^{15}$. The steady-state solution is depicted in Fig. 1. Tables 3 and 4 show the obtained convergence orders when $\ell = 1$ and $\ell = \min(\sqrt{|K_L|}, \sqrt{|K_R|})$, respectively.

For this test case, from Tables 3 and 4, we see that the diffusive length scale ℓ in the stabilization parameter has an effect on the order of convergence for the different variables. For degrees 1 and 2, if $\ell = \min(\sqrt{|K_L|}, \sqrt{|K_R|})$, we only obtain orders of convergence p for the auxiliary variables θ_1 and θ_2 and for degree $p = 1$, $\bar{u}$ converges with order $p+1$. On the other hand, u converges with orders $p+1$ for all considered p, the auxiliary variables converge with orders of at least $p+1$ for $p = 3$ and $\bar{u}$ superconverges with order $p+2$ for $p = 2, 3$.

Table 3 History of convergence for the steady-state boundary layer problem with $\ell = 1$

Degree	N_{cells}	$\lVert u - u_h \rVert_{L^2(\Omega)}$		$\lVert \theta_1 - \theta_1^h \rVert_{L^2(\Omega)}$		$\lVert \theta_2 - \theta_2^h \rVert_{L^2(\Omega)}$		$\lVert \bar{u} - \bar{u}_h \rVert_{L^2(\Omega)}$	
		Error	Order	Error	Order	Error	Order	Error	Order
1	8	8.46e–3	–	2.54e–3	–	2.54e–3	–	7.59e–4	–
	16	3.40e–3	1.3	1.09e–3	1.2	1.09e–3	1.2	2.26e–4	1.7
	32	1.16e–3	1.6	4.14e–4	1.4	4.14e–4	1.4	5.37e–5	2.1
	64	3.61e–4	1.7	1.47e–4	1.5	1.47e–4	1.5	1.10e–5	2.3
	128	1.06e–4	1.8	5.00e–5	1.6	5.00e–5	1.6	2.04e–6	2.4
2	8	7.08e–4	–	3.15e–4	–	3.15e–4	–	1.29e–4	–
	16	1.71e–4	2.0	5.62e–5	2.5	5.62e–5	2.5	7.96e–6	4.0
	32	3.61e–5	2.2	1.14e–5	2.3	1.14e–5	2.3	8.70e–7	3.2
	64	6.71e–6	2.4	2.38e–6	2.3	2.38e–6	2.3	5.26e–8	4.0
	128	1.13e–6	2.6	4.69e–7	2.3	4.69e–7	2.3	1.11e–9	5.6
3	8	4.14e–4	–	2.89e–4	–	2.89e–4	–	1.48e–4	–
	16	5.44e–5	2.9	3.84e–5	2.9	3.84e–5	2.9	1.21e–5	3.6
	32	3.72e–6	3.9	2.63e–6	3.9	2.63e–6	3.9	3.60e–7	5.1
	64	1.29e–7	4.9	9.17e–8	4.8	9.17e–8	4.8	4.06e–9	6.5
	128	5.80e–9	4.5	4.10e–9	4.5	4.10e–9	4.5	1.97e–11	7.7

Table 4 History of convergence for the steady-state boundary layer problem with $\ell = \min(\sqrt{|K_L|}, \sqrt{|K_R|})$

Degree	N_{cells}	$\lVert u - u_h \rVert_{L^2(\Omega)}$		$\lVert \theta_1 - \theta_1^h \rVert_{L^2(\Omega)}$		$\lVert \theta_2 - \theta_2^h \rVert_{L^2(\Omega)}$		$\lVert \bar{u} - \bar{u}_h \rVert_{L^2(\Omega)}$	
		Error	Order	Error	Order	Error	Order	Error	Order
1	8	7.57e–3	–	2.54e–3	–	2.54e–3	–	7.85e–4	–
	16	2.83e–3	1.4	1.13e–3	1.2	1.13e–3	1.2	2.42e–4	1.7
	32	8.78e–4	1.7	4.82e–4	1.2	4.82e–4	1.2	6.53e–5	1.9
	64	2.42e–4	1.9	2.27e–4	1.0	2.27e–4	1.0	1.77e–5	1.9
	128	6.42e–5	1.9	1.20e–4	0.9	1.20e–4	0.9	5.29e–6	1.7
2	8	7.97e–4	–	3.15e–4	–	3.15e–4	–	1.27e–4	–
	16	1.59e–4	2.3	5.95e–5	2.4	5.95e–5	2.4	9.00e–6	3.8
	32	2.69e–5	2.6	1.47e–5	2.0	1.47e–5	2.0	1.44e–6	2.6
	64	4.09e–6	2.7	4.39e–6	1.7	4.39e–6	1.7	9.69e–8	3.9
	128	5.55e–7	2.9	1.42e–6	1.6	1.42e–6	1.6	2.72e–9	5.2
3	8	4.29e–4	–	2.96e–4	–	2.96e–4	–	1.54e–4	–
	16	5.76e–5	2.9	4.03e–5	2.9	4.03e–5	2.9	1.31e–5	3.6
	32	4.06e–6	3.8	2.86e–6	3.8	2.86e–6	3.8	4.40e–7	4.9
	64	1.44e–7	4.8	1.02e–7	4.8	1.02e–7	4.8	5.96e–9	6.2
	128	6.06e–9	4.6	4.29e–9	4.6	4.29e–9	4.6	3.55e–11	7.4

Taking $\ell = 1$, for degree 1 and 2, it seems that u has difficulty converging with the expected orders of convergence $p+1$, while the auxiliary variables and the mean show better orders of convergence than for the case $\ell = \min(\sqrt{|K_L|}, \sqrt{|K_R|})$. For degree $p = 3$, u, θ_1, θ_2 converge with orders of at least $p+1$ and $\bar{u}$ converges with order of at least $p+2$!

4.3 A Rotating Gaussian Pulse on a Moving/Deforming Mesh

Finally, we consider the transport of a two-dimensional rotating Gaussian pulse, a test case that was presented in [14]. We, however, consider a moving and deforming space-time domain $\mathcal{E}$. Let the rotating velocity field be prescribed as $a = (-4x_2, 4x_1)$. We consider the solution at final time $T = \pi/4$, which is the time period for one-half rotation of the Gaussian pulse. The initial condition is given by

$$u_0(x_1, x_2) = \exp\left(-\frac{(x_1 - x_{1c})^2 + (x_2 - x_{2c})^2}{2\sigma^2}\right),$$

where (x_{1c}, x_{2c}) is the center and σ is the standard deviation. The exact solution with constant diffusivity constant κ is given by

$$u(x_1, x_2) = \frac{2\sigma^2}{2\sigma^2 + 4\kappa t} \exp\left(-\frac{(\tilde{x}_1 - x_{1c})^2 + (\tilde{x}_2 - x_{2c})^2}{2\sigma^2 + 4\kappa t}\right),$$

where $\tilde{x}_1 = x_1 \cos(4t) + x_2 \sin(4t)$ and $\tilde{x}_2 = -x_1 \sin(4t) + x_2 \cos(4t)$. The Dirichlet boundary condition g is deduced from the exact solution. As in [14], we choose $(x_{1c}, x_{2c}) = (-0.2, 0)$ and take $\sigma = 0.1$. As diffusivity constant, we take $\kappa = 0.01$.

The deformation of the space-time domain $\mathcal{E}$ is based on the following transformation of a uniform mesh of the space-time domain $[t, t + \Delta t] \times [-0.5, 0.5]^2$. Let (x_0^u, x_1^u, x_2^u) be the coordinates on the uniform mesh. Then we consider the following mapping:

$$x_i = x_i^u + A\left(\frac{1}{2} - x_i^u\right) \sin\left(2\pi\left(\frac{1}{2} - x_*^u + t^*\right)\right),$$

$$t^* = \begin{bmatrix} t & \text{if } x_0^u = t, \\ t + \Delta t & \text{if } x_0^u = t + \Delta t, \end{bmatrix}$$

$$x_*^u = \begin{bmatrix} x_2 & \text{if } i = 1, \\ x_1 & \text{if } i = 2, \end{bmatrix}$$

where A is the amplitude. We set $A = 0.1$. Furthermore, for the diffusivity constant we take $\kappa = 0.01$. We consider the convergence properties of the space-time HDG method for two given *CFL* numbers, namely $CFL = 1$ and $CFL = 10$. The history of convergence for the given *CFL* numbers is given in, respectively, Tables 5 and 6. In Fig. 2, we show some snapshots of the solution and mesh at different time levels.

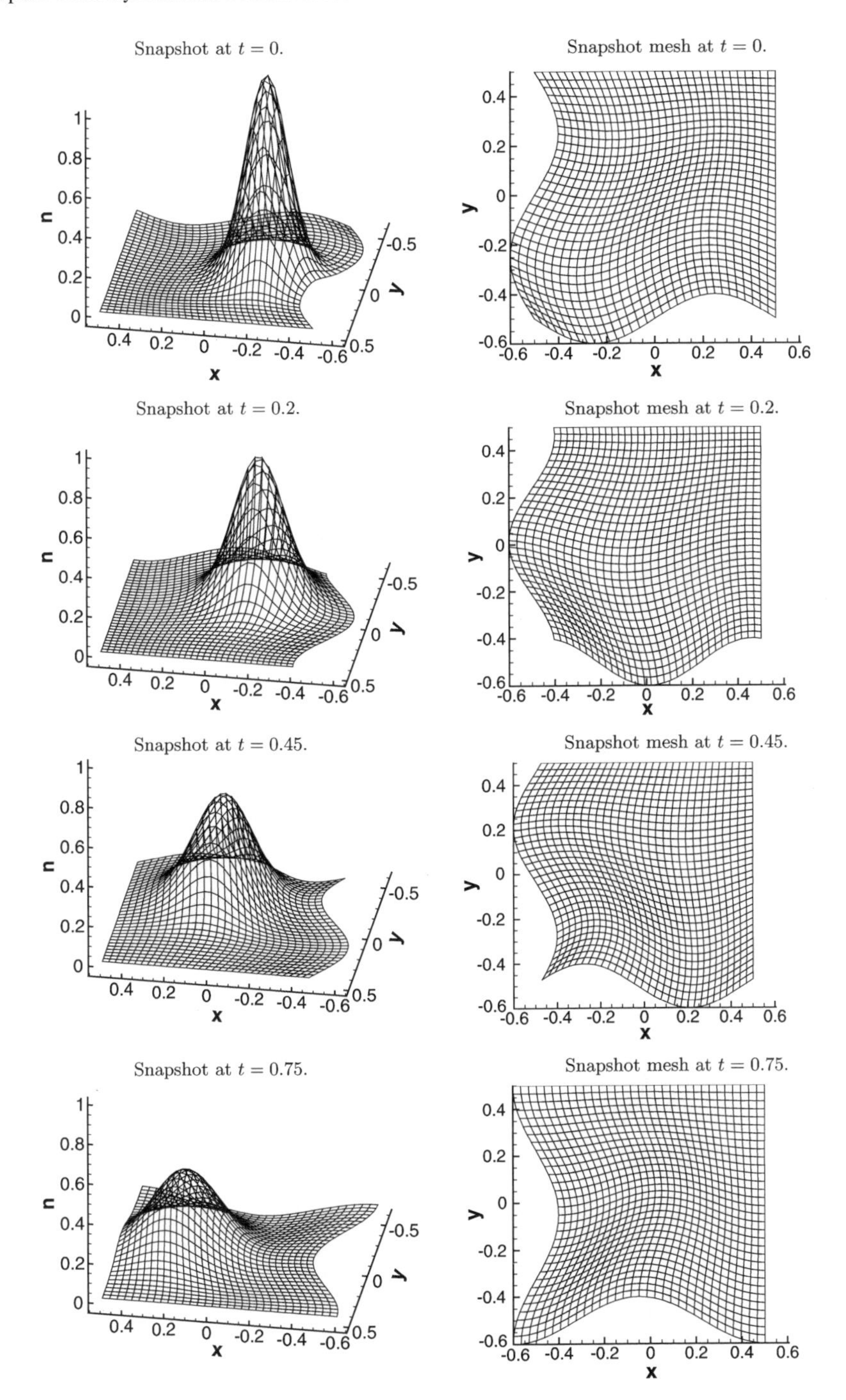

Fig. 2 Snapshots of the rotating Gaussian pulse on a moving/deforming mesh with $CFL = 1$ on a grid with 32×32 elements using a cubic polynomial approximation

Table 5 History of convergence for the rotating Gaussian pulse on a moving/deforming mesh with $CFL = 1$

Degree	N_{cells}	$\|u - u_h\|_{L^2(\Omega)}$		$\|\theta_1 - \theta_1^h\|_{L^2(\Omega)}$		$\|\theta_2 - \theta_2^h\|_{L^2(\Omega)}$		$\|\bar{u} - \bar{u}_h\|_{L^2(\Omega)}$	
		Error	Order	Error	Order	Error	Order	Error	Order
1	8	1.50e–2	–	1.26e–3	–	1.23e–3	–	1.31e–2	–
	16	3.26e–3	2.2	3.48e–4	1.9	3.57e–4	1.8	2.93e–3	2.2
	32	5.46e–4	2.6	8.30e–5	2.1	9.31e–5	1.9	4.47e–4	2.7
	64	9.78e–5	2.5	2.19e–5	1.9	2.60e–5	1.8	5.91e–5	2.9
2	8	1.61e–3	–	2.09e–4	–	2.22e–4	–	1.22e–3	–
	16	1.19e–4	3.8	2.93e–5	2.8	3.46e–5	2.7	6.19e–5	4.3
	32	1.22e–5	3.3	4.78e–6	2.6	5.71e–6	2.6	2.25e–6	4.8
	64	1.48e–6	3.0	7.30e–7	2.7	8.81e–7	2.7	7.61e–8	4.9
3	8	1.49e–4	–	2.91e–5	–	3.34e–5	–	8.67e–5	–
	16	6.72e–6	4.5	2.46e–6	3.6	2.82e–6	3.6	1.21e–6	6.2
	32	3.94e–7	4.1	2.06e–7	3.6	2.35e–7	3.6	1.41e–8	6.4
	64	2.38e–8	4.0	1.60e–8	3.7	1.81e–8	3.7	2.07e–10	6.0

Table 6 History of convergence for the rotating Gaussian pulse on a moving/deforming mesh with $CFL = 10$

Degree	N_{cells}	$\|u - u_h\|_{L^2(\Omega)}$		$\|\theta_1 - \theta_1^h\|_{L^2(\Omega)}$		$\|\theta_2 - \theta_2^h\|_{L^2(\Omega)}$		$\|\bar{u} - \bar{u}_h\|_{L^2(\Omega)}$	
		Error	Order	Error	Order	Error	Order	Error	Order
1	8	4.44e–2	–	2.76e–3	–	3.51e–3	–	4.04e–2	–
	16	1.79e–2	1.3	1.25e–3	1.1	1.56e–3	1.2	1.69e–2	1.3
	32	4.81e–3	1.9	3.44e–4	1.9	4.84e–4	1.7	4.57e–3	1.9
	64	9.43e–4	2.4	7.67e–5	2.2	1.05e–4	2.2	8.64e–4	2.4
2	8	2.01e–2	–	1.70e–3	–	1.90e–3	–	1.69e–2	–
	16	3.44e–3	2.5	3.57e–4	2.3	3.85e–4	2.3	3.09e–3	2.5
	32	2.83e–4	3.6	3.42e–5	3.4	4.46e–5	3.1	2.25e–4	3.8
	64	2.37e–5	3.6	4.69e–6	2.9	6.03e–6	2.9	9.71e–6	4.5
3	8	7.64e–3	–	9.59e–4	–	8.82e–4	–	6.03e–3	–
	16	5.08e–4	3.9	7.67e–5	3.6	8.38e–5	3.4	4.13e–4	3.9
	32	2.47e–5	4.4	5.49e–6	3.8	6.87e–6	3.6	7.97e–6	5.7
	64	1.64e–6	3.9	3.99e–7	3.8	5.40e–7	3.7	1.36e–7	5.9

For this test case, from Tables 5 and 6, we consider the effect of the CFL number on the convergence orders. Even though the mesh is moving/deforming, the results are very good. Indeed, for $CFL = 1$ for degree p, we achieve for u, θ_1, θ_2 the order of convergence $p+1$ while $\bar{u}$ seems to superconverge with order at least $p+2$! Moreover, for $CFL = 10$, we also find orders of convergence $p+1$ for u, θ_1, θ_2. For $p = 1$, it seems that $\bar{u}$ only converges with order $p+1$, but for $p = 2, 3$, we find again that $\bar{u}$ superconvergence with an order of at least $(p+2)$!

5 Conclusions

We have introduced and numerically tested the first space-time HDG method for time-dependent advection–diffusion problems. We have showed that, when the stabilization function is suitably defined, the method provides optimally convergent approximations, even in the advection-dominated regime and with highly-deformed and moving meshes. Moreover, the superconvergence of the local averages seems to be, to the knowledge of the authors, a new phenomenon whose theoretical study constitutes the subject of ongoing research.

Acknowledgements All test cases were implemented using hpGEM [23] for which we thank V.R. Ambati for technical support. Sander Rhebergen gratefully acknowledges funding by a Rubicon Fellowship from the Netherlands Organisation for Scientific Research (NWO) and the Marie Curie Cofund Action. Bernardo Cockburn was supported in part by the National Science Foundation (Grant DMS-0712955) and by the University of Minnesota Supercomputing Institute.

References

1. Ambati, V.R., Bokhove, O.: Space-time discontinuous Galerkin finite element method for shallow water flows. J. Comput. Appl. Math. **204**, 452–462 (2007)
2. Ambati, V.R., Bokhove, O.: Space-time discontinuous Galerkin discretization of rotating shallow water equations. J. Comput. Phys. **225**, 1233–1261 (2007)
3. Cockburn, B., Dong, B., Guzmán, J.: A superconvergent LDG-hybridizable Galerkin method for second-order elliptic problems. Math. Comput. **77**, 1887–1916 (2008)
4. Cockburn, B., Dong, B., Guzmán, J., Restelli, M., Sacco, R.: Superconvergent and optimally convergent LDG-hybridizable discontinuous Galerkin methods for convection-diffusion-reaction problems. SIAM J. Sci. Comput. **31**, 3827–3846 (2009)
5. Cockburn, B., Gopalakrishnan, J.: A characterization of hybridized mixed methods for the Dirichlet problem. SIAM J. Numer. Anal. **42**, 283–301 (2004)
6. Cockburn, B., Gopalakrishnan, J., Lazarov, R.: Unified hybridization of discontinuous Galerkin, mixed, and continuous Galerkin methods for second order elliptic problems. SIAM J. Numer. Anal. **47**, 1319–1365 (2009)
7. Cockburn, B., Gopalakrishnan, J., Nguyen, N.C., Peraire, J., Sayas, F.J.: Analysis of an HDG method for Stokes flow. Math. Comput. **80**, 723–760 (2011)
8. Cockburn, B., Gopalakrishnan, J., Sayas, F.-J.: A projection-based error analysis of HDG methods. Math. Comput. **79**, 1351–1367 (2010)
9. Gopalakrishnan, J., Tan, S.: A convergent multigrid cycle for the hybridized mixed method. Numer. Linear Algebra Appl. **16**, 689–714 (2009)
10. Hueber, B., Walhorn, E., Dinkler, D.: A monolithic approach to fluid-structure interaction using space-time finite elements. Comput. Methods Appl. Mech. Eng. **193**, 2087–2104 (2004)

11. Klaij, C.M., van der Vegt, J.J.W., van der Ven, H.: Space-time discontinuous Galerkin method for the compressible Navier–Stokes equations. J. Comput. Phys. **217**, 589 (2006)
12. Klaij, C.M., van Raalte, M.H., van der Ven, H., van der Vegt, J.J.W.: Vegt, h-multigrid for space-time discontinuous Galerkin discretizations of the compressible Navier–Stokes equations. J. Comput. Phys. **227**, 1024–1045 (2007)
13. Lesoinne, M., Farhat, C.: Geometric conservation laws for flow problems with moving boundaries and deformable meshes, and their impact on aeroelastic computations. Comput. Methods Appl. Mech. Eng. **134**, 71–90 (1996)
14. Nguyen, N.C., Peraire, J., Cockburn, B.: An implicit high-order hybridizable discontinuous Galerkin method for linear convection–diffusion equations. J. Comput. Phys. **228**, 3232–3254 (2009)
15. Nguyen, N.C., Peraire, J., Cockburn, B.: An implicit high-order hybridizable discontinuous Galerkin method for nonlinear convection–diffusion equations. J. Comput. Phys. **228**, 8841–8855 (2009)
16. Nguyen, N.C., Peraire, J., Cockburn, B.: Hybridizable discontinuous Galerkin methods. In: Proceedings of the International Conference on Spectral and High Order Methods, June 2009, Trondheim, Norway. LNCSE. Springer, Berlin (2009)
17. Nguyen, N.C., Peraire, J., Cockburn, B.: A hybridizable discontinuous Galerkin method for Stokes flow. Comput. Methods Appl. Mech. Eng. **193**, 2087–2104 (2010). I have: **199**, 582–597 (2010)
18. Nguyen, N.C., Peraire, J., Cockburn, B.: An implicit high-order hybridizable discontinuous Galerkin method for the incompressible Navier–Stokes equations. J. Comput. Phys. **230**, 1147–1170 (2011)
19. Peraire, J., Nguyen, N.C., Cockburn, B.: A hybridizable discontinuous Galerkin method for the incompressible Navier–Stokes equations. In: AIAA, Orlando, Florida, p. 362 (2010)
20. Peraire, J., Nguyen, N.C., Cockburn, B.: A hybridizable discontinuous Galerkin finite method for the compressible Euler and Navier–Stokes equations. In: AIAA, Orlando, Florida, p. 363 (2010)
21. Persson, P.-O., Bonet, J., Peraire, J.: Discontinuous Galerkin solution of the Navier–Stokes equations on deformable domains. Comput. Methods Appl. Mech. Eng. **198**, 1585–1595 (2009)
22. Persson, P.-O., Peraire, J.: Newton-GMRES preconditionning for discontinuous Galerkin discretizations of the Navier–Stokes equations. SIAM J. Sci. Comput. **30**, 2709–2733 (2008)
23. Pesch, L., Bell, A., Sollie, W.E.H., Ambati, V.R., Bokhove, O., van der Vegt, J.J.W.: hpGEM—a software framework for discontinuous Galerkin finite element methods. ACM Trans. Math. Softw. **33** (2007)
24. Pesch, L., van der Vegt, J.J.W.: A discontinuous Galerkin finite element discretization of the Euler equations for compressible and incompressible fluids. J. Comput. Phys. **227**, 5426–5446 (2008)
25. Reed, W.H., Hill, T.R.: Triangular mesh methods for the neutron transport equation. Los Alamos Scientific Laboratory, Report LA-UR-73-479 (1973)
26. Rhebergen, S., Bokhove, O., van der Vegt, J.J.W.: Discontinuous Galerkin finite element methods for hyperbolic nonconservative partial differential equations. J. Comput. Phys. **227**, 1887 (2008)
27. Rhebergen, S., Bokhove, O., van der Vegt, J.J.W.: Discontinuous Galerkin finite element method for shallow two-phase flows. Comput. Methods Appl. Mech. Eng. **198**, 819–830 (2009)
28. Rhebergen, S., van der Vegt, J.J.W., van der Ven, H.: Multigrid optimization for space-time discontinuous Galerkin discretizations of advection dominated flows. In: Kroll, N., Bieler, H., Deconinck, H., Couallier, V., Van der Ven, H., Sorensen, K. (eds.) ADIGMA—A European Initiative on the Development of Adaptive Higer-Order Variational Methods for Aerospace Applications. Notes on Numerical Fluid Mechanics and Multidisciplinary Design, vol. 113, pp. 257–269. Springer, Berlin (2010)

29. Sollie, W.E.H., Bokhove, O., van der Vegt, J.J.W.: Space-time discontinuous Galerkin finite element method for two-fluid flows. J. Comput. Phys. **230**, 789–817 (2011)
30. Sudirham, J.J., van der Vegt, J.J.W., van Damme, R.M.J.: Space-time discontinuous Galerkin method for advection-diffusion problems on time-dependent domains. Appl. Numer. Math. **56**, 1491–1518 (2006)
31. Tassi, P.A., Rhebergen, S., Vionnet, C.A., Bokhove, O.: A discontinuous Galerkin finite element model for river bed evolution under shallow flows. Comput. Methods Appl. Mech. Eng. **197**, 2930–2947 (2008)
32. van der Vegt, J.J.W., Sudirham, J.J.: A space-time discontinuous Galerkin method for the time-dependent Oseen equations. Appl. Numer. Math. **58**, 1892–1917 (2008)
33. van der Vegt, J.J.W., van der Ven, H.: Space-time discontinuous Galerkin finite element method with dynamic grid motion for inviscid compressible flows I. General formulation. J. Comput. Phys. **182**, 546–585 (2002)
34. van der Vegt, J.J.W., Rhebergen, S.: hp-Multigrid as smoother algorithm for higher order discontinuous Galerkin discretizations of advection dominated flows. Part I. Multilevel analysis. J. Comput. Phys. **231**(22), 7537–7563 (2012)
35. van der Vegt, J.J.W., Rhebergen, S.: hp-Multigrid as smoother algorithm for higher order discontinuous Galerkin discretizations of advection dominated flows. Part II. Optimization of the Runge-Kutta smoother. J. Comput. Phys. **231**(22), 7564–7583 (2012)

A Numerical Algorithm for Ambrosetti–Prodi Type Operators

José Teixeira Cal Neto and Carlos Tomei

Abstract We consider the numerical solution of the equation $-\Delta u - f(u) = g$, for the unknown u satisfying Dirichlet conditions in a bounded domain Ω. The nonlinearity f has bounded, continuous derivative. The algorithm uses the finite element method combined with a global Lyapunov–Schmidt decomposition.

Keywords Semilinear elliptic equations · Finite element method · Lyapunov–Schmidt decomposition

1 Introduction

We consider the partial differential equation

$$F(u) = -\Delta u - f(u) = g, \qquad u|_{\partial\Omega} = 0,$$

on domains $\Omega \in \mathbb{R}^n$, taken to be open, bounded, connected subsets of $\mathbb{R}^n$ with piecewise smooth boundary $\partial\Omega$, assumed to be at least Lipschitz at all points. There is a vast literature concerning the number of solutions for general and positive solutions for different kinds of nonlinearity f and right-hand side g (to cite a few, [1–9]).

Here we assume that the nonlinearity $f : \mathbb{R} \to \mathbb{R}$ has a bounded, continuous derivative, $a \leq f'(y) \leq b$. We show how a global Lyapunov–Schmidt decomposition introduced by Berger and Podolak [10] in their proof of the Ambrosetti–Prodi theorem (see [3, 4]) gives rise to a satisfactory solution algorithm using the finite element method. The decomposition was rediscovered by Smiley [11], who realized its potential for numerics: our results advance along these lines.

Write Δ_D for the Dirichlet Laplacian in Ω. The algorithm is especially convenient when the number d of eigenvalues of $-\Delta_D$ in the range of f' is small: the infinite dimensional equation reduces to the inversion of a map from $\mathbb{R}^d$ to itself.

J.T. Cal Neto
DME, UNIRIO, Rio de Janeiro, Brazil
e-mail: jose.calneto@uniriotec.br

C. Tomei (✉)
Departamento de Matemática, PUC-Rio, Rio de Janeiro, Brazil
e-mail: carlos@mat.puc-rio.br

C.A. de Moura, C.S. Kubrusly (eds.), *The Courant–Friedrichs–Lewy (CFL) Condition*, DOI 10.1007/978-0-8176-8394-8_5,

The subject of semilinear elliptic equations is sufficiently mature that algorithms should stand side by side with theory. The situation may be compared to the study of functions of one variable in a basic calculus course. Some functions, like parabolas, may be handled without substantial computational effort, but understanding increases with graphs, which are obtained by following a standard procedure.

We do not handle the difficulties and opportunities related to the finite dimensional inversion: a generic solver (as in [12] and, for $d = 2$, [13]) should be replaced by an algorithm which makes use of features inherited by the original map F. Here we only deal with examples for which $d = 1$ and 2, and there is some craftsmanship in handling the 2-dimensional example. It is in this step of the PDE solver that delicate issues like nonresonance and lack of properness come up.

2 The Basic Estimate

We consider the semilinear elliptic equation presented in the introduction for a nonlinearity $f(y) : \mathbb{R} \to \mathbb{R}$ with bounded, continuous derivative.

With these hypotheses, it is not hard to see that $F(u) = -\Delta u - f(u)$ is a C^1 map between the Sobolev spaces $H_0^2(\Omega)$ and $L^2(\Omega) = H^0(\Omega)$ and between $H_0^1(\Omega)$ and $H^{-1}(\Omega) \simeq H_0^1(\Omega)$. We concentrate on the second scenario, which is natural for the weak formulation of the problem. Still, the geometric statements below hold in both cases. To fix notation, set $F : X \to Y$, where $X = H_0^1(\Omega)$ and $Y = H^{-1}(\Omega)$.

The basic estimate is given in Proposition 1. Its proof is a simple extension of the argument in [10].

Define $\overline{f'(\mathbb{R})} = [a, b]$ (a allowed to be $-\infty$) and a larger interval $[\tilde{a}, \tilde{b}] \supset [a, b]$. Label the eigenvalues of $-\Delta_D$ in non-decreasing order. The *index set* J associated to $[\tilde{a}, \tilde{b}]$ is the collection of indices of eigenvalues of $-\Delta_D$ in that interval. The set J is associated to the nonlinearity f if $[\tilde{a}, \tilde{b}] = \overline{f'(\mathbb{R})}$. An index set defined this way is *complete*: it contains all indices labeling an eigenvalue in the interval.

Denote the *vertical subspaces* by $V_X \subset X$ and $V_Y \subset Y$, the spans of the normalized eigenfunctions ϕ_j, $j \in J$ in X and Y, respectively, with orthogonal complements W_X and W_Y. Let P and Q be the orthogonal projections on V and W. Clearly, the dimension of the vertical subspaces equals $|J|$, the cardinality of J. Let $v + W_X \subset X$ be the *horizontal affine subspace* of vectors $v + w$, $w \in W_X$ and consider a *projected restriction* $F_v : v + W_X \to W_Y$, the restriction of $P_Y F$ to $v + W_X$.

Proposition 1 *Let J be the index set associated to the nonlinearity f (or to any interval $[\tilde{a}, \tilde{b}]$ containing $\overline{f'(\mathbb{R})}$). Then the derivatives $DF_v : v + W_X \to W_Y$ are uniformly bounded from below. More precisely, there exists $C > 0$ such that*

$$\forall v \in V_X \; \forall w \in v + W_X \; \forall h \in W_X, \quad \left\| DF_v(w)h \right\|_Y \geq C \|h\|_X. \tag{1}$$

All such maps are invertible.

A direct application of Hadamard globalization theorem [15] implies that the projected restrictions are diffeomorphisms, for each $v \in V_X$.

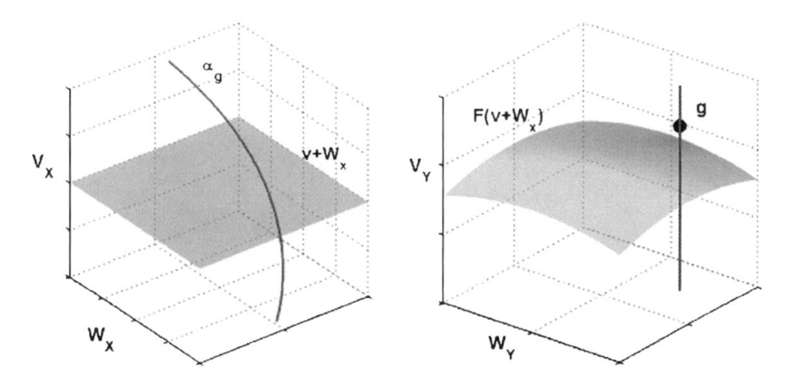

Fig. 1 Horizontal affine subspace, fiber; sheet, vertical affine subspace

3 The Underlying Picture

The geometric implications are very natural. The image under F of each horizontal affine subspace $v + W_X$ is a *sheet*, i.e., a surface which projects under P_Y diffeomorphically to the horizontal subspace W_Y. In particular, every vertical affine subspace $w + V_Y$ intercepts each sheet exactly at a single point. It is not hard to see that the intersection is transversal: tangent spaces of sheet and affine subspace form a direct sum decomposition of Y.

A *fiber* is the inverse image of a vertical affine subspace: see Fig. 1. In a similar fashion, fibers are surfaces of dimension $|J|$ which meet every horizontal affine subspace $v + W_X$ at a single point—again, the intersection is transversal. Thus, a vertical subspace parameterizes diffeomorphically each fiber, or, said differently, each fiber has a single point of a given *height*.

Recall a key idea in [10] and [12]. It is clear that X and Y are respectively foliated by fibers and vertical affine subspaces. By definition, all the solutions of $F(u) = g$ must lie in the fiber $\alpha_g = F^{-1}(g + V_Y)$. So, in principle, one might solve the equation by first identifying $\alpha_g \simeq \mathbb{R}^{|J|}$ and then facing the finite dimensional inversion of $F : \alpha_g \to g + V_Y$.

Horizontal affine subspaces are taken diffeomorphically to sheets, but fibers are not taken diffeomorphically to vertical affine subspaces. In a sense, the nonlinearity of the problem was reduced to a finite dimensional issue.

4 Finding the Fiber

Recall that each horizontal affine subspace $v + W_X$ contains exactly one element of each fiber. So, to identify α_g, choose $v + W_X$ and search in it for an element of α_g. Said differently, one may think of $F_v : v + W_X \to W_Y$ as being a diffeomorphism between fibers (represented by points in $v + W_X$) and vertical affine subspaces (represented by points in W_Y). The situation is ideal for an application of Newton's method: local improvements are performed by linearization of the diffeomorphism.

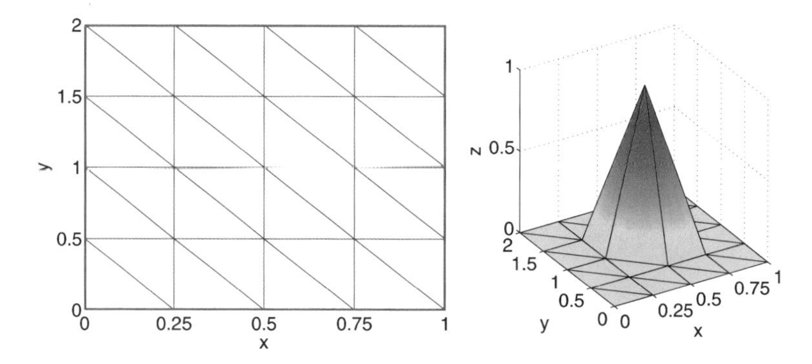

Fig. 2 Uniform triangulations of $[0, 1] \times [0, 2]$ and a nodal function

There is one difficulty, however, related to implementation issues. The functional spaces X and Y give rise to finite dimensional vector spaces generated by *finite elements*. We provide some detail; an excellent reference is [16]. First of all, *triangulate* the domain Ω, i.e., split it into disjoint simplices in $\mathbb{R}^n$. In the examples of Sect. 7, Ω is the uniformly triangulated rectangle $[0, 1] \times [0, 2]$. A *nodal function* is a continuous function that is affine linear on each simplex and has value one at a given vertex and zero at the remaining vertices. These functions form a *nodal basis* which spans a finite dimensional subspace of $H_0^1(\Omega)$. Figure 2 shows an example of a triangulation of Ω and one nodal function.

Inner products of nodal functions, both in H^1 and L^2, are often zero, a fact which simplifies the numerics associated to the weak formulation of the equation $F(u) = g$. The vertical subspaces V_X and V_Y are spanned by eigenfunctions ϕ_j, $j \in J$ and are well approximated by a few linear combinations on the nodal basis. On the other hand, obtaining a similar basis for the approximation of the orthogonal subspaces W_X and W_Y requires much more numerical effort and should be avoided.

To circumvent this problem, extend the Jacobian of $F_v : v + W_X \to W_Y$ at a point u to an invertible operator $L_u : X \to Y$ which is easy to handle and apply Newton's method to L_u instead. Setting

$$L_u z = -\Delta z - P_Y f'(u) P_X z,$$

it is clear that L_u has the required properties: it takes W_X to W_Y and V_X to V_Y and the restriction to $v + W_X$ equals DF_v, which is invertible. Moreover, the restriction to V_X coincides with $-\Delta$. This map is no longer a differential operator, due to the integrals needed to compute the projections P. But those new terms are innocuous in the finite element formulation—the sparsity of the underlying matrices is preserved, together with the possibility of standard preconditioning routines.

We search for a point of a horizontal affine subspace $v + W_X$ which belongs to α_g, $g \in Y$. The algorithm is straightforward; see Fig. 3. Choose a starting point u_0 and consider its image F_0. All would be well if the projections of F_0 and g on the horizontal subspace W_Y were equal or at least very near. When this does not happen, proceed by a continuation method to join both projections. Notice that the

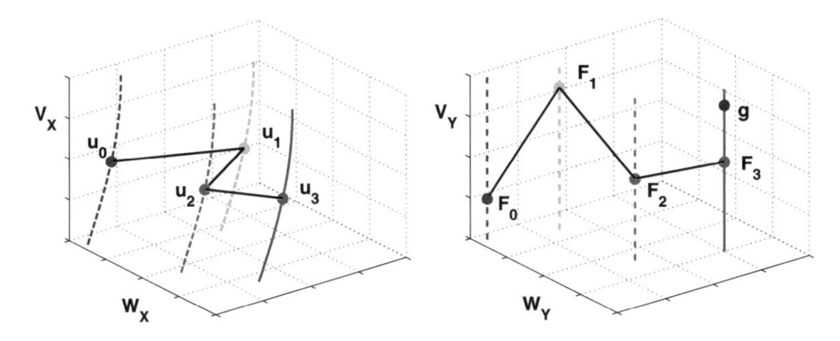

Fig. 3 Finding the right fiber

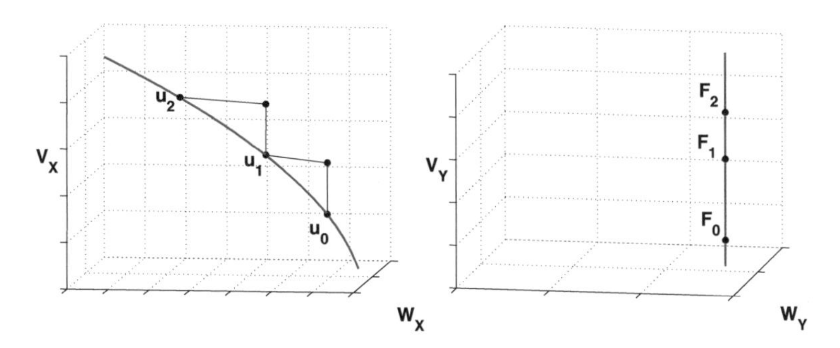

Fig. 4 Mapping a 1-D fiber

algorithm searches for the fiber (i.e., for a point in the fiber) by moving horizontally in the domain. A direct Newton iteration does not necessarily work: think of finding the (trivial) root of $\arctan(x) = 0$ starting sufficiently far from the origin.

5 Moving Along the Fiber

The necessary ingredients for a simple predictor–corrector method to move along a fiber are now available. Say $u \in \alpha_g$ and we want to find another point in α_g. Recall that fibers are parameterized by height $v \in V_X$. Take $u + v$, which is probably not in α_g, as a starting point for the algorithm in Sect. 4 to obtain the point of α_g in the same horizontal affine subspace of $u + v$ (see Fig. 4 for two such steps).

We don't know much about the behavior of F restricted to a fiber: the hypotheses on the nonlinearity f are not sufficient to imply properness of F, for example. In particular, it is not clear that the restrictions of F to a fiber are also proper.

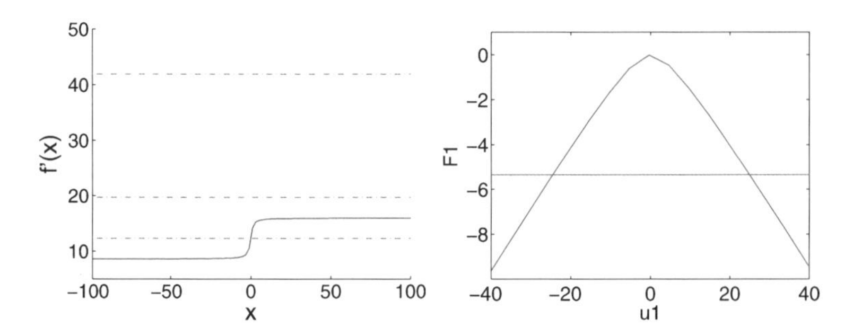

Fig. 5 The derivative of f and the image of α_g

6 Stability Issues

Proposition 1 in Sect. 2 ensures geometric stability, in the sense that the global Lyapunov–Schmidt decompositions preserve their properties under perturbations. This is convenient when replacing the vertical subspaces spanned by eigenfunctions by their finite elements counterparts.

As for the algorithm itself, the identification of the fiber is robust, being a standard continuation method associated to a diffeomorphism between horizontal affine spaces. The numerical analysis along a fiber is a different matter, and the fundamental issue was addressed by Smiley and Chun [12]: they showed that the finite element approximations to the restriction of the function F to (compact sets of) the fiber can be made arbitrarily close to the original map in the appropriate Sobolev norm. Here one must proceed with caution: small metric perturbation may induce variations in the number of solutions, as when changing from $x \mapsto x^2$, $x \in \mathbb{R}$ to $x \mapsto x^2 - \epsilon$, which is a perturbation of order ϵ for arbitrary C^k norms. Still, solutions of F which are regular points are stable: they correspond to nearby solutions of sufficiently good approximations F^h.

A different approach might be to interpret the algorithm as a provider of good starting points for Newton's iteration or at least a continuation method. As stated in [17], computer assisted arguments require good approximations for the eventual validation of solutions.

7 Some Examples

For the examples that follow, $F(u) = -\Delta u - f(u) = g$, with Dirichlet conditions on $\Omega = [0,1] \times [0,2]$. Here, $-\Delta_D$ has simple eigenvalues and $\lambda_1 = \frac{5}{4}\pi^2 \approx 12.34$, $\lambda_2 = 2\pi^2 \approx 19.74$, $\lambda_3 = \frac{17}{4}\pi^2 \approx 41.95$. Denote by ϕ_k^X and ϕ_k^Y the eigenfunctions of $-\Delta_D$ normalized in X and Y.

The first example is a nonlinearity f satisfying the hypotheses of Ambrosetti–Prodi theorem with $f(0) = 0$ and derivative $f'(x) = \alpha \arctan(x) + \beta$ with

$$\operatorname{Ran}(f') = \left(\frac{3\lambda_1 - \lambda_2}{2}, \frac{\lambda_1 + \lambda_2}{2} \right) > 0.$$

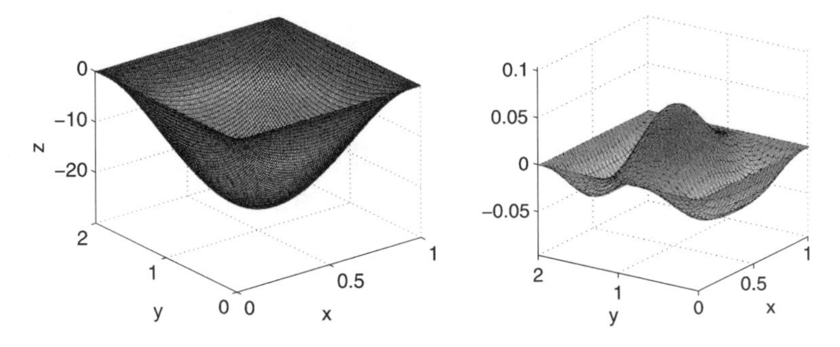

Fig. 6 A right-hand side and a function on its fiber

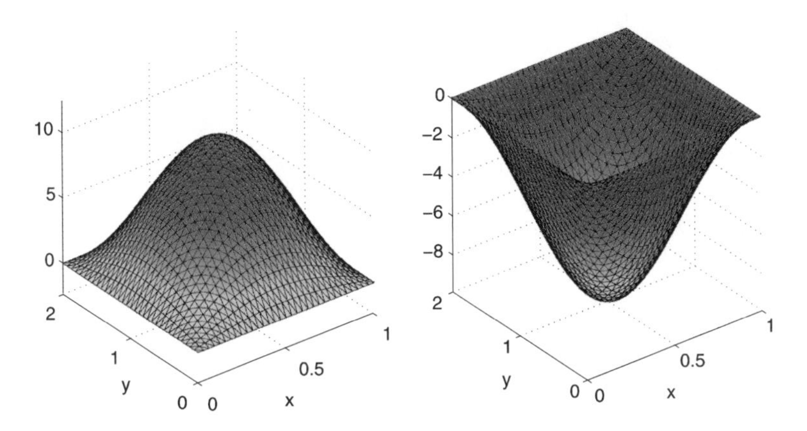

Fig. 7 Ambrosetti–Prodi solutions

The graph of f' is shown on the left of Fig. 5. Here, the index set associated to f is $J = \{1\}$: V_X and V_Y are spanned by $\phi_1^X, \phi_1^Y \geq 0$. For right-hand side set $g(x) = -100x(x-1)y(y-2)$, shown on the left of Fig. 6, which has a large negative component along the ground state.

We search for an element of the fiber α_g in the horizontal subspace W_X, starting from $u_0 = 0$, in the notation of Sect. 4. The result is the function on the right of Fig. 6. Now move along α_g, as in Sect. 5. The graph on the right of Fig. 5 plots $\langle u, \phi_1^X \rangle_X$ (the height of $u \in \alpha_g$) versus $\langle F(u), \phi_1^Y \rangle_Y$ (the height of $F(u)$). The horizontal line indicates the height of g: the solutions of the original PDE correspond to the intersections between the curve and this line. The two solutions found in this case are presented in Fig. 7.

For the next example, $J = \{1\}$ but f is a nonconvex function whose derivative is depicted on the left of Fig. 8. We consider the fiber through $u_0(x) = -50\phi_1^X(x) + 10\phi_2^X(x)$, i.e., $\alpha_{F(u_0)}$. According to Fig. 8(right), moving up the fiber yields three distinct solutions.

As a concluding example, we take a nonlinearity f for which $J = \{1, 2\}$: here the vertical spaces are spanned by the first two eigenfunctions. The function f is of

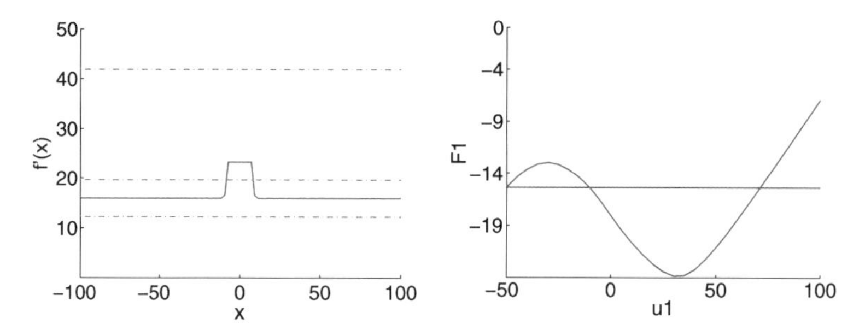

Fig. 8 Non-convex f

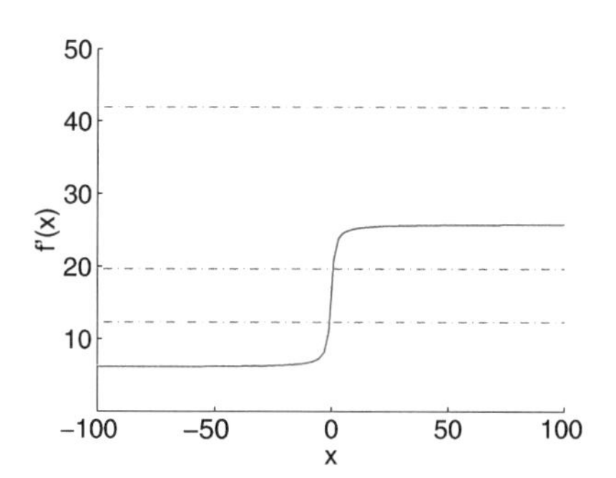

Fig. 9 The range of f' contains λ_1 and λ_2

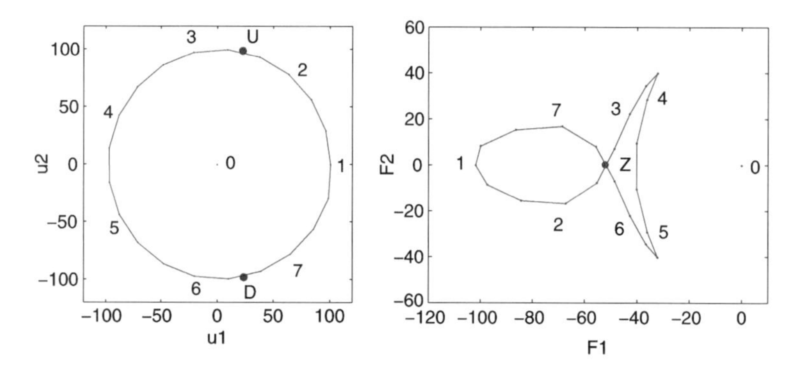

Fig. 10 Two preimages on the circle

the same form as the first example and its derivative is shown in Fig. 9. We study the fiber through the point $u_0 = 0$, which is α_0, since $F(0) = 0$. Recall from Sect. 4 that there is exactly one point of α_0 for each height, i.e., given a point $u \in V_X$, there is a unique point $\zeta(u) \in \alpha_0$ in the same horizontal affine subspace as u. For a circle C centered at the origin in V_X, $\zeta(C) \in \alpha_0$. The image $F(\zeta(C))$ is shown in the right side of Fig. 10: here, we must project $F(\zeta(u))$ along directions ϕ_1^Y and ϕ_2^Y. Clearly,

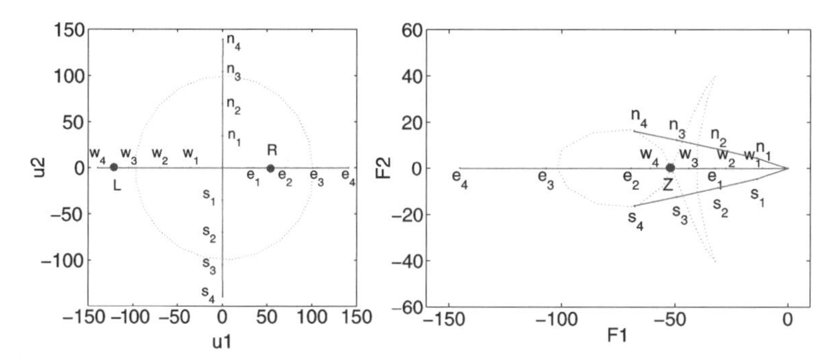

Fig. 11 Two preimages along the horizontal axis

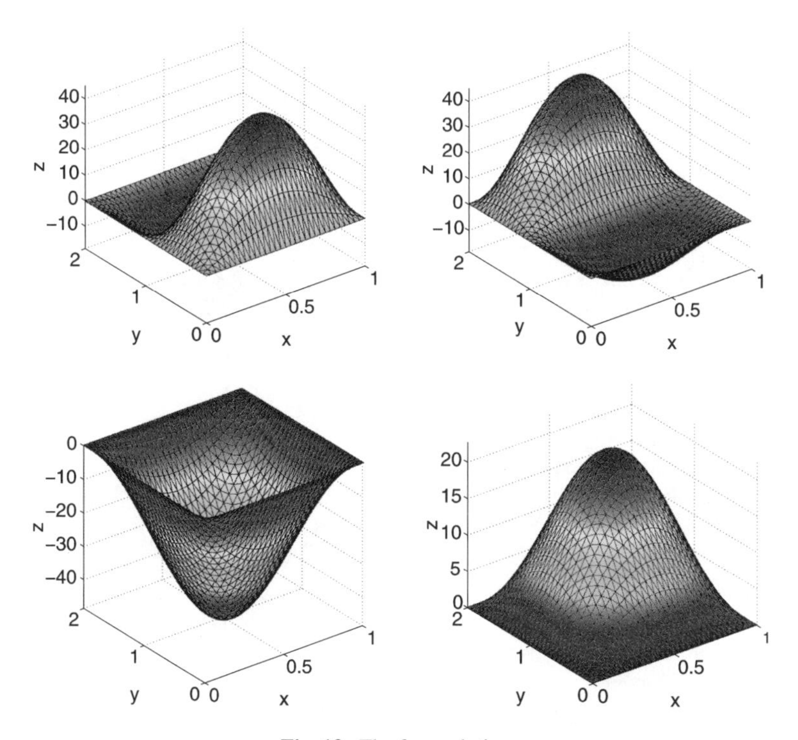

Fig. 12 The four solutions

there is a double point Z in $F(\zeta(C))$ and it is not hard to identify in C its two pre-images, U and D marked in the left of Fig. 10.

We now obtain two additional preimages of Z in a rather naive fashion. The images under $F \circ \zeta$ of the four half-axes of V_X are drawn on the right of Fig. 11. It is clear, then, that the horizontal axis contains two preimages L and R of Z, which are easily computed.

For the sake of completeness, Fig. 12 displays the four solutions.

Acknowledgements The results are part of the PhD thesis of the first author [14]. Complete proofs are presented elsewhere. The authors are grateful to CAPES, CNPq, and Faperj for support.

References

1. Hammerstein, A.: Nichtlineare Integralgleichungen nebst Anwendungen. Acta Math. **54**(1), 117–176 (1930)
2. Dolph, C.L.: Nonlinear integral equations of the Hammerstein type. Trans. Am. Math. Soc. **66**, 289–307 (1949)
3. Ambrosetti, A., Prodi, G.: On the inversion of some differentiable mappings with singularities between Banach spaces. Ann. Mat. Pura Appl. **93**, 231–246 (1972)
4. Manes, A., Micheletti, A.M.: Un'estensione della teoria variazionale classica degli autovalori per operatori ellittici del secondo ordine. Boll. Unione Mat. Ital. **7**, 285–301 (1973)
5. Lazer, A.C., McKenna, P.J.: On the number of solutions of a nonlinear Dirichlet problem. J. Math. Anal. Appl. **84**(1), 282–294 (1981)
6. Lupo, D., Solimini, S., Srikanth, P.N.: Multiplicity results for an ODE problem with even nonlinearity. Nonlinear Anal. **12**(7), 657–673 (1988)
7. Dancer, E.N.: A counterexample to the Lazer–McKenna conjecture. Nonlinear Anal. **13**(1), 19–21 (1989)
8. Costa, D.G., de Figueiredo, D.G., Srikanth, P.N.: The exact number of solutions for a class of ordinary differential equations through Morse index computation. J. Differ. Equ. **96**(1), 185–199 (1992)
9. Breuer, B., McKenna, P.J., Plum, M.: Multiple solutions for a semilinear boundary value problem: a computational multiplicity proof. J. Differ. Equ. **195**(1), 243–269 (2003)
10. Berger, M.S., Podolak, E.: On the solutions of a nonlinear Dirichlet problem. Indiana Univ. Math. J. **24**, 837–846 (1974)
11. Smiley, M.W.: A finite element method for computing the bifurcation function for semilinear elliptic BVPs. J. Comput. Appl. Math. **70**(2), 311–327 (1996)
12. Smiley, M.W., Chun, C.: Approximation of the bifurcation function for elliptic boundary value problems. Numer. Methods Partial Differ. Equ. **16**(2), 194–213 (2000)
13. Malta, I., Saldanha, N.C., Tomei, C.: The numerical inversion of functions from the plane to the plane. Math. Comput. **65**(216), 1531–1552 (1996)
14. Cal Neto, J.: Numerical analysis of Ambrosetti–Prodi type operators. PhD thesis, Departamento de Matemática–Pontifícia Universidade Católica do Rio de Janeiro (PUC-Rio) (2010)
15. Berger, M.S.: Nonlinearity and Functional Analysis. Academic Press [Harcourt Brace Jovanovich Publishers], New York (1977). Lectures on nonlinear problems in mathematical analysis, pure and applied mathematics
16. Ciarlet, P.G.: The Finite Element Method for Elliptic Problems. Classics in Applied Mathematics, vol. 40. Society for Industrial and Applied Mathematics (SIAM), Philadelphia (2002). Reprint of the 1978 original, North-Holland, Amsterdam, MR0520174 (58 #25001)
17. Plum, M.: Computer-assisted proofs for semilinear elliptic boundary value problems. Jpn. J. Ind. Appl. Math. **26**(2–3), 419–442 (2009)

On the Quadratic Finite Element Approximation of 1D Waves: Propagation, Observation, Control, and Numerical Implementation

Aurora Marica and Enrique Zuazua

Abstract In arXiv:1112.4297, we studied the propagation, observation, and control properties of the 1D wave equation on a bounded interval semi-discretized in space using the *quadratic classical finite element* approximation. It was shown that the discrete wave dynamics consisting of the interaction of *nodal* and *midpoint components* leads to the existence of two different eigenvalue branches in the spectrum: an *acoustic* one, of physical nature, and an *optic* one, of spurious nature. The fact that both *dispersion relations* have critical points where the corresponding *group velocities* vanish produces numerical wave packets whose energy is concentrated in the interior of the domain, without propagating, and for which the *observability constant* blows up as the mesh size goes to zero. This extends to the *quadratic finite element* setting the fact that the classical property of continuous waves being observable from the boundary fails for the most classical approximations on *uniform meshes* (*finite differences*, *linear finite elements*, etc.). As a consequence, the numerical controls of minimal norm may blow up as the mesh size parameter tends to zero. To cure these high-frequency pathologies, in arXiv:1112.4297 we designed a *filtering mechanism* consisting in taking *piecewise linear and continuous initial data* (so that the curvature component vanishes at the initial time) with nodal components given by a *bi-grid algorithm*. The aim of this article is to implement this filtering technique and to show numerically its efficiency.

Keywords Linear and quadratic finite element method · Uniform mesh · Vanishing group velocity · Observability/controllability property · Acoustic/optic mode · Bi-grid algorithm · Conjugate gradient algorithm

---

A. Marica · E. Zuazua
BCAM—Basque Center for Applied Mathematics, Alameda de Mazarredo 14, 48009, Bilbao, Basque Country, Spain

A. Marica
e-mail: marica@bcamath.org

E. Zuazua (✉)
Ikerbasque—Basque Foundation for Science, Alameda Urquijo 36-5, Plaza Bizkaia, 48011, Bilbao, Basque Country, Spain
e-mail: zuazua@bcamath.org

C.A. de Moura, C.S. Kubrusly (eds.), *The Courant–Friedrichs–Lewy (CFL) Condition*, DOI 10.1007/978-0-8176-8394-8_6,

1 Preliminaries on the Continuous Model and Problem Formulation

Consider the 1D wave equation with non-homogeneous boundary conditions:

$$\begin{cases} y_{tt}(x,t) - y_{xx}(x,t) = 0, & x \in (0,1),\ t > 0, \\ y(0,t) = 0, \quad y(1,t) = v(t), & t > 0, \\ y(x,0) = y^0(x), \quad y_t(x,0) = y^1(x), & x \in (0,1). \end{cases} \tag{1}$$

System (1) is said to be *exactly controllable* in time $T \geq 2$ if, for all $(y^0, y^1) \in L^2 \times H^{-1}(0,1)$, there exists a control function $v \in L^2(0,T)$ such that the solution of (1) can be driven to rest at time T, i.e. $y(x,T) = y_t(x,T) = 0$.

We also introduce the adjoint 1D wave equation with homogeneous boundary conditions:

$$\begin{cases} u_{tt}(x,t) - u_{xx}(x,t) = 0, & x \in (0,1),\ t > 0, \\ u(0,t) = u(1,t) = 0, & t > 0, \\ u(x,T) = u^0(x), \quad u_t(x,T) = u^1(x), & x \in (0,1). \end{cases} \tag{2}$$

This system is well known to be well posed in the energy space $\mathcal{V} := H_0^1 \times L^2(0,1)$ and the energy below is conserved in time:

$$\mathcal{E}(u^0,u^1) = \frac{1}{2}\left(\|u(\cdot,t)\|_{H_0^1}^2 + \|u_t(\cdot,t)\|_{L^2}^2\right) = \frac{1}{2}\left(\|u^0\|_{H_0^1}^2 + \|u^1\|_{L^2}^2\right).$$

The Hilbert Uniqueness Method (HUM) introduced in [8] allows showing that the property above of *exact controllability* for (1) is equivalent to the *boundary observability property* of (2). The *observability property* ensures that the following *observability inequality* holds for all solutions of (2), provided $T \geq 2$:

$$\mathcal{E}(u^0,u^1) \leq C(T) \int_0^T |u_x(1,t)|^2\, dt. \tag{3}$$

The best constant $C(T)$ in (3) is the so-called *observability constant*. The observability time T has to be larger than the characteristic one, $T^\star := 2$, needed by any initial data (u^0,u^1) supported in a very narrow neighborhood of $x = 1$ to travel along the characteristic rays parallel to $x(t) = x - t$, touch the boundary $x = 0$ and bounce back to the boundary $x = 1$ along the characteristics parallel to $x(t) = x + t$.

The HUM control v, the one of minimal $L^2(0,T)$-norm, for which the solution of (1) fulfills $y(x,T) = y_t(x,T) = 0$, has the explicit form

$$v(t) = \tilde{v}(t) := \tilde{u}_x(1,t), \tag{4}$$

where $\tilde{u}(x,t)$ is the solution of (2) corresponding to the minimum $(\tilde{u}^0, \tilde{u}^1) \in \mathcal{V}$ of the quadratic functional

$$\mathcal{J}(u^0,u^1) = \frac{1}{2}\int_0^T |u_x(1,t)|^2\, dt - \left\langle (y^1, -y^0), \left(u(\cdot,0), u_t(\cdot,0)\right)\right\rangle_{\mathcal{V}',\mathcal{V}}. \tag{5}$$

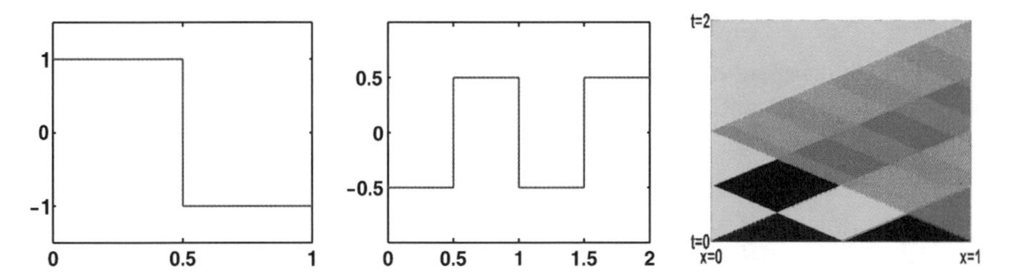

Fig. 1 The initial position $H(x)$ (*left*) versus the HUM control $\tilde{v}_H$ (*middle*) versus the solution y of the control problem (1) (*right*) (*red* $=1$, *orange* $=1/2$, *green* $=0$, *cyan* $=-1/2$, and *blue* $=-1$)

Here, $\mathcal{V}' = H^{-1} \times L^2(0,1)$ and $\langle \cdot, \cdot \rangle_{\mathcal{V}',\mathcal{V}}$ is the duality product between $\mathcal{V}$ and $\mathcal{V}'$.

In this paper, in order to analyze the efficiency of the various models under consideration, we shall run the simulations on a specific example. We consider the particular case of the characteristic control time $T=2$ and of initial data (y^0, y^1) in (1) given by $y^1 \equiv 0$ and the Heaviside function H as initial position:

$$y^0(x) = H(x) := \begin{cases} 1, & x \in [0, 1/2), \\ -1, & x \in [1/2, 1]. \end{cases} \tag{6}$$

The initial position, having discontinuities, involves significant high frequency components that will be the source of instabilities for the numerical methods under consideration. This example allows us to highlight the high-frequency pathologies of the numerical approximations of the controlled wave problem (1) and the effects of the filtering techniques we propose. In this particular case, the HUM control can be explicitly computed by Fourier expansions, using the periodicity with time period $\tau = 2$ of the solutions (cf. Sect. 3.3 in [3]), and it is given by (see Fig. 1):

$$\tilde{v}(t) = \tilde{v}_H(t) = \begin{cases} -1/2, & t \in (0, 1/2] \cup (1, 3/2], \\ 1/2, & t \in (1/2, 1] \cup (3/2, 2). \end{cases} \tag{7}$$

The *discrete approach* to the numerical approximation of this kind of control problems has been intensively studied during the last years, starting from some simple models on uniform meshes like *finite differences* or *linear finite element methods* in [7] and, more recently, more complex schemes like the *discontinuous Galerkin* ones in [10]. The problem consists in analyzing whether the controls of a numerical approximation scheme of (1), obtained in a similar manner, i.e., by minimizing a suitable discrete version of (5), converge to the control v of the wave equation (1) as the mesh size parameter tends to zero. In all these cases, the convergence of the approximation scheme in the classical sense of the numerical analysis does not suffice to guarantee that the sequence of discrete controls converges to the continuous ones, as one could expect. This is due to the fact that there are classes of initial data for the discrete adjoint problem generating high-frequency wave packets propagating at a very low group velocity and that, consequently, cannot be observed from the boundary of the domain during a finite time uniformly with the mesh size parameter. This leads to the divergence of the discrete observability constant as the mesh size tends to zero.

Similar high-frequency pathological phenomena have also been observed for numerical approximation schemes of other models, like the linear Schrödinger equation (cf. [5]), in which one is also interested in the uniformity of the so-called *dispersive estimates*, which play an important role in the study of the well-posedness of some nonlinear models.

The rest of the paper is organized as follows. In Sect. 2, we summarize some well-known results on the boundary controllability of the classical finite element space semi-discretizations, especially the linear and the quadratic ones, emphasizing the high frequency pathologies and their remedies based on the *bi-grid algorithm*. In Sect. 3, we present in detail the implementation of the conjugate gradient algorithm giving the numerical HUM controls, together with its two-grid adaptation, and we show some numerical results to illustrate the validity of the theoretical ones. In Sect. 4, we will summarize the conclusions of our paper and some related open problems.

Before starting, let us give some basic notation. All vectors we deal with will be considered as being column vectors and will be denoted by bold capital letters. We will use capital letters for the components of the vectors and for matrices and calligraphic capital letters for the discrete spaces. We denote: by h—the mesh size and it will be the first superscript; by p—the degree of the numerical approximation and it will be the first subscript; by the superscript $*$—the transposition of a matrix; and by the overline symbol—the complex conjugation.

2 Preliminaries on Numerical Controls Using P_1 and P_2 Finite Element Approximations

Let us now introduce *the quadratic P_2 finite element* approximation method and recall the main existent results, taken essentially from [11]. We consider $N \in \mathbb{N}$, $h = 1/(N+1)$, and $0 = x_0 < x_j < x_{N+1} = 1$ to be the *nodes* of a *uniform grid* of the interval $[0, 1]$, with $x_j = jh, 0 \le j \le N+1$, constituted by the subintervals $I_j = (x_j, x_{j+1})$, with $0 \le j \le N$. On this grid, we also define the *midpoints* $x_{j+1/2} = (j+1/2)h$, with $0 \le j \le N$. Let us introduce the space $\mathcal{P}_p(a,b)$ of polynomials of order p on the interval (a,b) and the *space of piecewise quadratic and continuous functions* $\mathcal{U}_2^h := \{u \in H_0^1(0,1) \text{ s.t. } u|_{I_j} \in \mathcal{P}_2(I_j),\ 0 \le j \le N\}$. The space $\mathcal{U}_2^h$ can be written as

$$\mathcal{U}_2^h = \text{span}\{\phi_{2,j}^h, 1 \le j \le N\} \oplus \text{span}\{\phi_{2,j+1/2}^h, 0 \le j \le N\},$$

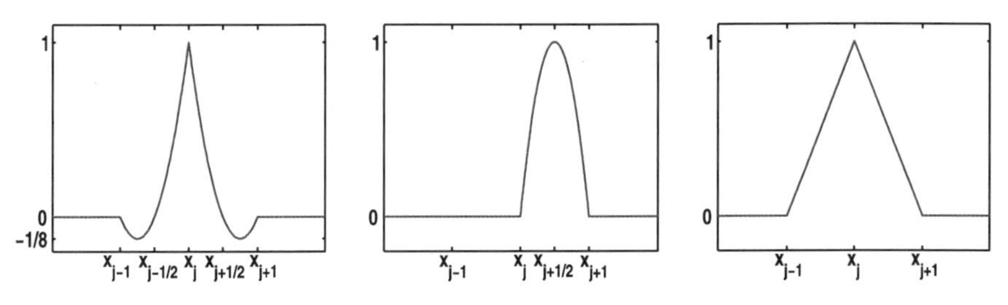

Fig. 2 The basis functions: $\phi_{2,j}^h$ (*left*), $\phi_{2,j+1/2}^h$ (*middle*), and $\phi_{1,j}^h$ (*right*)

where the two classes of basis functions are represented in Fig. 2 and are explicitly given by

$$\phi_{2,j}^h(x) = \begin{cases} \frac{2}{h^2} Q(x, x_{j-1}, x_{j-1/2}), & x \in I_{j-1}, \\ \frac{2}{h^2} Q(x, x_{j+1/2}, x_{j+1}), & x \in I_j, \\ 0, & \text{otherwise}, \end{cases} \tag{8}$$

$$\phi_{2,j+1/2}^h(x) = \left[-\frac{4}{h^2} Q(x, x_j, x_{j+1}) \right]^+$$

with $Q(x,a,b) = (x-a)(x-b)$ and $[f]^+$—the positive part of f.

We will compare the results obtained when numerically approximating the controls on this basis with the ones obtained by the *linear* P_1 *finite element* approximation. In order to do this, in the same *uniform grid* of size h defined by the *nodal points* x_j, $0 \le j \le N+1$, we introduce the *space of piecewise linear and continuous functions* $\mathcal{U}_1^h := \{u \in H_0^1(0,1) \text{ s.t. } u|_{I_j} \in \mathcal{P}_1(I_j),\ 0 \le j \le N\}$, which can be written as $\mathcal{U}_1^h = \text{span}\{\phi_{1,j}^h,\ 1 \le j \le N\}$, where $\phi_{1,j}^h(x) = [1 - (x - x_j)/h]^+$. The linear/quadratic approximation of the adjoint problem (2) is

$$\begin{cases} \text{Find } u_p^h(\cdot,t) \in \mathcal{U}_p^h \text{ s.t. } \frac{d^2}{dt^2}(u_p^h(\cdot,t),\varphi)_{L^2} + (u_p^h(\cdot,t),\varphi)_{H_0^1} = 0, \quad \forall \varphi \in \mathcal{U}_p^h, \\ u_p^h(x,T) = u_p^{h,0}(x), \qquad u_{p,t}^h(x,T) = u_p^{h,1}(x), \quad x \in (0,1). \end{cases} \tag{9}$$

The solution $u_p^h(\cdot,t) \in \mathcal{U}_p^h$ admits the decomposition

$$u_p^h(x,t) = \sum_{j=1}^{pN+p-1} U_{p,j/p}(t)\phi_{p,j/p}^h(x).$$

Consequently, the function $u_p^h(\cdot,t)$ can be identified with the vector of its coefficients, $\mathbf{U}_p^h(t) = (U_{p,j/p}(t))_{1 \le j \le pN+p-1}$. Thus, problem (9) can be written as a system of second-order linear ordinary differential equations (ODEs):

$$M_p^h \mathbf{U}_{p,tt}^h(t) + S_p^h \mathbf{U}_p^h(t) = 0, \quad \mathbf{U}_p^h(T) = \mathbf{U}_p^{h,0}, \quad \mathbf{U}_{p,t}^h(T) = \mathbf{U}_p^{h,1}, \ p = 1,2, \tag{10}$$

where M_1^h and S_1^h are the following $N \times N$ *tri-diagonal mass* and *stiffness* matrices

$$M_1^h = \begin{pmatrix} \frac{2h}{3} & \frac{h}{6} & 0 & \cdots & 0 & 0 \\ \frac{h}{6} & \frac{2h}{3} & \frac{h}{6} & \cdots & 0 & 0 \\ \vdots & \ddots & \ddots & \ddots & \vdots & \vdots \\ 0 & 0 & 0 & \cdots & \frac{h}{6} & \frac{2h}{3} \end{pmatrix},$$

$$S_1^h = \begin{pmatrix} \frac{2}{h} & -\frac{1}{h} & 0 & \cdots & 0 & 0 \\ -\frac{1}{h} & \frac{2}{h} & -\frac{1}{h} & \cdots & 0 & 0 \\ \vdots & \ddots & \ddots & \ddots & \vdots & \vdots \\ 0 & 0 & 0 & \cdots & -\frac{1}{h} & \frac{2}{h} \end{pmatrix}$$

and M_2^h and S_2^h are the following $(2N+1) \times (2N+1)$ *pentha-diagonal mass* and *stiffness* matrices

$$M_2^h = \begin{pmatrix} \frac{8h}{15} & \frac{h}{15} & 0 & 0 & 0 & 0 & \cdots & 0 & 0 & 0 & 0 \\ \frac{h}{15} & \frac{4h}{15} & \frac{h}{15} & -\frac{h}{30} & 0 & 0 & \cdots & 0 & 0 & 0 & 0 \\ 0 & \frac{h}{15} & \frac{8h}{15} & \frac{h}{15} & 0 & 0 & \cdots & 0 & 0 & 0 & 0 \\ 0 & -\frac{h}{30} & \frac{h}{15} & \frac{4h}{15} & \frac{h}{15} & -\frac{h}{30} & \cdots & 0 & 0 & 0 & 0 \\ \vdots & \vdots & \ddots & \ddots & \ddots & \ddots & \ddots & \vdots & \vdots & \vdots & \vdots \\ 0 & 0 & 0 & 0 & 0 & 0 & \cdots & -\frac{h}{30} & \frac{h}{15} & \frac{4h}{15} & \frac{h}{15} \\ 0 & 0 & 0 & 0 & 0 & 0 & \cdots & 0 & 0 & \frac{h}{15} & \frac{8h}{15} \end{pmatrix}$$

and

$$S_2^h = \begin{pmatrix} \frac{16}{3h} & -\frac{8}{3h} & 0 & 0 & 0 & 0 & \cdots & 0 & 0 & 0 & 0 \\ -\frac{8}{3h} & \frac{14}{3h} & -\frac{8}{3h} & \frac{1}{3h} & 0 & 0 & \cdots & 0 & 0 & 0 & 0 \\ 0 & -\frac{8}{3h} & \frac{16}{3h} & -\frac{8}{3h} & 0 & 0 & \cdots & 0 & 0 & 0 & 0 \\ 0 & \frac{1}{3h} & -\frac{8}{3h} & \frac{14}{3h} & -\frac{8}{3h} & \frac{1}{3h} & \cdots & 0 & 0 & 0 & 0 \\ \vdots & \vdots & \ddots & \ddots & \ddots & \ddots & \ddots & \vdots & \vdots & \vdots & \vdots \\ 0 & 0 & 0 & 0 & 0 & 0 & \cdots & \frac{1}{3h} & -\frac{8}{3h} & \frac{14}{3h} & -\frac{8}{3h} \\ 0 & 0 & 0 & 0 & 0 & 0 & \cdots & 0 & 0 & -\frac{8}{3h} & \frac{16}{3h} \end{pmatrix}.$$

For $p = 1, 2$ (corresponding to the linear/quadratic approximation), let us introduce some notations for the discrete analogues of $H_0^1(0,1)$, $L^2(0,1)$, and $H^{-1}(0,1)$,

$$\mathcal{H}_p^{h,i} := \left\{ \mathbf{F}_p^h = (F_{p,j/p})_{1 \le j \le pN+p-1} \in \mathbb{C}^{pN+p-1} \text{ s.t. } \left\| \mathbf{F}_p^h \right\|_{\mathcal{H}_p^{h,i}} < \infty \right\},$$
$$i = 1, 0, -1.$$

The elements of $\mathcal{H}_p^{h,1}$ verify the additional requirement $F_{p,0} = F_{p,N+1} = 0$. The inner products defining the discrete spaces $\mathcal{H}_p^{h,i}$, $i = 1, 0, -1$, are given by

$$(\mathbf{E}_p^h, \mathbf{F}_p^h)_{\mathcal{H}_p^{h,i}} := \big((M_p^h (S_p^h)^{-1})^{1-i} S_p^h \mathbf{E}_p^h, \mathbf{F}_p^h\big)_{p,e}, \quad i = 1, 0, -1, \tag{11}$$

and the norms are given by $\|\mathbf{F}_p^h\|^2_{\mathcal{H}_p^{h,i}} := (\mathbf{F}_p^h, \mathbf{F}_p^h)_{\mathcal{H}_p^{h,i}}$, for all $i = 1, 0, -1$. Here, $(\cdot,\cdot)_{p,e}$ is the inner product in the Euclidean space $\mathbb{C}^{pN+p-1}$, defined by

$$(\mathbf{F}_p^h, \mathbf{G}_p^h)_{p,e} := \sum_{k=1}^{pN+p-1} F_{p,k/p} \overline{G}_{p,k/p}.$$

Set $\mathcal{V}_p^h := \mathcal{H}_p^{h,1} \times \mathcal{H}_p^{h,0}$ and its dual $\mathcal{V}_p^{h,'} := \mathcal{H}_p^{h,-1} \times \mathcal{H}_p^{h,0}$, the duality product $\langle \cdot, \cdot \rangle_{\mathcal{V}_p^{h,'}, \mathcal{V}_p^h}$ between $\mathcal{V}_p^{h,'}$ and $\mathcal{V}_p^h$ being defined as

$$\big\langle (\mathbf{F}_{p,1}^h, \mathbf{G}_{p,1}^h), (\mathbf{F}_{p,2}^h, \mathbf{G}_{p,2}^h) \big\rangle_{\mathcal{V}_p^{h,'}, \mathcal{V}_p^h} := (\mathbf{F}_{p,1}^h, \mathbf{F}_{p,2}^h)_{\mathcal{H}_p^{h,0}} + (\mathbf{G}_{p,1}^h, \mathbf{G}_{p,2}^h)_{\mathcal{H}_p^{h,0}}.$$

Problem (10) is well posed in $\mathcal{V}_p^h$. The total energy of its solutions defined below is conserved in time:

$$\begin{aligned} \mathcal{E}_p^h(\mathbf{U}_p^{h,0}, \mathbf{U}_p^{h,1}) &= \frac{1}{2}\big(\|\mathbf{U}_p^h(t)\|^2_{\mathcal{H}_p^{h,1}} + \|\mathbf{U}_{p,t}^h(t)\|^2_{\mathcal{H}_p^{h,0}}\big) \\ &= \frac{1}{2}\big(\|\mathbf{U}_p^{h,0}\|^2_{\mathcal{H}_p^{h,1}} + \|\mathbf{U}_p^{h,1}\|^2_{\mathcal{H}_p^{h,0}}\big). \end{aligned} \tag{12}$$

In [7] and [11], the following discrete version of the observability inequality (3) for the linear ($p = 1$) and for the quadratic ($p = 2$) approximation was analyzed:

$$\mathcal{E}_p^h(\mathbf{U}_p^{h,0}, \mathbf{U}_p^{h,1}) \le C_p^h(T) \int_0^T \|B_p^h \mathbf{U}_p^h(t)\|^2_{p,e}\, dt, \tag{13}$$

where B_p^h is a $(pN + p - 1) \times (pN + p - 1)$ *observability matrix operator*. Within this paper we focus on the particular case of *boundary observation operators* B_p^h, in the sense that they approximate the normal derivative $u_x(x,t)$ of the solution of the continuous adjoint problem (2) at $x = 1$ as $h \to 0$. One of the simplest examples of such boundary matrix operators B_h^p that will be used throughout this paper is:

$$B_{p,ij} := \begin{cases} -\frac{1}{h}, & (i,j) = (pN + p - 1, pN), \\ 0, & \text{otherwise}. \end{cases} \tag{14}$$

The only non-trivial component of $B_p^h \mathbf{U}_p^h(t)$ is the last one which equals to $u_{h,x}(x_{N+(p-1)/p}, t)$ and is a first-order approximation of $u_x(1,t)$, where u is a solution of (2).

As shown in [7] for $p=1$ and in [11] for $p=2$, the observability inequality (13) does not hold uniformly as $h\to 0$, meaning that the observability constant $C_p^h(T)$ in (13) blows up whatever $T>0$ is. This is due to the existence of solutions propagating very slowly concentrated on zones of the spectrum where the *spectral gap* or the *group velocity* tends to zero as $h\to 0$. To be more precise, for $\eta\in[0,\pi]$, let us introduce the *Fourier symbols*

$$\Lambda_1(\eta):=\frac{6(1-\cos(\eta))}{2+\cos(\eta)},$$

$$\Lambda_2^{\alpha}(\eta):=\frac{22+8\cos^2(\eta/2)+2\mathrm{sign}(\alpha)\sqrt{\Delta(\eta)}}{1+\sin^2(\eta/2)},\quad \text{for } \alpha\in\{\mathrm{a},\mathrm{o}\},$$

where $\mathrm{sign}(\mathrm{a})=-1,\ \mathrm{sign}(\mathrm{o})=1,$ and

$$\Delta(\eta):=1+268\cos^2(\eta/2)-44\cos^4(\eta/2).$$

Define $\lambda_1(\eta):=\sqrt{\Lambda_1(\eta)}$ and $\lambda_2^{\alpha}(\eta):=\sqrt{\Lambda_2^{\alpha}(\eta)}$, $\alpha\in\{\mathrm{a},\mathrm{o}\}$. Set $\Lambda_1^k:=\Lambda_1(k\pi h)$ and $\Lambda_2^{\alpha,k}:=\Lambda_2^{\alpha}(k\pi h)$, $\alpha\in\{\mathrm{a},\mathrm{o}\}$, and consider the following spectral problem:

$$S_p^h\boldsymbol{\varphi}_p^h=\Lambda_p^h M_p^h\boldsymbol{\varphi}_p^h. \tag{15}$$

We take L^2-normalized eigenvectors, i.e., $\|\boldsymbol{\varphi}_p^h\|_{\mathcal{H}_p^{h,0}}=1$. The eigenvalues are explicitly given by

$$\Lambda_1^{h,k}=\Lambda_1^k/h^2,\qquad \Lambda_2^{h,\alpha,k}=\Lambda_2^{\alpha,k}/h^2,$$

with $\alpha\in\{\mathrm{a},\mathrm{o}\}$ and $1\le k\le N$. The superscripts a, o entering in the notation of the P_2-eigenvalues stand for *acoustic/optic*, respectively, to distinguish these two main branches of the spectrum. In the quadratic case, $p=2$, additionally to the $2N$ modes $\Lambda_2^{h,\alpha,k}$, with $1\le k\le N$ and $\alpha\in\{\mathrm{a},\mathrm{o}\}$, there is also the so-called *resonant* mode, given by $\Lambda_2^{h,\mathrm{r}}=10/h^2$. In Fig. 3, we represent $\lambda_p^h:=\sqrt{\Lambda_p^h}$ for different values of p.

The solutions of (10) admit the following Fourier representation:

$$\mathbf{U}_p^h(t)=\sum_{\pm}\sum_{(\Lambda_p^h,\boldsymbol{\varphi}_p^h)}\widehat{u}_{p,\pm}\exp\big(\pm it\lambda_p^h\big)\boldsymbol{\varphi}_p^h,$$

where the second sum is taken over the all possible eigensolutions $(\Lambda_p^h,\boldsymbol{\varphi}_p^h)$ in (15). Here, $\widehat{u}_{p,\pm}=(\widehat{u}_p^0\pm\widehat{u}_p^1/i\lambda_p^h)/2$ and $\widehat{u}_p^i$ are the Fourier coefficients of the initial data $\mathbf{U}_p^{h,i}$ defined by $\widehat{u}_p^i:=(\mathbf{U}_p^{h,i},\varphi_p^h)_{\mathcal{H}_p^{h,0}}$.

Firstly, let us remark that as $kh\to 1$, $\Lambda_2^{\mathrm{a},k}\to 10$, Λ_1^k, $\Lambda_2^{\mathrm{o},k}\to 12$ and as $kh\to 0$, $\Lambda_2^{\mathrm{o},k}\to 60$. On the other hand, as we can see in Fig. 3, $\lambda_1^{h,k}$ and $\lambda_2^{h,\mathrm{a},k}$ are strictly increasing in k, while $\lambda_2^{h,\mathrm{o},k}$ is strictly decreasing. The *group velocities*, which are

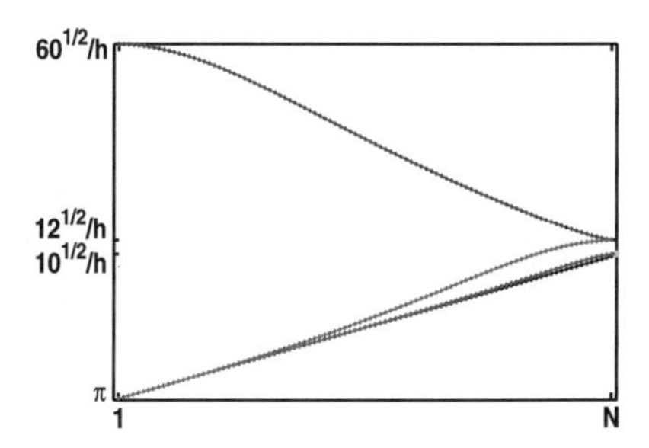

Fig. 3 The square roots of the eigenvalues, λ_p^h: the continuous (*black*), acoustic (*red*), optic (*blue*), resonant (*green*) modes for $p = 2$ and λ_1^h (*magenta*)

the first-order derivatives of the Fourier symbols $\lambda_p^{\cdot}$ and $\Lambda_p^{\cdot}$, verify

$$\partial_\eta \Lambda_1(\pi) = \partial_\eta \Lambda_1(0) = \partial_\eta \Lambda_2^{\mathrm{a}}(\pi) = \partial_\eta \Lambda_2^{\mathrm{o}}(\pi) = \partial_\eta \Lambda_2^{\mathrm{a}}(0) = \partial_\eta \Lambda_2^{\mathrm{o}}(0) = 0$$
$$\text{and} \quad \partial_\eta \lambda_1(\pi) = \partial_\eta \lambda_2^{\mathrm{a}}(\pi) = \partial_\eta \lambda_2^{\mathrm{o}}(\pi) = \partial_\eta \lambda_2^{\mathrm{o}}(0) = 0. \tag{16}$$

For all $\alpha \in \{\mathrm{a}, \mathrm{o}\}$ and all $1 \le k \le N$, the following spectral identities hold:

$$\|\boldsymbol{\varphi}_1^{h,k}\|^2_{\mathcal{H}_1^{h,1}} = \frac{6\|B_1^h \boldsymbol{\varphi}_1^{h,k}\|^2_{1,e}}{12 - \Lambda_1^k} \quad \text{and} \quad \|\boldsymbol{\varphi}_2^{h,\alpha,k}\|^2_{\mathcal{H}_2^{h,1}} = \frac{\|B_2^h \boldsymbol{\varphi}_2^{h,\alpha,k}\|^2_{2,e}}{W(\Lambda_2^{\alpha,k})}, \tag{17}$$

where

$$W(\Lambda) = \frac{24(\Lambda - 10)^2(\Lambda - 12)(\Lambda - 60)}{(-19\Lambda^2 - 120\Lambda + 3600)(\Lambda^2 + 16\Lambda + 240)}.$$

Thus, for a finite observability time T, by taking solutions of (10) of the form

$$\mathbf{U}_1^h(t) = \exp\bigl(i(T - t)\lambda_1^{h,N}\bigr)\boldsymbol{\varphi}_1^{h,N}$$

or

$$\mathbf{U}_2^h(t) = \exp\bigl(i(T - t)\lambda_2^{h,\alpha,k}\bigr)\boldsymbol{\varphi}_2^{h,\alpha,k},$$

with $(\alpha, k) \in \{(\mathrm{a}, N), (\mathrm{o}, N), (\mathrm{o}, 1)\}$, we obtain that the observability constant $C_p^h(T)$ blows up at least polynomially as $h \to 0$. In fact, by adapting the analysis in [10] based on the Stationary Phase Lemma, we can obtain a polynomial blow-up rate at any order. In [13], by arguments based on fine estimates on the family of bi-orthogonals that are expected to be adaptable to the approximations used in this paper, an exponential blow-up rate was proved for the finite difference semi-discretization scheme.

In Fig. 4(a), (e), we represent the solution of the continuous (abbreviated by c) adjoint system (2) with

$$u^0(x) = \exp\bigl(-\gamma(x - 1/2)^2/2\bigr)\exp(ix\xi_0) \quad \text{and} \quad u^1(x) = -u_x^0(x), \quad \gamma = h^{-0.9},$$

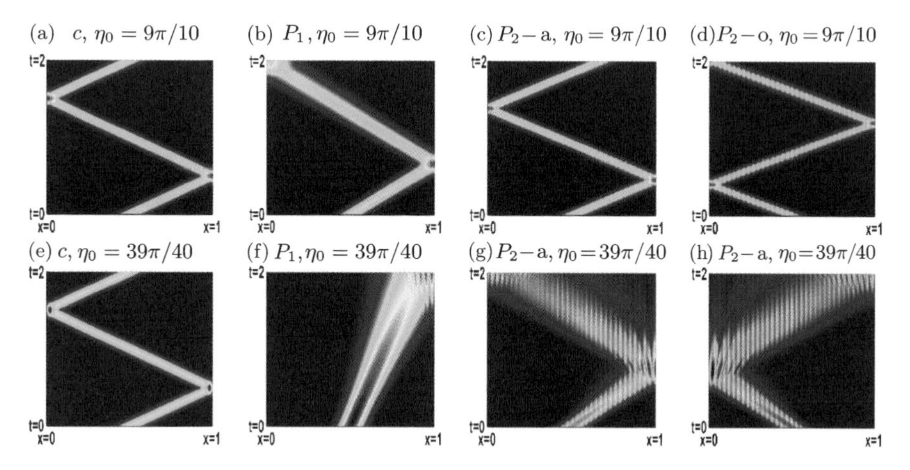

Fig. 4 Propagation along the rays of geometric optics of a Gaussian wave packet concentrated around the wave number $\xi_0 = \eta_0/h$ for $h = 1/1000$

for which the solution propagates at velocity one (the maximum amplitude for both initial time $t = 0$ and final one $t = 2$ is at $x = 1/2$ after two reflections on the boundary) along the generalized ray

$$x(t) = \begin{cases} 2k+1+1/2-t, & t \in (2k+1/2, 2k+1+1/2), \\ -(2k+1+1/2)+t, & t \in (2k+1+1/2, 2k+2+1/2), \end{cases} \quad k \in \mathbb{Z}.$$

Also no dispersive effect holds since the corresponding group acceleration is identically zero (see Fig. 5, the black curves), despite of the value of the wave number η_0. The presence of the dispersion effects due to the group acceleration is responsible for modifications on size of the support of the solutions as time evolves (cf. [12]), whereas their absence leads to the conservation of the support size.

In Fig. 4(b), (f), we represent the corresponding solution of the numerical adjoint problem (10) for $p = 1$; for both values of the wave number η_0, the solution propagates at a smaller group velocity than the continuous one since both $\eta_0 = 9\pi/10$ and $\eta_0 = 39\pi/40$ belong to the region where $\partial_\eta \lambda_1 < 1$; the dispersive effects are visible for both wave numbers, since the group acceleration $\partial_\eta^2 \lambda_1$ is non-trivial; however, they are more accentuated for $\eta_0 = 39\pi/40$ than for $\eta_0 = 9\pi/10$ since $|\partial_\eta^2 \lambda_1(39\pi/40)| > |\partial_\eta^2 \lambda_1(9\pi/10)|$, as we can see in Fig. 5, the blue curves.

In Fig. 4(c), (g), we represent the projection on the acoustic mode of the solution to the adjoint problem (10) for $p = 2$. For $\eta_0 = 9\pi/10$, the velocity of propagation is larger than one ($\partial_\eta \lambda_2^{\mathrm{a}}(9\pi/10) > 1$) (at the final time $t = 2$, the maximum amplitude is located at a space position $x > 1/2$, after two reflections on the border); almost no dispersive effect can be observed, since $\partial_\eta^2 \lambda_2^{\mathrm{a}}(\eta) \sim 0$, for all $\eta \in (0, 9\pi/10)$. On the other hand, for $\eta_0 = 39\pi/40$, the projection on the acoustic branch propagates at velocity $\partial_\eta \lambda_2^{\mathrm{a}}(39\pi/10) < 1$, so that it reflects only once on the boundary, but more rapidly than the corresponding wave packet for $p = 1$ (which even does not

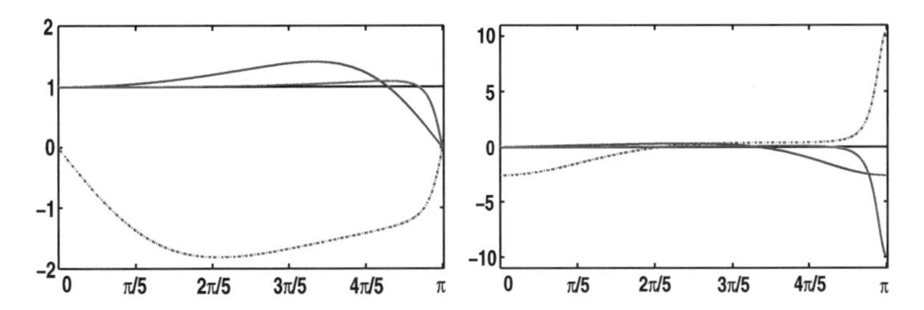

Fig. 5 Group velocities (*left*) versus group accelerations (*right*): continuous (*black*), $p=1$ (*blue*), $p=2$ acoustic (*red*), $p=2$ optic (*dotted red*)

reflect on the boundary) since $\partial_\eta\lambda_2^{\mathrm{a}}(39\pi/40) > \partial_\eta\lambda_1(39\pi/40)$. At the same time, the dispersive effects are much more accentuated for the projection on the acoustic branch than for $p=1$ since $|\partial_\eta^2\lambda_2^{\mathrm{a}}(39\pi/40)| > |\partial_\eta^2\lambda_1(39\pi/40)|$, as we can see in Fig. 5, the red curves.

In Fig. 4(d), (h), we represent the projection on the optic mode of the solution to the adjoint problem (10), which propagates in the opposite direction than the physical solution, due to the fact that $\partial_\eta\lambda_2^{\mathrm{o}}(\eta) < 0$, for all $\eta \in (0,\pi)$, while in the continuous case the group velocity is strictly positive ($\equiv 1$). For $\eta_0 = 9\pi/10$, the velocity of propagation is larger than the one for the corresponding acoustic projection (i.e., $|\partial_\eta\lambda_2^{\mathrm{o}}(9\pi/10)| > \partial_\eta\lambda_2^{\mathrm{a}}(9\pi/10)$), reflected in the fact that the maximum amplitude at $t=2$ is located next to $x=0$; almost no dispersive effects occur. For $\eta_0 = 39\pi/40$, the optic projection propagates almost at the same velocity as the acoustic one and almost with the same dispersive effects, the only visible change being the reverse direction (see Fig. 5, the dotted red lines).

Several *filtering techniques* have been designed to face these high frequency pathologies, all based on taking subclasses of initial data that filter them: the *Fourier truncation method* (cf. [7]), which simply eliminates all the Fourier components propagating non-uniformly, and the *bi-grid algorithm* (cf. [4]), rigorously studied in [6, 9] and [14] in the context of the finite differences semi-discretization of the 1D and 2D wave equation and of the Schrödinger equation (cf. [5]), which consists in taking initial data with slow oscillations obtained by linear interpolation from data given on a coarser grid. The interested reader is referred to the survey articles [3] and [15] for a more or less complete presentation of the development of this topic and the state of the art.

Let us describe how the bi-grid filtering acts for the linear and quadratic finite element approximations under consideration. To be more precise, for an *odd* N, let us define the *set of data on the fine grid obtained by linear interpolation from data on a twice coarser grid*,

$$\mathcal{B}_1^h := \Big\{ \mathbf{F}_1^h = (F_{1,j})_{1\le j\le N},\ \text{s.t. } F_{1,0} = F_{1,N+1} = 0,$$
$$\text{and } F_{1,2j+1} = \frac{1}{2}(F_{1,2j} + F_{1,2j+2}),\ \forall 0 \le j \le (N-1)/2 \Big\},$$

and the *set of linear data whose nodal components are given by a bi-grid algorithm*,

$$\mathcal{B}_2^h := \Big\{ \mathbf{F}_2^h = (F_{2,j/2})_{1\le j\le 2N+1} \text{ s.t. } F_{2,0} = F_{2,N+1} = 0,$$
$$F_{2,j+1/2} = \frac{1}{2}(F_{2,j} + F_{2,j+1}),\ \forall 0 \le j \le N,$$
$$\text{and } F_{2,2j+1} = \frac{1}{2}(F_{2,2j} + F_{2,2j+2}),\ \forall 0 \le j \le (N-1)/2 \Big\}.$$

The following result has been proved for the adjoint problem (10) for $p=1$ in [9] or [14] and for $p=2$ in [11]:

Theorem 1 *For all $T \ge 2$, the observability inequality* (13) *holds uniformly as $h \to 0$ within the class of initial data $(\mathbf{U}_p^{h,0}, \mathbf{U}_p^{h,1}) \in (\mathcal{B}_p^h \times \mathcal{B}_p^h) \cap \mathcal{V}_p^h$ in the adjoint problem* (10).

One of the possible proofs of this result is based on a *dyadic decomposition* argument like in [6]. For the case $p=1$, it reduces to showing that the total energy of solutions corresponding to initial data in $(\mathcal{B}_1^h \times \mathcal{B}_1^h) \cap \mathcal{V}_1^h$ can be uniformly bounded from above by the energy of their projection on the first half of the spectrum. The second step is to use the uniform observability inequality (13) in the class $\mathcal{T}_1^h \times \mathcal{T}_1^h$ consisting of discrete functions for which the second half of Fourier modes have been truncated; this result can be obtained by the *multiplier technique* (cf. [7]) or by *Ingham-type inequalities* (cf. [15]). For the quadratic case $p=2$, the projection on the first half of the acoustic mode has to be implemented to reduce the proof of Theorem 1 to the observability inequality (13) on the class $\mathcal{T}_2^h \times \mathcal{T}_2^h$ of functions for which the second half (the high frequency one) of the acoustic diagram and the whole optic diagram have been truncated. The fact that for $p=2$ the bi-grid algorithm in Theorem 1 essentially truncates $3/4$ of the spectrum versus only $1/2$ for $p=1$ can be intuitively seen in the fact that $\mathcal{B}_2^h$ involves two requirements on its elements versus only one requirement for $\mathcal{B}_1^h$. The observability time for these two bi-grid algorithms coincides with the continuous optimal one $T^\star = 2$, since the group velocities $\partial_\eta \lambda_1$ and $\partial_\eta \lambda_2^{\mathrm{a}}$ are increasing functions on $[0, \pi/2]$ and then $\partial_\eta \lambda_1(\eta) \ge \partial_\eta \lambda_1(0) = 1$ for all $\eta \in [0, \pi/2]$ and similarly for $\partial_\eta \lambda_2^{\mathrm{a}}$. Thus, the minimal velocity of propagation involving solutions with data in the class $\mathcal{T}_p^h \times \mathcal{T}_p^h$ for both $p=1$ and $p=2$ is equal to one.

In practice, one has to employ fully discrete schemes. In this respect, it is important to note that, using the results of Ervedoza–Zheng–Zuazua in [2] allowing to transfer observability results for time-continuous conservative semigroups on time-discrete conservative schemes, we see that our observability results in Theorem 1 are also valid for any conservative fully discrete finite element approximation, like, for example, the implicit *midpoint* time-discretization scheme

$$\mathbf{U}_p^{h,k+1} - 2\mathbf{U}_p^{h,k} + \mathbf{U}_p^{h,k-1} + (\delta t)^2 \big(M^h\big)^{-1} S^h \big(\mathbf{U}_p^{h,k+1} + \mathbf{U}_p^{h,k-1}\big)/2 = 0,$$

where δt is the time step and $\mathbf{U}_p^{h,k} \approx \mathbf{U}_p^h(k\delta t)$. Note, however, that the results in [2] do not yield the optimal observability time, a subject that needs further investigation.

Once the observability problem is well understood, we are in conditions to address the discrete control problem. For a particular solution $\tilde{\mathbf{U}}_p^h(t)$ of the adjoint problem (10), let us consider the following non-homogeneous discrete problem

$$\begin{aligned} &M_p^h \mathbf{Y}_{p,tt}^h(t) + S_p^h \mathbf{Y}_p^h(t) = -\big(B_p^h\big)^* B_p^h \tilde{\mathbf{U}}_p^h(t), \qquad \mathbf{Y}_p^h(0) = \mathbf{Y}_p^{h,0}, \\ &\mathbf{Y}_{p,t}^h(0) = \mathbf{Y}_p^{h,1}. \end{aligned} \tag{18}$$

Multiplying system (18) by any solution $\mathbf{U}_p^h(t)$ of the adjoint problem (10), integrating in time and imposing that at $t = T$ the solution is at rest, i.e.,

$$\big\langle \big(\mathbf{Y}_{p,t}^h(T), -\mathbf{Y}_p^h(T)\big), \big(\mathbf{U}_p^{h,0}, \mathbf{U}_p^{h,1}\big)\big\rangle_{\mathcal{V}_p^{h,\prime}, \mathcal{V}_p^h} = 0, \quad \forall \big(\mathbf{U}_p^{h,0}, \mathbf{U}_p^{h,1}\big) \in \mathcal{V}_p^h, \tag{19}$$

we obtain the identity,

$$\int_0^T \big(B_p^h \tilde{\mathbf{U}}_p^h(t), B_p^h \mathbf{U}_p^h(t)\big)_{p,e}\, dt = \big\langle \big(\mathbf{Y}_p^{h,1}, -\mathbf{Y}_p^{h,0}\big), \big(\mathbf{U}_p^h(0), \mathbf{U}_{p,t}^h(0)\big)\big\rangle_{\mathcal{V}_p^{h,\prime}, \mathcal{V}_p^h}, \tag{20}$$

for all $(\mathbf{U}_p^{h,0}, \mathbf{U}_p^{h,1}) \in \mathcal{V}_p^h$. This is the Euler–Lagrange equation corresponding to the quadratic functional, the discrete analogue of $\mathcal{J}$ in (5):

$$\begin{aligned} &\mathcal{J}_p^h\big(\mathbf{U}_p^{h,0}, \mathbf{U}_p^{h,1}\big) \\ &\quad = \frac{1}{2}\int_0^T \|B_p^h \mathbf{U}_p^h(t)\|_{p,e}^2\, dt - \big\langle \big(\mathbf{Y}_p^{h,1}, -\mathbf{Y}_p^{h,0}\big), \big(\mathbf{U}_p^h(0), \mathbf{U}_{p,t}^h(0)\big)\big\rangle_{\mathcal{V}_p^{h,\prime}, \mathcal{V}_p^h}, \end{aligned}$$

$\mathbf{U}_p^h(t)$ being the solution of the adjoint problem (10) with initial data $(\mathbf{U}_p^{h,0}, \mathbf{U}_p^{h,1})$ and $(\mathbf{Y}_p^{h,1}, \mathbf{Y}_p^{h,0}) \in \mathcal{V}_p^{h,\prime}$ the initial data to be controlled in (18). Actually, (18) and (20) are completely equivalent so that, in practice, it is sufficient to prove the existence of a critical point for $\mathcal{J}_p^h$ to deduce the existence of a control for (18). The uniform observability inequality (13) within the class of initial data $\mathcal{B}_p^h \times \mathcal{B}_p^h$ guarantees the uniform coercivity of $\mathcal{J}_p^h$ and the convergence of the last component $\tilde{v}_p^h$ of $B_p^h \tilde{\mathbf{U}}_p^h(t)$, the discrete control, to the optimal control $\tilde{v}$ for the continuous wave equation given by (4) when the initial data $(\mathbf{Y}_p^{h,0}, \mathbf{Y}_p^{h,1})$ in (18) approximates well the initial data (y^0, y^1) in the continuous problem (1). Here $\tilde{\mathbf{U}}_p^h(t)$ is the solution of the discrete adjoint system (10) corresponding to the minimizer $(\tilde{\mathbf{U}}_p^{h,0}, \tilde{\mathbf{U}}_p^{h,1}) \in \mathcal{B}_p^h \times \mathcal{B}_p^h$ of $\mathcal{J}_p^h$.

3 Implementation of the Bi-grid Algorithm and Numerical Results

The aim of this section is to numerically illustrate the three high frequency pathologies for the quadratic approximation of the control problem (18) and the way in

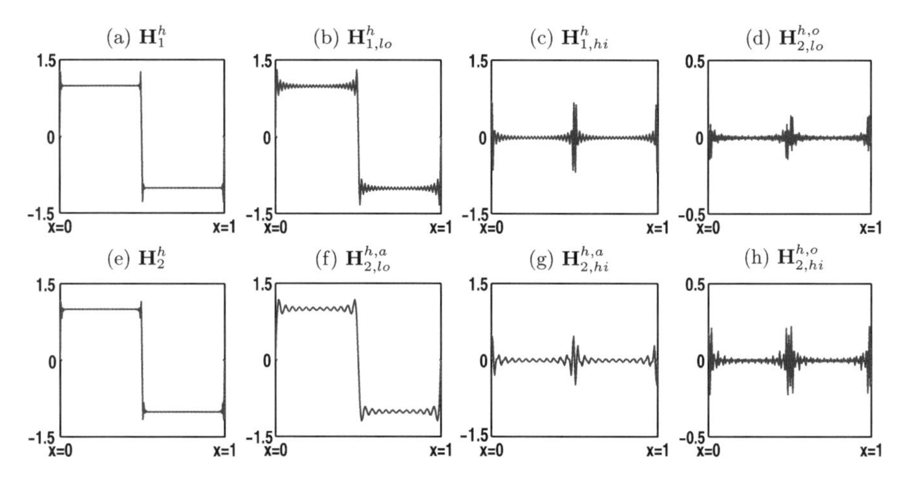

Fig. 6 The discrete Heaviside functions $\mathbf{H}_p^h$ and their projections

which the bi-grid filtering leads to the convergence of the solution of (18) to the continuous one. We will also compare the numerical results obtained for the quadratic case $p = 2$ and for the linear one $p = 1$.

In order to simplify the presentation, we will take as discrete initial data $(\mathbf{Y}_p^{h,0}, \mathbf{Y}_p^{h,1})$ in (18) $\mathbf{Y}_p^{h,1} = 0$, $\mathbf{Y}_p^{h,0}$ being an approximation of the Heaviside function $H(x)$ in (6). Firstly, let us define the vectors $\tilde{\mathbf{H}}_p^h = (\tilde{H}_{p,j/p})_{1\le j\le pN+p-1}$, where $\tilde{H}_{p,j/p} = (H, \phi_{p,j/p}^h)_{L^2}$, for all $1 \le j \le pN + p - 1$. The numerical approximation of $H(x)$ we consider is

$$\mathbf{Y}_p^{h,0} = \mathbf{H}_p^h := \left(M_p^h\right)^{-1} \tilde{\mathbf{H}}_p^h. \tag{21}$$

For all $\alpha \in \{\mathrm{a}, \mathrm{o}\}$ and all $\beta \in \{\mathrm{lo}, \mathrm{hi}\}$, (lo/hi standing for *low*/*high*-frequency), we also define the projections of $\mathbf{H}_p^h$ on some parts of the spectrum as follows:

$$\mathbf{H}_{1,\beta}^h = \sum_{k_\beta^-}^{k_\beta^+} \left(\mathbf{H}_1^h, \boldsymbol{\varphi}_1^{h,k}\right)_{\mathcal{H}_1^{h,0}} \boldsymbol{\varphi}_1^{h,k} \quad \text{and} \quad \mathbf{H}_{2,\beta}^{h,\alpha} = \sum_{k_\beta^-}^{k_\beta^+} \left(\mathbf{H}_2^h, \boldsymbol{\varphi}_2^{h,\alpha,k}\right)_{\mathcal{H}_2^{h,0}} \boldsymbol{\varphi}_2^{h,\alpha,k},$$

where $(k_\beta^-, k_\beta^+) = (1, (N-1)/2)$ if $\beta = \mathrm{lo}$ and $(k_\beta^-, k_\beta^+) = ((N+1)/2, N)$ if $\beta = \mathrm{hi}$. More precisely, $\mathbf{H}_{1,\mathrm{lo}}^h$ is the projection of $\mathbf{H}_1^h$ on the first half of the spectrum and $\mathbf{H}_{2,\mathrm{lo}}^{h,\mathrm{a}}$ that of $\mathbf{H}_2^h$ on the first half of the acoustic diagram (see Fig. 6).

Since the datum $H(x)$ in (6) is irregular due to the presence of the jump, it involves high-frequency eigenfunctions. This also happens with its numerical approximations $\mathbf{H}_p^h$, as it can be easily observed in Fig. 7. These high-frequency components will lead to the divergence of the corresponding numerical controls.

In order to find the minimum of the discrete functional $\mathcal{J}_p^h$, we will apply the *Conjugate Gradient* (CG) algorithm (see [1, 4]) to iteratively solve the Euler–

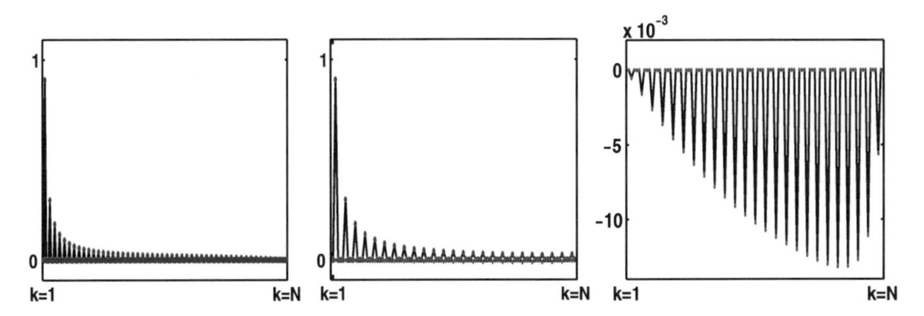

Fig. 7 The Fourier coefficients of $\mathbf{H}_p^h$ for $p = 1$ (*left*), $p = 2$ (*center*, *blue* = acoustic, *red* = optic), $p = 2$—the optic branch (*right*)

Lagrange equation (20). Let us briefly recall it when no-filtering technique is applied.

Firstly, fix the initial data to be controlled $(\mathbf{Y}_{p,0}^{h,0}, \mathbf{Y}_{p,0}^{h,1})$, a tolerance ϵ $(= 0.001$ in our particular case) and a maximum number of iterations $n_{\max}$ $(= 200)$, aimed to be a stopping criterium. In order to better follow the CG algorithm, we divide it into several steps as follows:

Step 1. We initialize the algorithm solving the adjoint problem (10) with arbitrary data $(\mathbf{U}_p^{h,0}, \mathbf{U}_p^{h,1}) = (\mathbf{U}_{p,0}^{h,0}, \mathbf{U}_{p,0}^{h,1}) \in \mathcal{V}_p^h$, for example, the trivial one. This step yields the solution $\mathbf{U}_{p,0}^h(t)$.

Step 2. Compute the first gradient $(\mathbf{G}_{p,0}^{h,0}, \mathbf{G}_{p,0}^{h,1}) := \nabla \mathcal{J}_p^h(\mathbf{U}_{p,0}^{h,0}, \mathbf{U}_{p,0}^{h,1})$ by solving the non-homogeneous problem (18) with initial data $(\mathbf{Y}_{p,0}^{h,0}, \mathbf{Y}_{p,0}^{h,1})$ and $\tilde{\mathbf{U}}_p^h(t) = \mathbf{U}_{p,0}^h(t)$. This produces the solution $\mathbf{Y}_{p,0}^h(t)$. Then

$$\mathbf{G}_{p,0}^{h,0} = -\left(S_p^h\right)^{-1} M_p^h \mathbf{Y}_{p,0,t}^h(T) \quad \text{and} \quad \mathbf{G}_{p,0}^{h,1} = \mathbf{Y}_{p,0}^h(T).$$

Step 3. If $\|\mathbf{G}_{p,0}^{h,0}\|_{\mathcal{H}_p^{h,1}}^2 + \|\mathbf{G}_{p,0}^{h,1}\|_{\mathcal{H}_p^{h,0}}^2 \geq \epsilon^2$, compute the first descent direction

$$\left(\mathbf{D}_{p,0}^{h,0}, \mathbf{D}_{p,0}^{h,1}\right) = -\left(\mathbf{G}_{p,0}^{h,0}, \mathbf{G}_{p,0}^{h,1}\right).$$

Step 4. Given $(\mathbf{U}_{p,n}^{h,0}, \mathbf{U}_{p,n}^{h,1})$, $(\mathbf{G}_{p,n}^{h,0}, \mathbf{G}_{p,n}^{h,1})$ and $(\mathbf{D}_{p,n}^{h,0}, \mathbf{D}_{p,n}^{h,1})$ in $\mathcal{V}_p^h$, we compute these quantities at the next iteration $n + 1$ as follows:

Step 4.a. Solve (10) for $(\mathbf{U}_p^{h,0}, \mathbf{U}_p^{h,1}) = (\mathbf{D}_{p,n}^{h,0}, \mathbf{D}_{p,n}^{h,1})$. Denote the solution by $\mathbf{D}_{p,n}^h(t)$.

Step 4.b. Solve (18) with trivial initial data and $\tilde{\mathbf{U}}_p^h(t) = \mathbf{D}_{p,n}^h(t)$ and denote the solution by $\mathbf{Y}_{p,n+1}^h(t)$. Take

$$\mathbf{Z}_{p,n}^{h,0} = -\left(S_p^h\right)^{-1} M_p^h \mathbf{Y}_{p,n+1,t}^h(T) \quad \text{and} \quad \mathbf{Z}_{p,n}^{h,1} = \mathbf{Y}_{p,n+1}^h(T).$$

Step 4.c. Set

$$\rho_{p,n} := -\frac{\|\mathbf{G}_{p,n}^{h,0}\|_{\mathcal{H}_p^{h,1}}^2 + \|\mathbf{G}_{p,n}^{h,1}\|_{\mathcal{H}_p^{h,0}}^2}{(\mathbf{Z}_{p,n}^{h,0}, \mathbf{D}_{p,n}^{h,0})_{\mathcal{H}_p^{h,1}} + (\mathbf{Z}_{p,n}^{h,1}, \mathbf{D}_{p,n}^{h,1})_{\mathcal{H}_p^{h,0}}}.$$

Step 4.d. Compute the next iteration

$$\left(\mathbf{U}_{p,n+1}^{h,0}, \mathbf{U}_{p,n+1}^{h,1}\right) := \left(\mathbf{U}_{p,n}^{h,0}, \mathbf{U}_{p,n}^{h,1}\right) + \rho_{p,n}\left(\mathbf{D}_{p,n}^{h,0}, \mathbf{D}_{p,n}^{h,1}\right).$$

Step 4.e. Compute the next gradient

$$\left(\mathbf{G}_{p,n+1}^{h,0}, \mathbf{G}_{p,n+1}^{h,1}\right) := \nabla \mathcal{J}_p^h\left(\mathbf{U}_{p,n+1}^{h,0}, \mathbf{U}_{p,n+1}^{h,1}\right)$$

by

$$\left(\mathbf{G}_{p,n+1}^{h,0}, \mathbf{G}_{p,n+1}^{h,1}\right) := \left(\mathbf{G}_{p,n}^{h,0}, \mathbf{G}_{p,n}^{h,1}\right) + \rho_{p,n}\left(\mathbf{Z}_{p,n}^{h,0}, \mathbf{Z}_{p,n}^{h,1}\right).$$

Step 4.f. Compute the next descent direction

$$\begin{aligned}&\left(\mathbf{D}_{p,n+1}^{h,0}, \mathbf{D}_{p,n+1}^{h,1}\right)\\ &\quad := -\left(\mathbf{G}_{p,n+1}^{h,0}, \mathbf{G}_{p,n+1}^{h,1}\right) + \frac{\|\mathbf{G}_{p,n+1}^{h,0}\|_{\mathcal{H}_p^{h,1}}^2 + \|\mathbf{G}_{p,n+1}^{h,1}\|_{\mathcal{H}_p^{h,0}}^2}{\|\mathbf{G}_{p,n}^{h,0}\|_{\mathcal{H}_p^{h,1}}^2 + \|\mathbf{G}_{p,n}^{h,1}\|_{\mathcal{H}_p^{h,0}}^2}\left(\mathbf{D}_{p,n}^{h,0}, \mathbf{D}_{p,n}^{h,1}\right).\end{aligned}$$

The algorithm ends up when for some $n < n_{\max}$ we obtain

$$\left\|\mathbf{G}_{p,n}^{h,0}\right\|_{\mathcal{H}_p^{h,1}}^2 + \left\|\mathbf{G}_{p,n}^{h,1}\right\|_{\mathcal{H}_p^{h,0}}^2 < \epsilon^2$$

or when $n \geq n_{\max}$. When the second stopping criterium holds, we understand that the CG algorithm does not converge (due to the fact that $\mathcal{J}_p^h$ looses coercivity). For both stopping criteria, we take the minimizer of $\mathcal{J}_p^h$ to be $(\tilde{\mathbf{U}}_p^{h,0}, \tilde{\mathbf{U}}_p^{h,1}) := (\mathbf{U}_{p,n}^{h,0}, \mathbf{U}_{p,n}^{h,1})$, where n is the last iteration number before stopping.

Let us now describe the changes we have to do in the CG algorithm to implement the bi-grid filtering we propose in Theorem 1. The linear case $p = 1$ has been implemented in [3]. For this reason, we restrict ourselves to the quadratic case $p = 2$. However, whenever we have to implement any filtering technique, the only steps we have to modify are Steps 2 and 4.b above. In order to simplify the presentation, we describe only the modifications to be done on Step 2, the ones on Step 4.b being similar. Firstly, set

$$\mathbf{F}_2^{h,0} = -M_2^h \mathbf{Y}_{2,0,t}^h(T) \quad \text{and} \quad \mathbf{F}_2^{h,1} = M_2^h \mathbf{Y}_{2,0}^h(T),$$

and observe that, for all test function $(\mathbf{U}_2^{h,0}, \mathbf{U}_2^{h,1}) \in \mathcal{V}_2^h$, the Gateaux derivative of $\mathcal{J}_2^h$ at $(\mathbf{U}_{2,0}^{h,0}, \mathbf{U}_{2,0}^{h,1})$ has the following expressions:

$$(\mathcal{J}_2^h)'(\mathbf{U}_{2,0}^{h,0},\mathbf{U}_{2,0}^{h,1})(\mathbf{U}_2^{h,0},\mathbf{U}_2^{h,1}) = (\mathbf{F}_2^{h,0},\mathbf{U}_2^{h,0})_{2,e} + (\mathbf{F}_2^{h,1},\mathbf{U}_2^{h,1})_{2,e}$$
$$= (\mathbf{G}_{2,0}^{h,0},\mathbf{U}_2^{h,0})_{\mathcal{H}_2^{h,1}} + (\mathbf{G}_{2,0}^{h,1},\mathbf{U}_2^{h,1})_{\mathcal{H}_2^{h,0}}.$$

Let us observe that the linear functions with nodal components given by a bi-grid algorithm in $\mathcal{B}_2^h$ are in fact linear functions on a grid of size $2h$. We consider that both the test functions $(\mathbf{U}_2^{h,0},\mathbf{U}_2^{h,1})$ and the gradient $(\mathbf{G}_{2,0}^{h,0},\mathbf{G}_{2,0}^{h,1})$ belong to $\mathcal{B}_2^h \times \mathcal{B}_2^h$. Consider the *restriction operator* Π that associates to any quadratic function of coefficients $\mathbf{E}_2^h = (E_{2,j/2})_{1\le j\le 2N+1}$ the linear function on the mesh of size $2h$ of coefficients $(\Pi\mathbf{E}_2^h)_j = E_{2,2j}$, for all $1 \le j \le (N-1)/2$. When both $(\mathbf{U}_2^{h,0},\mathbf{U}_2^{h,1})$ and $(\mathbf{G}_{2,0}^{h,0},\mathbf{G}_{2,0}^{h,1})$ belong to $\mathcal{B}_2^h \times \mathcal{B}_2^h$, then

$$(\mathbf{G}_{2,0}^{h,0},\mathbf{U}_2^{h,0})_{\mathcal{H}_2^{h,1}} + (\mathbf{G}_2^{h,1},\mathbf{U}_2^{h,1})_{\mathcal{H}_2^{h,0}}$$
$$= (\Pi\mathbf{G}_{2,0}^{h,0},\Pi\mathbf{U}_2^{h,0})_{\mathcal{H}_1^{2h,1}} + (\Pi\mathbf{G}_{2,0}^{h,1},\Pi\mathbf{U}_2^{h,1})_{\mathcal{H}_1^{2h,0}}.$$

Consider another *restriction operator* Γ defined as

$$(\Gamma\mathbf{E}_2^h)_j = E_{2,2j} + \frac{3}{4}(E_{2,2j+1/2} + E_{2,2j-1/2})$$
$$+ \frac{1}{2}(E_{2,2j+1} + E_{2,2j-1}) + \frac{1}{4}(E_{2,2j+3/2} + E_{2,2j-3/2}).$$

Then

$$(\mathbf{F}_2^{h,0},\mathbf{U}_2^{h,0})_{2,e} + (\mathbf{F}_2^{h,1},\mathbf{U}_2^{h,1})_{2,e} = (\Gamma\mathbf{F}_2^{h,0},\Pi\mathbf{U}_2^{h,0})_{1,e} + (\Gamma\mathbf{F}_2^{h,1},\Pi\mathbf{U}_2^{h,1})_{1,e}$$

and the two components of the gradient are explicitly given by

$$\mathbf{G}_{2,0}^{h,0} = \Pi^{-1}(S_1^{2h})^{-1}\Gamma\mathbf{F}_2^{h,0} \quad \text{and} \quad \mathbf{G}_{2,0}^{h,1} = \Pi^{-1}(M_1^{2h})^{-1}\Gamma\mathbf{F}_2^{h,1},$$

where Π^{-1} is the inverse of the restriction operator Π defined as the linear interpolation on a grid of size $h/2$ of a function defined on a grid of size $2h$. Therefore, our filtering mechanism in Theorem 1 for $p = 2$ acts in fact like a classical bi-grid algorithm of mesh ratio $1/4$. This is very similar to the bi-grid algorithm designed in [5] to ensure discrete dispersive estimates for the finite difference semi-discretization of the Schrödinger equation uniformly in the mesh size parameter h. In that case, the bi-grid algorithm has to face the two singularities of the Fourier symbol $p(\eta) = 4\sin^2(\eta/2)$ defined on $\eta \in [0,\pi]$: the *vanishing group velocity* at $\eta = \pi$, yielding the non-uniform gain of $1/2$-derivative, and the *vanishing group acceleration* at $\eta = \pi/2$, related to the non-uniform L_x^p–L_t^q-integrability (see Fig. 8, right). In our case, by ordering in an increasing way the eigenvalues on the two dispersion curves and constructing $\lambda_2(\eta) = \lambda_2^a(\eta)$, for $\eta \in [0,\pi]$ and $\lambda_2(\eta) = \lambda_2^o(2\pi - \eta)$, for $\eta \in [\pi, 2\pi]$, we formally obtain a discrete wave equation on the grid $h/2$ whose dispersion relation $\lambda_2(\eta)$, $\eta \in [0, 2\pi]$, has *vanishing group*

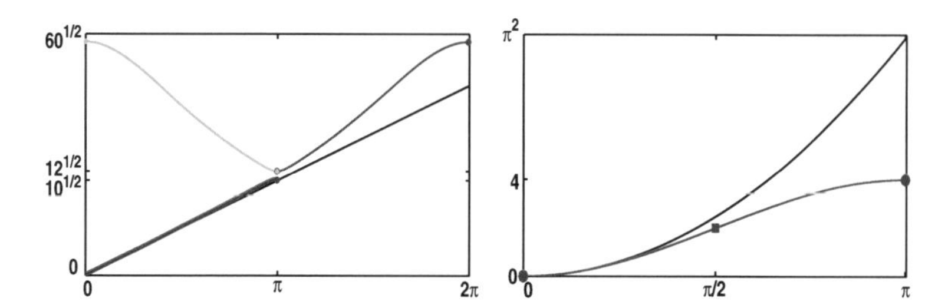

Fig. 8 Dispersion relations for the P_2-approximation of the wave equation (*left*): continuous (*black*), λ_2^a (*red*), λ_2^o (*green*), λ_2 (*blue*), versus Fourier symbols of the finite difference semi-discretization of the Schrödinger equation (*right*): continuous (*black*) and discrete (*blue*). At the marked points, the symbols have vanishing group velocity (*circles*) or vanishing group acceleration (*squares*)

velocity at $\eta = \pi^{\pm}$ and at $\eta = 2\pi$ (see Fig. 8, left). In order to remedy the pathologies associated to both singular points $\pi^{\pm}$ and 2π, a bi-grid of mesh ratio $1/4$ should suffice, despite of the discontinuity of λ_2 at $\eta = \pi$.

Remark 1 In practice, one has to reduce the semi-discrete problem to be solved, $M^h \mathbf{U}_{tt}^h(t) + S^h \mathbf{U}^h(t) = \mathbf{F}^h(t)$, to a fully discrete system with time-step of size δt and to take $\mathbf{U}^{h,k} \approx \mathbf{U}^h(k\delta t)$. Set $\mu := \delta t / h$ to be the *Courant number*. When using an *explicit time scheme*, for example, the *leap-frog* one,

$$\mathbf{U}^{h,k+1} - 2\mathbf{U}^{h,k} + \mathbf{U}^{h,k-1} + \mu^2 h^2 \left(M^h\right)^{-1} S^h \mathbf{U}^{h,k} = (\delta t)^2 \left(M^h\right)^{-1} \mathbf{F}^h(k\delta t),$$

a careful *von Neumann analysis* shows that the *Courant–Friedrichs–Lewy (CFL) condition* for μ is $\mu \leq \min_{\Lambda} \sqrt{4/\Lambda}$, where the minimum is taken over all the eigenvalues Λ of the matrix $h^2 (M^h)^{-1} S^h$. When dealing with (10) or (18) for the linear approximation $p = 1$, this analysis gives $\mu \leq 1/\sqrt{3}$. For the case $p = 2$, we obtain $\mu \leq 1/\sqrt{15}$ if we work with solutions involving both modes or $\mu \leq \sqrt{2/5}$ if the numerical solution involves only the acoustic mode. We observe that, globally, the quadratic scheme requires smaller Courant numbers than the linear one, whereas the resolution of the homogeneous problem (10) with data concentrated only on the acoustic mode admits larger μ's than in the linear case.

We end up this section by discussing the numerical results in Figs. 9, 10, 11, 12. For the P_1-approximation, we take $h = 1/200$ and for the P_2-one, $h = 1/100$, in order to have the same number of degrees of freedom in both approximations.

- Without restricting the space where the functional $\mathcal{J}_p^h$ is minimized, the numerical controls are highly oscillatory and diverge (see Figs. 9 and 10(a)–(b)). This is due to the fact that the initial data $\mathbf{H}_p^h$ involves the critical modes on the high-frequency regime of the dispersion relations for which the numerical controls diverge. These pathological effects can be seen separately by controlling the corresponding projections of the data $\mathbf{H}_p^h$ on the high frequency modes (see Fig. 9(f)

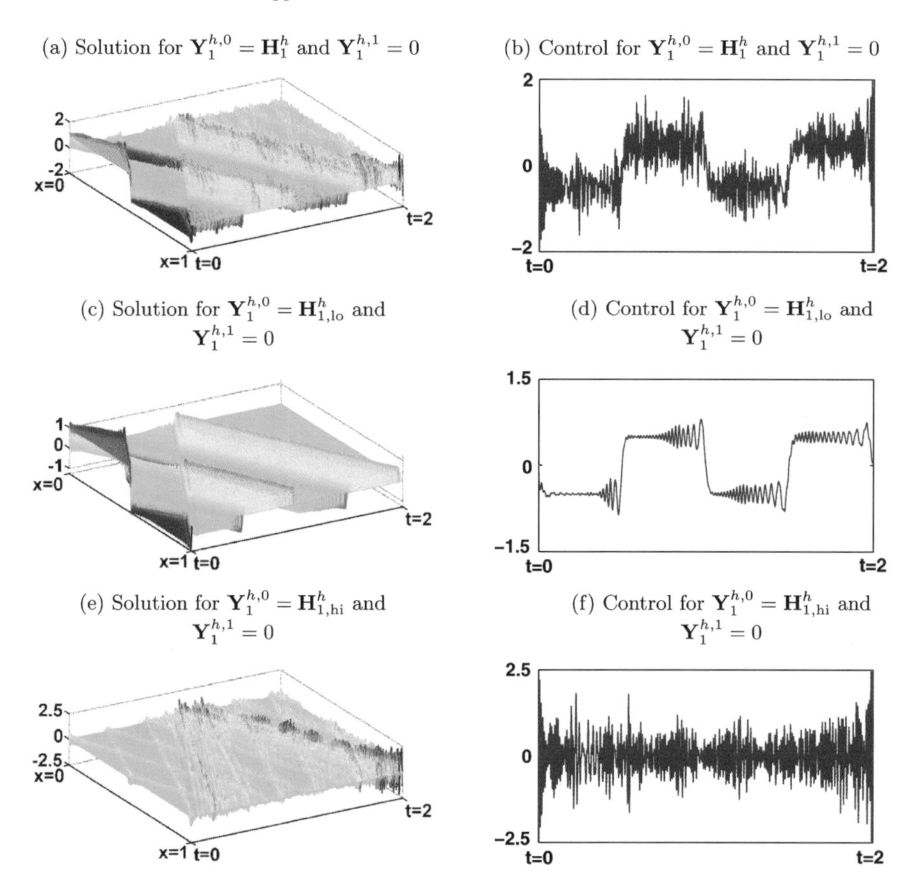

Fig. 9 Solutions of the control problem (18) versus their numerical controls for $p = 1$ arising by minimizing $\mathcal{J}_1^h$ over the whole space $\mathcal{V}_1^h$

for $p = 1$ and Fig. 10(f), (h), (j) for $p = 2$). As long as the initial data $\mathbf{H}_p^h$ is projected on the first half of the acoustic mode for $p = 2$ or on the first half of the spectrum for $p = 1$, the CG algorithm and the numerical controls converge (see Figs. 9(d) and 10(d)). The numerical controls obtained for these projections $\mathbf{H}_{1,\mathrm{lo}}^h$ and $\mathbf{H}_{2,\mathrm{lo}}^{h,\mathrm{a}}$ as initial positions in the control problem (18) without filtering are approximately the same as the ones obtained by the bi-grid filtering mechanism taking as initial position the whole $\mathbf{H}_p^h$ (see also Fig. 11(b) and 12(b)). This is due to the fact that the controls obtained by the bi-grid algorithm damp out the high-frequency effects and for this reason they act mainly on the eigenmodes involved in $\mathbf{H}_{1,\mathrm{lo}}^h$ or $\mathbf{H}_{2,\mathrm{lo}}^{h,\mathrm{a}}$.

- Without filtering, the high-frequency modes produce instabilities in the form of oscillations of larger and larger amplitude which accumulate as time evolves in the solutions of the control problem (18) (see Figs. 9(a), (e) and 10(a), (e), (g), (i)). These high frequency effects are larger in the P_2 case than in the P_1 one, due

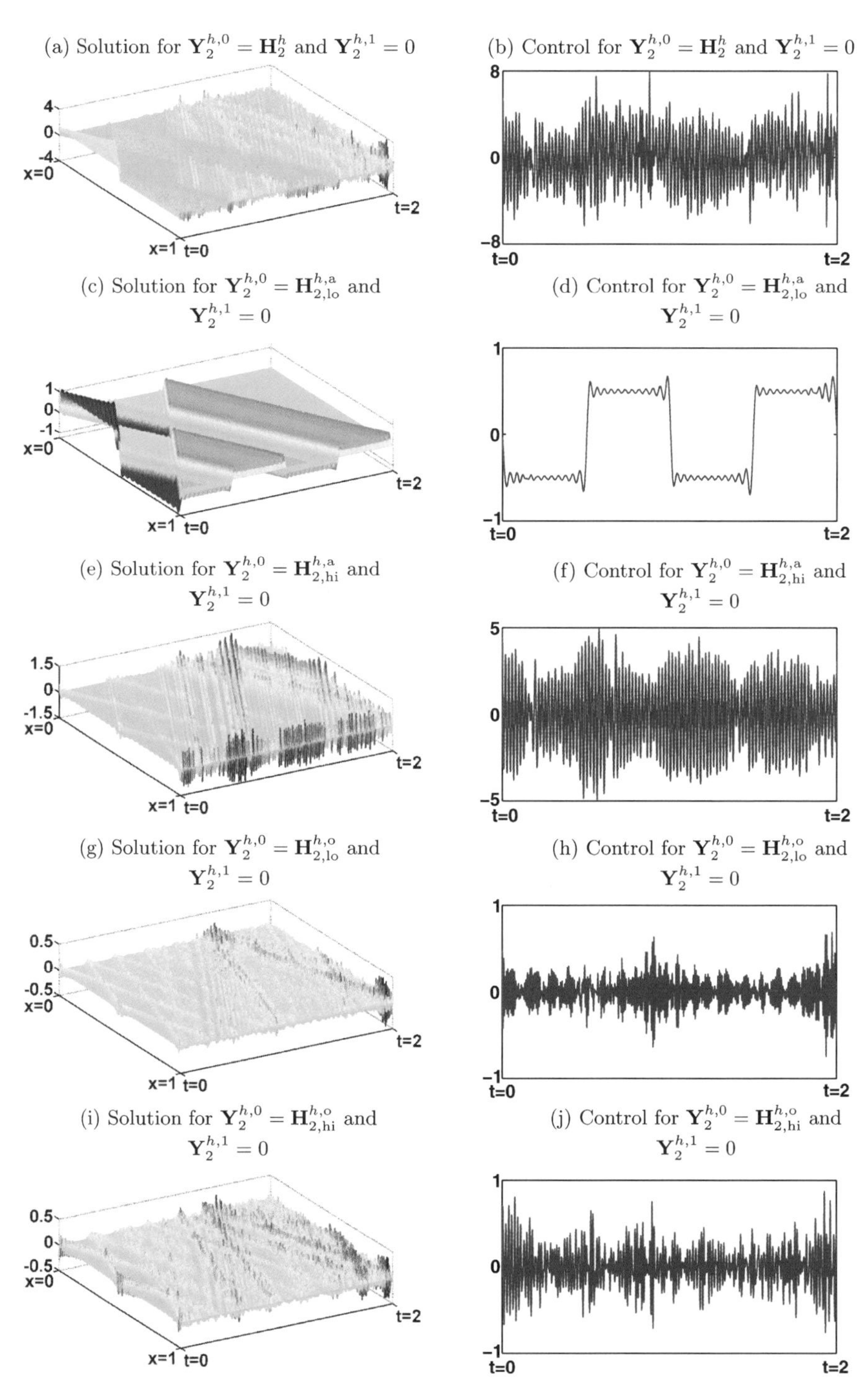

Fig. 10 Solutions of the control problem (18) versus their numerical controls for $p = 2$ arising by minimizing $\mathcal{J}_2^h$ over the whole space $\mathcal{V}_2^h$

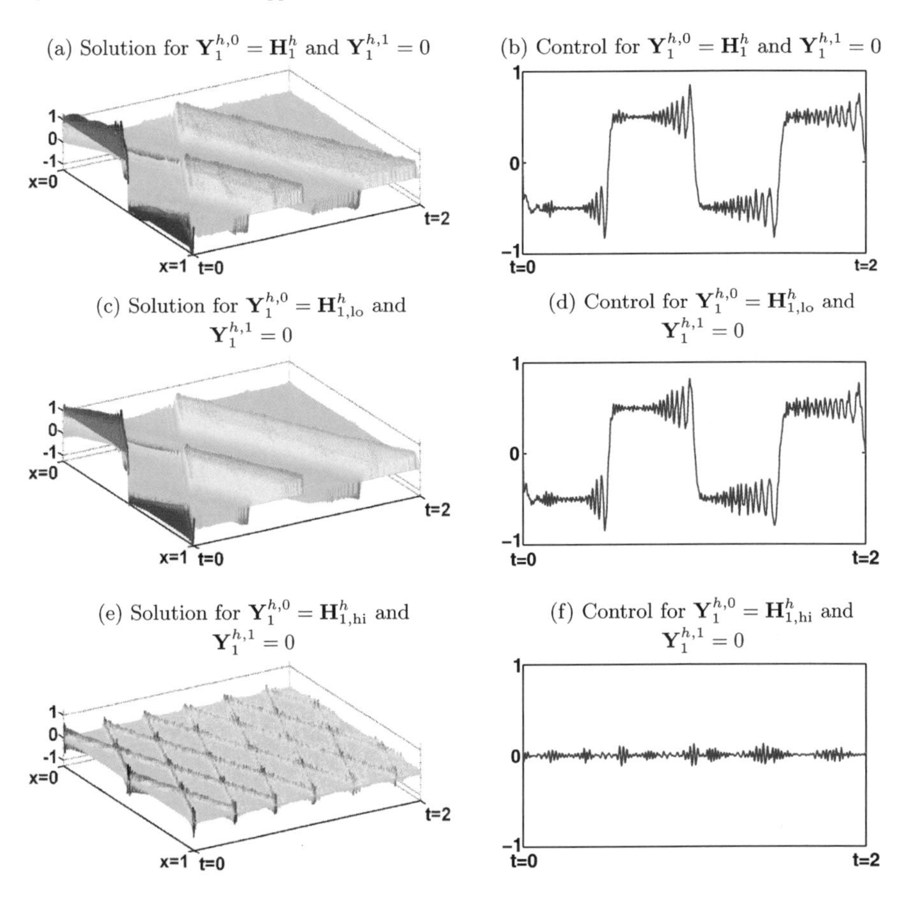

Fig. 11 Solutions of the control problem (18) versus their numerical controls for $p = 1$ arising by minimizing $\mathcal{J}_1^h$ over the restricted space $(\mathcal{B}_1^h \times \mathcal{B}_1^h) \cap \mathcal{V}_1^h$

to the presence of the optic mode whose largest eigenvalues are much above the largest ones for the linear approximation ($60/h^2$ versus $12/(h/2)^2 = 48/h^2$). The solutions of the adjoint problem (10) corresponding to the minimizer $(\tilde{\mathbf{U}}_p^{h,0}, \tilde{\mathbf{U}}_p^{h,1})$ of $\mathcal{J}_p^h$ over $\mathcal{V}_p^h$ are typically highly oscillatory wave packets whose energy is concentrated far from the boundary $x = 1$ at any time $t \in [0, 2]$ (see Fig. 13, left).

- When the space over which the functional $\mathcal{J}_p^h$ is minimized is restricted to the bi-grid class $(\mathcal{B}_p^h \times \mathcal{B}_p^h) \cap \mathcal{V}_p^h$, the high-frequency modes diminish in time for both the linear and the quadratic approximation as it can be observed in Figs. 11(e) and 12(e), (g), (i). For the case $p = 2$, the optic modes are more dissipated than the acoustic ones. However, by comparing Figs. 11(a)–(b) and 12(a)–(b), we observe that the numerical controls and the solutions of the discrete control problem (18) under filtering are much more accurate in the quadratic case than in the linear one. As we made it precise before, for $p = 2$, the bi-grid filtering acts mainly like a Fourier truncation of the whole optic mode λ_2^{o} and of the second

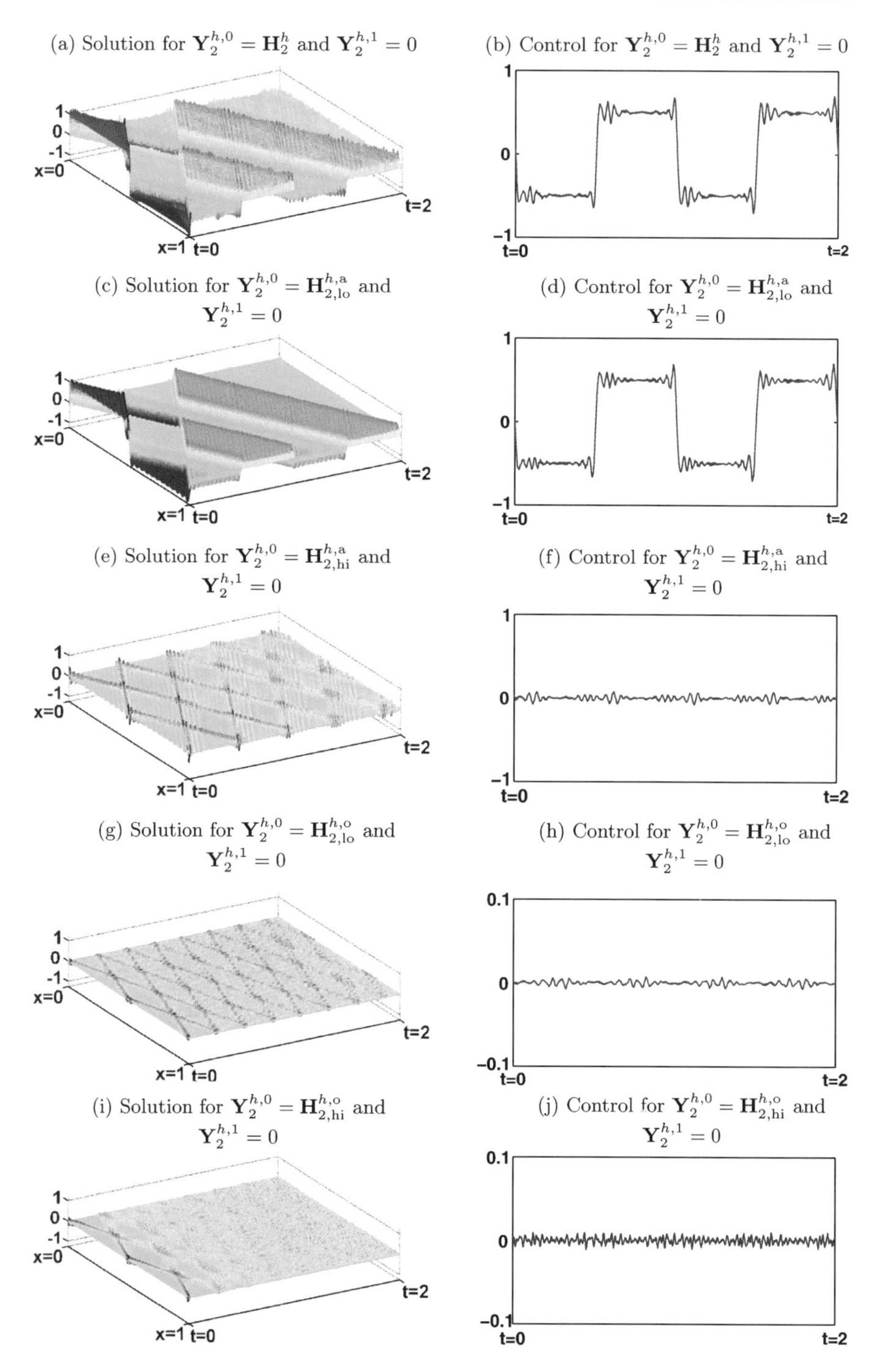

Fig. 12 Solutions of the control problem (18) versus their numerical controls for $p = 2$ arising by minimizing $\mathcal{J}_2^h$ over the restricted space $(\mathcal{B}_2^h \times \mathcal{B}_2^h) \cap \mathcal{V}_2^h$

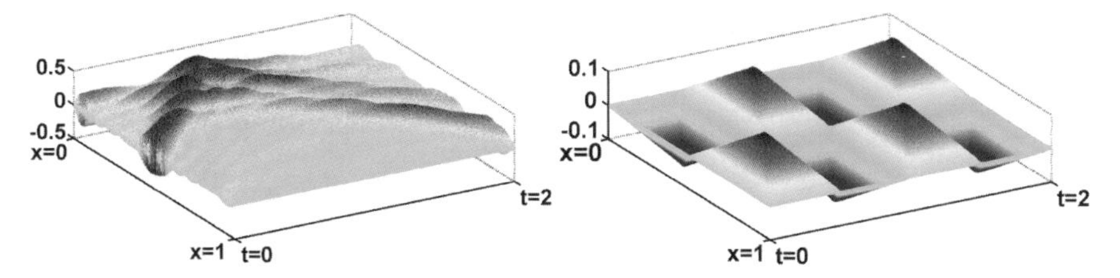

Fig. 13 Typical solution of the adjoint problem (10) corresponding to the minimizer $(\tilde{\mathbf{U}}_p^{h,0}, \tilde{\mathbf{U}}_p^{h,1})$ of $\mathcal{J}_p^h$ over $\mathcal{V}_p^h$ (*left*) or over $(\mathcal{B}_p^h \times \mathcal{B}_p^h) \times \mathcal{V}_p^h$ (*right*)

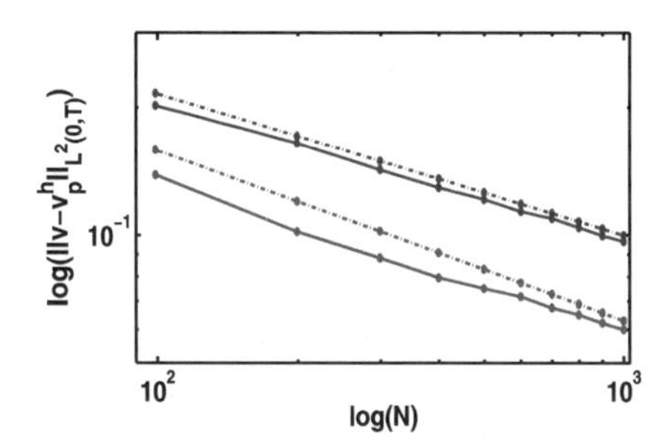

Fig. 14 The error $\|\tilde{v}_p^h - \tilde{v}\|_{L^2(0,T)}$ for $p = 1$ (*blue*) and $p = 2$ (*red*) versus the number of degrees of freedom N at the logarithmic scale. In *dotted blue/red*, we represent $N^{-1/3}$ and $N^{-2/5}$ also at the logarithmic scale. Here N takes values from 99 to 999 with increments of 100

half of the acoustic one λ_2^{a}, whereas for $p = 1$, it behaves like a Fourier truncation of the second half of the dispersion diagram λ_1. But the low frequencies of the acoustic mode approximate much better the continuous dispersion relation $\lambda(\eta) = \eta$, $\eta \in \mathbb{R}$, than the dispersion diagram of the linear approximation. Indeed, as $\eta \sim 0$, $\lambda_1(\eta) \sim \eta + \eta^3/24 + \eta^5/1920$, whereas $\lambda_2^{\mathrm{a}}(\eta) \sim \eta + \eta^5/1440$. According to the results in [3], this improves the convergence rate $h^{2/3}$ of the numerical controls towards the continuous ones corresponding to the case $p = 1$ for initial data (y^0, y^1) in the continuous control problem (1) belonging to the more regular space $H_0^1 \times L^2(0, 1)$, so that a convergence order $h^{4/5}$ is obtained for $p = 2$ under the same regularity assumptions. In Fig. 14, we represent the errors in the numerical controls obtained by the bi-grid filtering at the logarithmic scale for both approximations $p = 1$ and $p = 2$ for the initial position $\mathbf{H}_p^h$ approximating the Heaviside function H. The continuous initial data $(y^0, y^1) = (H, 0) \in H^{1/2-\epsilon} \times H^{-1/2-\epsilon}(0, 1)$, for any $\epsilon > 0$, which is less than the regularity imposed in [3]. Consequently, by interpolation, a natural sharp bound for the convergence orders of the numerical controls should be $h^{1/3}$ for $p = 1$ versus $h^{2/5}$ for $p = 2$. This is confirmed by our numerical results.

All the numerical simulations in this paper are realized under the Matlab environment. The corresponding numerical codes can be found following the link

www.bcamath.org/projects/MTM2008-03541/sim/images/.

4 Conclusions and Open Problems

In this paper, we have discussed and illustrated numerically the high frequency pathological effects of the P_2 approximation of the 1D wave equation, previously analyzed rigorously in [11] in what concerns the boundary observation and control problems. We have also illustrated the efficiency of the bi-grid filtering algorithm in recovering the convergence of the numerical controls and compared the results obtained by this quadratic finite element method with those one recovers by means of the P_1-approximation. Our conclusion is that, after applying the bi-grid filtering, the quadratic approximation leads to more accurate controls than the linear one.

This filtering technique can be easily generalized to higher order finite element approximation methods of waves ($p \geq 3$) on uniform meshes, a higher and higher accuracy of the numerical controls being expected. However, the high-frequency effects of the numerical approximations on irregular meshes is a completely open problem.

Acknowledgements Both authors were partially supported by ERC Advanced Grant FP7-246775 NUMERIWAVES, Grants MTM2008-03541 and MTM2011-29306 of MICINN Spain, Project PI2010-04 of the Basque Government, and ESF Research Networking Programme OPT-PDE. Additionally, the first author was supported by Grant PN-II-ID-PCE-2011-3-0075 of CNCS-UEFISCDI Romania.

References

1. Castro, C., Cea, M., Micu, S., Münch, A., Negreanu, M., Zuazua, E.: Wavecontrol: a program for the control and stabilization of waves. Manual. http://www.bcamath.org/documentos_public/archivos/personal/conferencias/manual200805.pdf
2. Ervedoza, S., Zheng, C., Zuazua, E.: On the observability of time-discrete conservative linear systems. J. Funct. Anal. **254**(12), 3037–3078 (2008)
3. Ervedoza, S., Zuazua, E.: The wave equation: control and numerics. In: Cannarsa, P.M., Coron, J.M. (eds.) Control and stabilization of PDEs. Lecture Notes in Mathematics, CIME Subseries, vol. 2048, pp. 245–339. Springer, Berlin (2012)
4. Glowinski, R., Li, C.H., Lions, J.L.: A numerical approach to the exact boundary controllability of the wave equation. I. Dirichlet controls: description of the numerical methods. Jpn. J. Appl. Math. **7**(1), 1–76 (1990)
5. Ignat, L., Zuazua, E.: Numerical dispersive schemes for the nonlinear Schrödinger equation. SIAM J. Numer. Anal. **47**(2), 1366–1390 (2009)
6. Ignat, L., Zuazua, E.: Convergence of a two-grid algorithm for the control of the wave equation. J. Eur. Math. Soc. **11**(2), 351–391 (2009)
7. Infante, J.-A., Zuazua, E.: Boundary observability for the space semidiscretization of the 1D wave equation. Modél. Math. Anal. Numér. **33**, 407–438 (1999)
8. Lions, J.L.: Contrôlabilité Exacte, Perturbations et Stabilisation des Systèmes Distribués, vol. 1. Masson, Paris (1988)
9. Loreti, P., Mehrenberger, M.: An Ingham type proof for a two-grid observability theorem. ESAIM Control Optim. Calc. Var. **14**(3), 604–631 (2008)
10. Marica, A., Zuazua, E.: Localized solutions and filtering mechanisms for the discontinuous Galerkin semi-discretizations of the 1D wave equation. C. R. Acad. Sci. Paris Ser. I **348**, 1087–1092 (2010)

11. Marica, A., Zuazua, E.: On the quadratic finite element approximation of 1D waves: propagation, observation and control. SIAM J. Numer. Anal. (accepted). arXiv:1112.4297
12. Marica, A., Zuazua, E.: High frequency wave packets for the Schrödinger equation and its numerical approximations. C. R. Acad. Sci. Paris Ser. I **349**, 105–110 (2011)
13. Micu, S.: Uniform boundary controllability of a semi-discrete 1D wave equation. Numer. Math. **91**, 723–768 (2002)
14. Negreanu, M., Zuazua, E.: Convergence of a multigrid method for the controllability of a 1D wave equation. C. R. Math. Acad. Sci. Paris **338**, 413–418 (2004)
15. Zuazua, E.: Propagation, observation, control and numerical approximations of waves. SIAM Rev. **47**(2), 197–243 (2005)

Space-Time Adaptive Multiresolution Techniques for Compressible Euler Equations

Margarete O. Domingues, Sônia M. Gomes, Olivier Roussel, and Kai Schneider

Abstract This paper considers space-time adaptive techniques for finite volume schemes with explicit time discretization. The purpose is to reduce memory and to speed-up computations by a multiresolution representation of the numerical solution on adaptive grids which are introduced by suitable thresholding of its wavelet coefficients. Further speed-up is obtained by the combination of the multiresolution scheme with an adaptive strategy for time integration, which is classical for ODE simulations. It considers variable time steps, controlled by a given precision, using embedded Runge–Kutta schemes. As an alternative to the celebrated CFL condition, the aim in the application of such an time-adaptive scheme for PDE simulations is to obtain accurate and safe integrations. The efficiency of this adaptive space-time method is analyzed in applications to typical Riemann–Lax test problems for the compressible Euler equations in one and two space dimensions. The results show that the accuracy properties of the reference finite volume scheme on the finest regular grid, where the time step is determined by the CFL condition, is preserved. Nevertheless, both CPU time and memory requirements are considerably reduced, thanks to the efficient self-adaptive grid refinement and controlled time-stepping.

Keywords Wavelets · Multiresolution · Partial differential equations · Finite volume · Runge–Kutta · Adaptivity · Time step control

M.O. Domingues
Laboratório Associado de Computação e Matemática Aplicada (LAC), Instituto Nacional de Pesquisas Espaciais (INPE), São José dos Campos, Brazil
e-mail: margarete@lac.inpe.br

S.M. Gomes
IMECC, Universidade Estadual de Campinas, Campinas, Brazil
e-mail: soniag@ime.unicamp.br

O. Roussel
Eurobios, 94234 Cachan, France
e-mail: o_roussel@yahoo.fr

K. Schneider (✉)
M2P2-CNRS, Aix-Marseille Université, 38 rue Joliot-Curie, 13451 Marseille cedex 20, France
e-mail: kschneid@cmi.univ-mrs.fr

C.A. de Moura, C.S. Kubrusly (eds.), *The Courant–Friedrichs–Lewy (CFL) Condition*,
DOI 10.1007/978-0-8176-8394-8_7,

1 Introduction

Multiresolution techniques have become an efficient tool, not only for data compression, but also for adaptive simulations of partial differential equations, since the seminal work by Ami Harten [14]. The main purpose of this paper is to show results illustrating the efficiency of a fully adaptive space-time multiresolution finite volume method for solving the compressible Euler equations.

The Finite Volume method (FV) is one of the most robust and versatile discretization techniques used in computational fluid dynamics [18]. Given a partition of the computational domain, the quantities (such as mass, momentum, energy, and species concentrations) are represented by their cell averages over the grid cells, which are modified in each time step by the flux through the cells edges. Such a procedure provides conservation of the quantities, which is valid locally, for any grid cell, as well as globally, for the whole computational domain.

FV calculations can be accelerated significantly by the use of adaptive grids, motivated by the presence of inhomogeneous singular features in the numerical solution (as interfaces, shocks or reaction zones). Here we consider multiresolution (MR) adaptive strategies to speed up FV schemes for time dependent partial differential equations in Cartesian geometries.

The MR technique is known to yield an appropriate framework to construct adaptive FV schemes for hyperbolic conservation laws since the work of Harten [15]. In this context, cell average discretizations are considered in a hierarchy of embedded partitions of the computational domain, at different scale levels. The principle of the multiresolution analysis is to represent a set of cell average data given on a fine grid as values on a coarser grid plus a series of differences at different levels, the wavelet contributions, containing the information of the solution when going from a coarse to a finer grid.

Using the wavelet coefficients as local regularity indicators—the idea of Harten, which was subsequently explored in [1, 4, 6]—is to reduce the number of costly flux evaluations to speed up the scheme, however, without reducing the memory requirements. Fully adaptive MR schemes with memory compression have been developed, where the representation of the numerical solutions is performed only by the cell averages on the adaptive partitions formed by cells corresponding to their significant wavelet coefficients [5, 7, 10, 11, 16, 20, 22–24]. Typically, little information is required in each time step, since fine grids are only used near the steep gradients, while coarser grids are sufficient to represent the solution in smooth regions.

To further speed-up space-adaptive simulations, adaptive strategies may be adopted for time integration. For instance, one strategy is the local scale-dependent time stepping, which is classical in the AMR (Adaptive Mesh Refinement) context [3]. The principle is to evolve the solution on large scale cells with larger time steps, which are determined locally by the CFL condition, according to each cell size. This technique has also been combined with MR schemes, as discussed in [9, 10, 21]. Another adaptive time integration, which is usual for ODE simulations, considers variable time steps, controlled by a given precision, using embedded Runge–Kutta schemes. The aim in the application of such schemes for PDE simulations is to obtain accurate and safe integrations without the requirement of a prescribed fixed CFL parameter. This controlled time-stepping technique, which can also be combined with local time stepping, has been applied for AMR schemes in [13], and for MR methods in [2, 5, 11, 12]. For the applications of the present paper, only the combination of MR with controlled time stepping is considered, which is herein referred as MR/CT scheme.

The text is organized as follows. In Sect. 2, we describe the reference FV scheme and the corresponding space-adaptive MR and space-time-adaptive MR/CT schemes. In Sect. 3, the MR/CT method is applied to the compressible Euler equations in one and two space dimensions. The results are compared with the exact solution in 1D, or with those obtained with the reference FV scheme on the finest regular grid in 2D. Their accuracy, CPU time, and memory compression are discussed taking into account two choices of the threshold parameters. Finally, conclusions from our results are drawn and some perspectives of this work are given in Sect. 4.

2 The Numerical Schemes

The compressible Euler equations can be written in the following conservation form,

$$\frac{\partial Q}{\partial t} + \nabla \cdot f(Q) = 0, \tag{1}$$

with $Q = (\rho, \rho \boldsymbol{v}, \rho e)^T$, where $\rho = \rho(\boldsymbol{x}, t)$ is the density, $\boldsymbol{v} = \boldsymbol{v}(\boldsymbol{x}, t)$ is the vector velocity with components (v_1, v_2, v_3), and $e = e(\boldsymbol{x}, t)$ is the energy per unit of mass, which are functions of time t and position $\boldsymbol{x} = (x_1, x_2) \in \Omega$. The flux function $f = (f_1, f_2)$ is given by

$$f_1 = \begin{pmatrix} \rho v_1 \\ \rho v_1^2 + p \\ \rho v_1 v_2 \\ (\rho e + p) v_1 \end{pmatrix}, \qquad f_2 = \begin{pmatrix} \rho v_2 \\ \rho v_1 v_2 \\ \rho v_2^2 + p \\ (\rho e + p) v_2 \end{pmatrix},$$

where $p = p(\boldsymbol{x}, t)$ denotes the pressure. The system is completed by an equation of state for a calorically ideal gas

$$p = \rho R T = (\gamma - 1)\rho \left(e - \frac{|\boldsymbol{v}|^2}{2} \right), \tag{2}$$

where $T = T(\boldsymbol{x}, t)$ is the temperature, γ the specific heat ratio, and R the universal gas constant. In dimensionless form, we obtain the same system of equations, but the equation of state becomes $p = \frac{\rho T}{\gamma M^2}$, where M denotes the Mach number. For the present applications, the physical parameters are $M = 1$ and $\gamma = 1.4$.

2.1 Reference Finite Volume Discretization

In the reference scheme for equations in the conservation form (1), the numerical solution is represented by the vector $\bar{Q}(t)$ of the approximated cell-averages $\bar{Q}_{k,m}(t)$

$$\bar{Q}_{k,m}(t) \approx \frac{1}{|\Omega_{k,m}|} \int_{\Omega_{k,m}} Q(\boldsymbol{x},t)\,d\boldsymbol{x}$$

on cells $\Omega_{k,m}$ of a grid with uniform spacing $\Delta x = \Delta y$. For space discretization, a finite volume method is chosen, which results in an ODE system of the form

$$\frac{d\bar{Q}}{dt} = F(\bar{Q}), \tag{3}$$

where $F(\bar{Q})$ denotes the vector of the numerical flux function. For time integration, approximate solutions $\bar{Q}^n$ at a sequence of time instants t^n are obtained using an explicit ODE solver, i.e., Runge–Kutta schemes. For stability, the time steps $\Delta t_n = t^{n+1} - t^n$ are determined by the CFL-condition [8]

$$CFL = \lambda_{\max} \frac{\Delta t}{\Delta x},$$

where $\lambda_{\max}$ is the maximum absolute value for the eigenvalues of the Jacobian matrix of $f(Q)$.

For the numerical tests of the present paper, the reference FV scheme uses a second order MUSCL scheme with an AUSM+ flux vector splitting scheme [19] and the van Albada limiter. For time integration, an explicit third-order Runge–Kutta (RK3) scheme is used.

2.2 Adaptive Multiresolution Methods

The adaptive methods of the present paper fall into the MR category, combined with a time adaptive strategy using controlled time-stepping.

MR Scheme The adaptive MR scheme belongs to a class of adaptive methods which are formed by two basic parts: the operational part and the representation part. The operational part consists of an accurate and stable discretization of the partial differential operators. In the representation part, multiresolution analysis tools of the discrete information are employed. The principle of the MR setting is to represent a set of function data as values on a coarser grid G_0 plus a series of differences at different levels of nested grids $G_j \subset G_{j+1}$, see Fig. 1. The information at consecutive scale levels are related by inter-level transformations: projection and prediction operators. The wavelet coefficients d_j are defined as prediction errors, and they retain the detail information when going from a coarse G_j to a finer grid G_{j+1} [23].

In MR schemes for the adaptive numerical solution of PDEs, the main idea is to use the decay of the wavelet coefficients to obtain information on the local regularity of the solution. Adaptive MR representations are obtained by stopping the

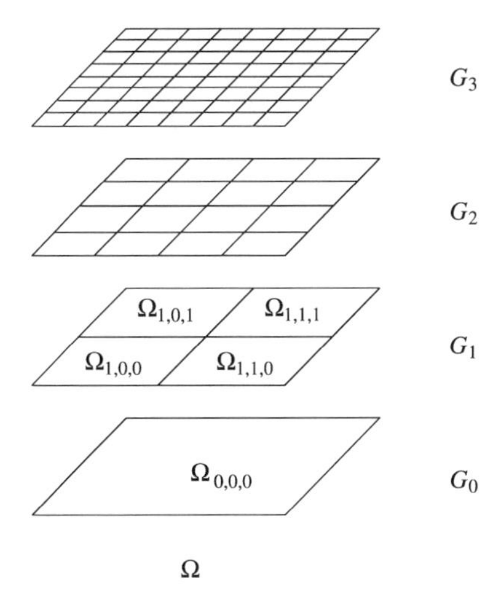

Fig. 1 Set of nested dyadic grids G_j, for $0 \leq j \leq 4$

refinement in a cell at a certain scale level where the wavelet coefficients are non-significant. In particular, these coefficients are small in regions where the solution is smooth and significant close to irregularities, e.g., steep gradients. In the finite volume context, the natural representation framework is the multiresolution analysis based on cell-averages. Instead of using the cell-average representation on the uniform fine grid, the MR scheme computes the numerical solution represented by its cell-averages on an adaptive sparse grid, which is formed by the cells whose wavelet coefficients are significant and above a given threshold.

An efficient way to store the reduced MR data is to use a tree data structure, where grid adaptivity is related with an incomplete tree, and where the refinement may be interrupted at intermediate scale levels. This means that, using the tree terminology, an MR grid is formed by *leaves*, which are nodes without children; for an illustration we refer to Fig. 2, left.

For the time evolution of the solution, three basic steps are considered: *refinement*, *evolution*, and *coarsening*. The refinement operator is a precautionary measure to account for possible translation of the solution or creation of finer scales in the solution between two subsequent time steps. Since the regions of smoothness or irregularities of the solution may change with time, the MR grid at t^n may not be convenient anymore at the next time step t^{n+1}. Therefore, before doing the time evolution, the representation of the solution should be interpolated onto an extended grid that is expected to be a refinement of the adaptive grid at t^n, and to contain the adaptive grid at t^{n+1}.

Then, the time evolution operator is applied on the leaves of the extended grid. To compute fluxes between leaves of different levels, we also add *virtual leaves* (Fig. 2, right). Conservation is ensured by the fact that the fluxes are always computed on the higher level, the value being reported on the leaves of a lower level.

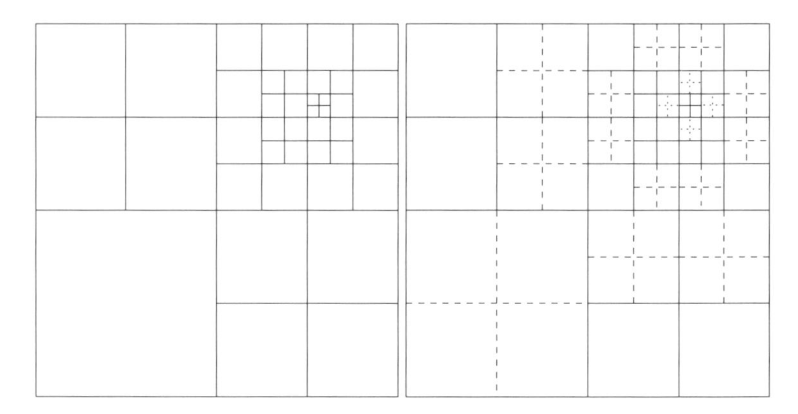

Fig. 2 *Left*: Sketch of a 2D tree structure. *Right*: corresponding sketch with its leaves (*plain*) and virtual leaves (*dashed*)

Finally, a wavelet thresholding operation (coarsening) is applied in order to unrefine the cells in the extended grid that are not necessary for an accurate representation of the solution at t^{n+1}. This data compression is based on the definition of deletable cells, where the wavelet coefficients are not significant, i.e., their magnitudes are bellow a threshold parameter ϵ_j, where j denotes the cell scale level. In order to control the L^1-norm, Harten's thresholding strategy is recommended, for which

$$\epsilon_j = \frac{\epsilon}{|\Omega|} 2^{d(j-J+1)}, \quad 0 \leq j \leq J-1,$$

where $d = 1, 2$, or 3 is the space dimension, and J is the finest scale level. For comparison, we shall also consider level independent threshold parameters: $\epsilon_j \equiv \epsilon$, for all $1 \leq j \leq J-1$.

For the applications of the present paper, the multiresolution analysis corresponds to a prediction operator based on a third order polynomial interpolation on the cell-averages. For further details on the adaptive MR scheme, we refer to [23].

MR/CT Scheme In the MR/CT scheme, the time integration is performed with a variable time step Δt, which size is chosen dynamically. It should be small enough to get a required precision and stability of the computed results, but sufficiently large to avoid unnecessary computational work.

Instead of determining the time step by the celebrated CFL condition, using a fixed CFL parameter determined a priori, the MR/CT scheme adopts a classical strategy from ODE simulations, where the time step size selection is based on estimated local truncation errors. The main reason of controlling the error in the solution is to obtain an accurate and safe integration in the whole interval. When the estimated local error is smaller than a given tolerance, denoted by δ_{desired}, the algorithm increases the step size to make the integration more efficient.

For the applications of the present paper, the MR/CT scheme is based on the embedded Runge–Kutta Fehlberg 2(3) ODE solver [26]. The initial time step is determined by an input $CFL(0)$ parameter provided by the user. From one time instant to the next, given the current Δt, the next Δt_{new} is determined to maintain the local truncation error below δ_{desired}. Precisely, it has the form $\Delta t_{\text{new}} = \Delta t \xi$, with

$$\xi = \left[\frac{\delta_{\text{desired}}}{|\bar{Q}_{(\text{low})} - \bar{Q}_{(\text{high})}|} \right]^{1/3},$$

where $\bar{Q}_{(\text{low})}$ and $\bar{Q}_{(\text{high})}$ stand for the solutions produced by RK2 and RK3 schemes, respectively. To prevent the time step of varying too abruptly or to be sure that Δt_{new} in fact will produce an error less than δ_{desired}, the time step variation is limited by a factor that decreases exponentially from 10 %, in the initial time step, to 1 % after a few iterations. For more details on the combined MR/CT scheme, we refer to [11].

3 Numerical Tests

To illustrate the accuracy and efficiency of the MR/CT method, we consider two test problems in one and two space dimensions. Since one of the purposes here is to evaluate the effect of different threshold strategies, we refer to MR/CT-ϵ and MR/CT-ϵ_j to distinguish between the constant and the Harten threshold parameters which are used.

The simulations presented in this section were performed using the multiresolution code Carmen 1.54, initially developed by Roussel et al. [23]. In this updated version, details are considered for averaged quantities of ρ, ρe, and $|\boldsymbol{v}|$.

3.1 *Lax Test-Case in 1D*

The Riemann problem for the unidimensional Euler equations with initial condition

$$Q(x,t=0) = \begin{pmatrix} 0.445 \\ 0.31061 \\ 8.928 \end{pmatrix}, \quad \text{if } x < 0, \quad \text{and}$$

$$Q(x,t=0) = \begin{pmatrix} 0.5 \\ 0 \\ 1.4275 \end{pmatrix}, \quad \text{otherwise}$$

is known as the Lax problem. Details on this test-case, and its exact solution can be found in [18, 27]. We compute the solution in the domain $\Omega = [-1, 1]$, with Neumann boundary conditions applied on both sides. The simulations are performed until physical time $t = 0.32$, and all errors are taken at this final time. We take the grid spacing $\Delta x = 2^{J-1}$ at the finest scale level, and the results are obtained for $J = 11$.

For this problem, the maximum absolute value for the eigenvalues of the Jacobian matrix is constant for $t > 0$. Therefore, for the FV simulations, we assume a constant time step Δt, which is obtained from the input $CFL(0)$. Within $t \leq 0.32$

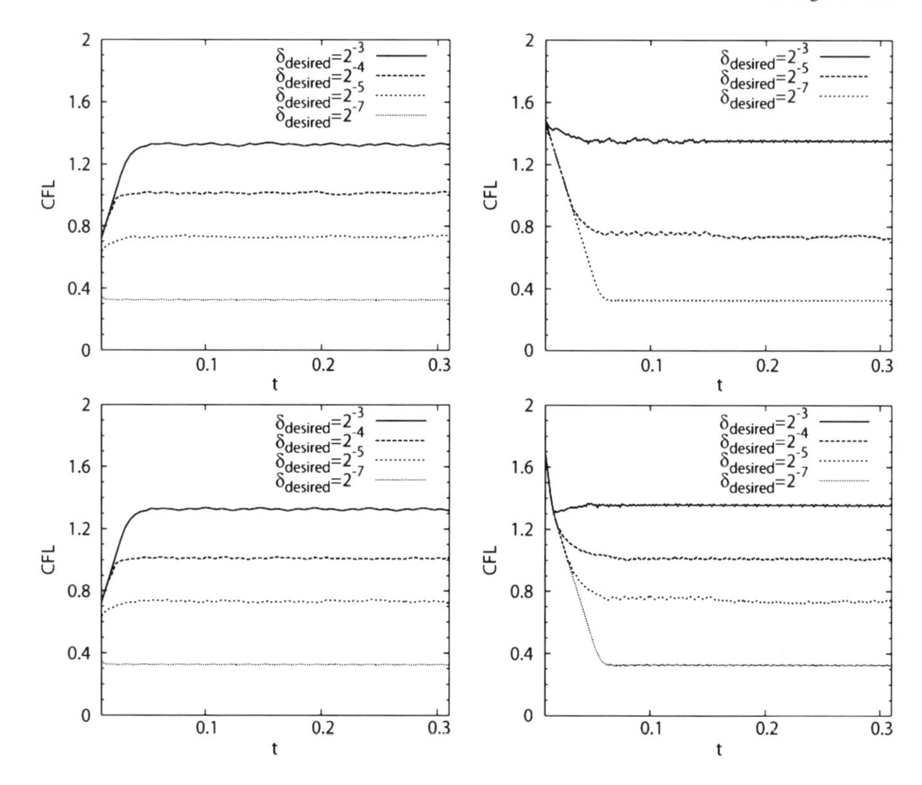

Fig. 3 1D Lax test-case: CFL evolution for MR/CT-ϵ (*top*) and MR/CT-ϵ_j (*bottom*) schemes, for different accuracy values δ_{desired}, with $J = 11$, $CFL(0) = 0.4$ (*left*) and $CFL(0) = 1.5$ (*right*)

using AUSM+ with van Albada limiter and an RK3 scheme, we found the stability limit $CFL \leq 1.33$.

For the adaptive MR/CT scheme, an input $CFL(0)$ parameter is provided by the user, and $CFL(t)$ evolves according to the new Δt obtained by the time step control. In [11], results are presented with initial $CFL(0)$ values below and above the CFL/RK3 stability limit. It was observed that initially there is a transient state, and then $CFL(t)$ becomes constant. The steady state value $CFL_\infty = CFL_\infty(\delta_{\text{desired}})$ decreases with δ_{desired}, remaining within the stability limit of the reference FV scheme. Nevertheless, it seems to be independent of the $CFL(0)$ input. The accuracy of the numerical solutions, measured in the L^1-norm, has the same behavior, which is almost insensitive to the CFL history, with significant gain in the amount of memory and CPU time required with respect to the reference FV scheme.

We revisit this test problem here to analyze the effect of the threshold strategy, having in mind the applications in two and three dimensions, where the savings in CPU time and memory are crucial. Therefore, we consider similar results using MR/CT-ϵ_j and MR/CT-ϵ schemes.

Figure 3 shows the CFL evolution for both threshold strategies, for $\delta_{\text{desired}} = 2^{-M}$, $M = 3, 4, 5$, and 7, with $J = 11$, confirming the conclusions presented in [11],

Table 1 1D Lax test-case: Comparison of L_1-errors for density ρ and kinetic energy E, speed-up and data compression of the numerical solutions at time $t = 0.32$ with $J = 11$ levels. For the MR/CT schemes, $\delta_{\text{desired}} = 2^{-4}$ and $CFL(0) = 0.4$ is used

Method	Error		CPU		
	L_1 ($\times 10^{-3}$)	E (%)	Time (%)	Memory (%)	Leaves (%)
FV (Ref.)	4.144	0.009	100	100	100
$\epsilon = 10^{-2}$					
MR/CT-ϵ	6.867	0.298	7.44	16.92	7.53
MR/CTS-ϵ_j	4.321	0.108	10	24	11
$\epsilon = 10^{-3}$					
MR/CT-ϵ	4.266	0.108	10	24	11
MR/CTS-ϵ_j	4.151	0.103	13	29	13
$\epsilon = 10^{-4}$					
MR/CT-ϵ	4.141	0.106	13	30	13
MR/CTS-ϵ_j	4.134	0.105	15	34	16

and showing that the influence of the threshold strategies seems to be insignificant in the choices of the Δt parameters in the MR/CT schemes.

Table 1 shows memory and CPU time compression effects of the adaptive schemes, for $J = 11$, $\delta_{\text{desired}} = 2^{-4}$, $CFL(0) = 0.4$, $\epsilon = 10^{-2}$, 10^{-3}, and 10^{-4}, together with L^1-errors on density and kinetic energy

$$E = \frac{1}{2}\int_{-1}^{1} \rho(x)\left|v(x)\right|^2 dx = 0.966568.$$

The reference FV scheme uses a constant time step $\Delta t = 9.694157 \times 10^{-5}$, which is obtained from $CFL(0) = 0.4$. In all the cases, concerning CPU time and compression gains with respect to the FV reference scheme, the effectiveness of the MR/CT schemes increases with increasing ϵ. In all the MR/CT cases, there is a slight variation of L_1 and the percentage of energy errors. As expected, with respect to the MR/CT-ϵ_j scheme, the MR/CT-ϵ scheme requires less memory, with gain in CPU time, but with a consequent increase in the L^1 and kinetic energy errors. However, these differences in efficiency become less important as ϵ decreases.

The plots for the exact density $\rho(x, t = 0.32)$ and its numerical approximations are shown in Fig. 4 (*top, left*), showing that the numerical solutions fit the exact one. The other three plots correspond to zooms onto the rarefaction boundary (*top, right*), the contact discontinuity (*bottom, left*), and the shock (*bottom, right*). We can observe that the MR/CT-ϵ solution loses resolution at the rarefaction part, where FV and MR/CT-ϵ_j solutions almost coincide with the exact solution. At the contact discontinuity and the shock, the three schemes have a comparable behavior.

In Fig. 5, the leaves of the MR adaptive meshes are represented in the position $\times$ level plane. As expected, the grid is refined close to irregularities, and the finest

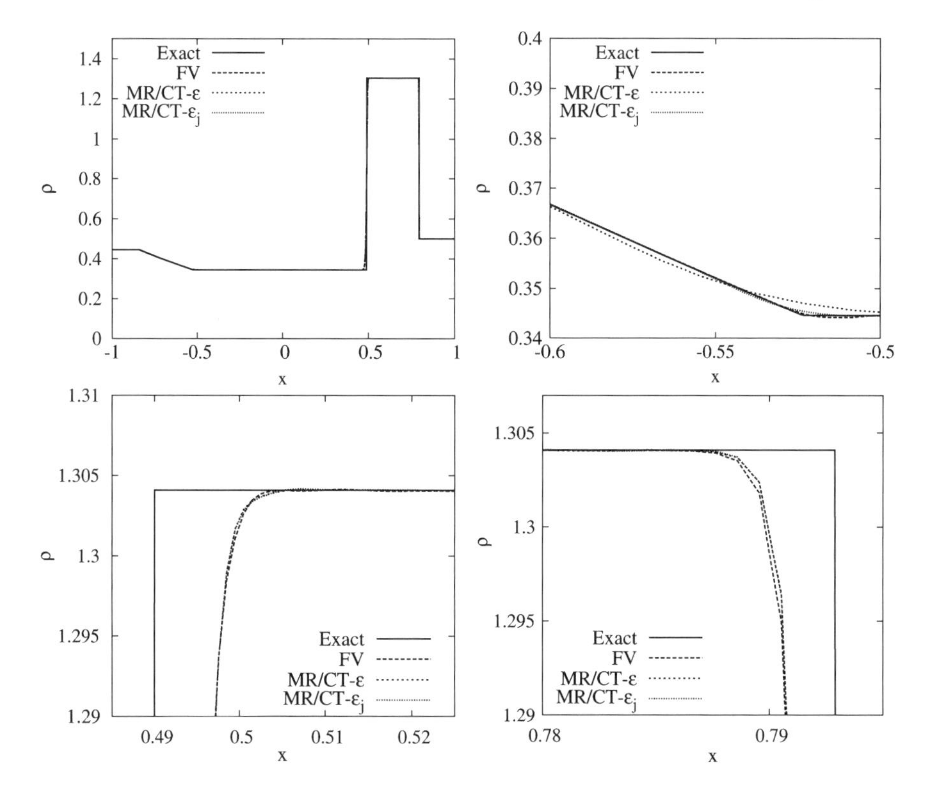

Fig. 4 1D Lax test-case: Exact and numerical density obtained with the FV and MR/CT-ϵ methods at $t = 0.32$ with $J = 11$ (*top, left*). For all MR cases $\epsilon = 10^{-3}$, and for the CTS cases $\delta_{\text{desired}} = 2^{-4}$, $CFL(0) = 0.4$. Zooms onto the rarefaction (*top, right*), the contact discontinuity (*bottom, left*) and the shock (*bottom, right*)

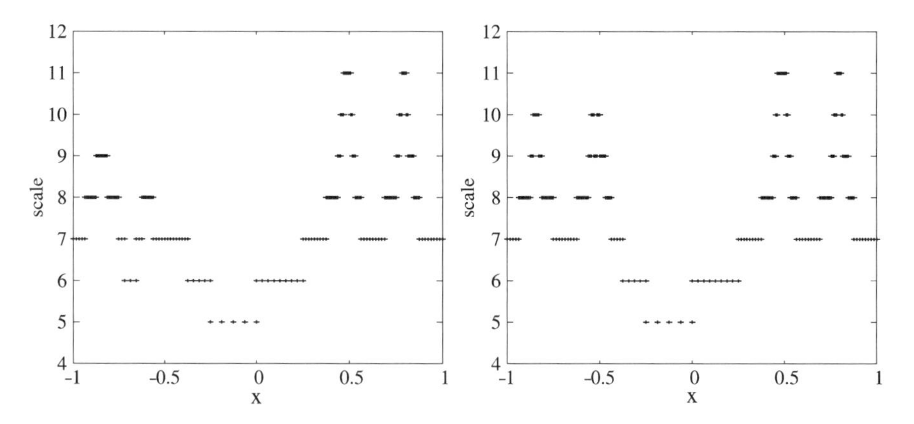

Fig. 5 1D Lax test-case: MR grids at time $t = 0.32$ with $J = 11$, $\delta_{\text{desired}} = 2^{-4}$, $CFL(0) = 0.4$ and $\epsilon = 10^{-3}$ for the MR/CT-ϵ scheme (*left*) and the MR/CT-ϵ_j scheme (*right*)

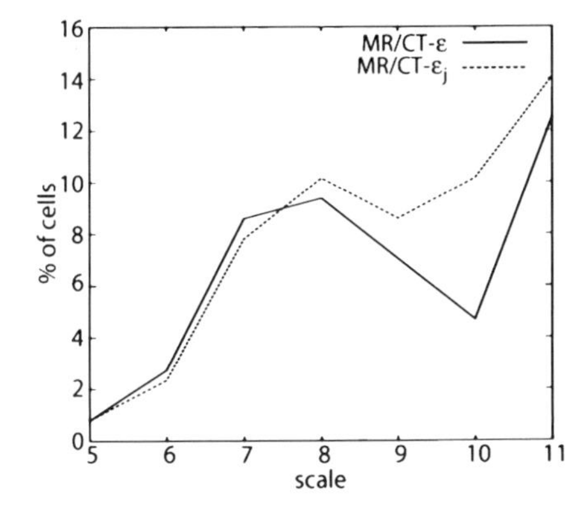

Fig. 6 1D Lax test-case: Distribution of the percentage of cells for each scale at time $t = 0.32$ with $J = 11$, $\delta_{\text{desired}} = 2^{-3}$, $CFL(0) = 0.4$ and $\epsilon = 10^{-3}$ for MR/CT-ϵ and MR/CT-ϵ_j

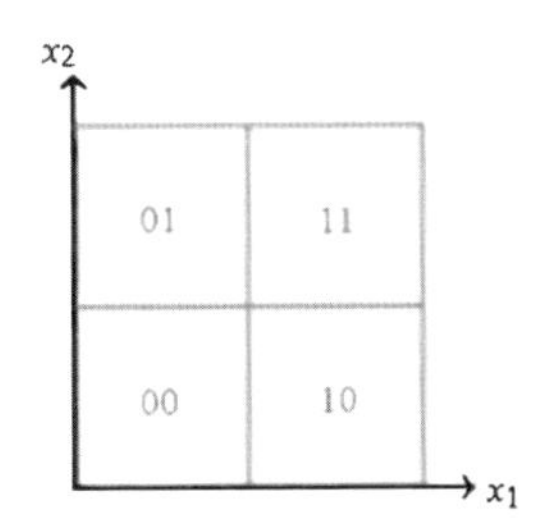

Fig. 7 2D Riemann problem: Domain decomposition for the definition of the initial condition

level is reached in the vicinity of the shock and the contact discontinuity. Close to the rarefaction boundaries the MR/CT-ϵ_j grid shows more refinement, explaining the better resolution of its solution at this location.

The percentage of the leaf cells on each scale level with respect to the full uniform grid at the finest resolution is presented in Fig. 6. The higher degree of refinement given by the MR/CT-ϵ_j scheme at higher levels, mainly at $j = 9$ and 10, is noticeable, which comes from the rarefaction zone, as it can be observed in Fig. 5.

3.2 *2D Test-Case: Lax–Liu Configuration #6*

The case study chosen here is a typical Riemann problem for 2D gas dynamics, and corresponds to the configuration #6 treated, e.g., in [17], and initially discussed in [25, 28]. The computational domain is the square $\Omega = [0, 1]^2$, with free boundary conditions. The domain is divided into four quadrants (Fig. 7), where the initial data are set constant in each quadrant, according to the values given in Table 2. The simulations are performed until $t = 0.3$, with $J = 10$ or 11. For the MR-CT schemes, we take the parameters $\epsilon = 10^{-3}$, $CFL(0) = 0.5$, and $\delta_{\text{desired}} = 2^{-4}$.

Table 2 Initial values for the 2D Lax–Liu configuration #6 [17]

Variables	Quadrant			
	00	01	10	11
Density (ρ)	1.0	2.0	3.0	1.0
Pressure (p)	1.0	1.0	1.0	1.0
Velocity component (v_1)	−0.75	0.75	−0.75	0.75
Velocity component (v_2)	0.5	0.5	−0.5	−0.5

Table 3 2D test-case: Speed-up and L_1-errors of density ρ and kinetic energy E of the MR/CT numerical solutions at time $t = 0.3$ with $J = 10$ and 11 levels, $\epsilon = 10^{-3}$, $\delta_{\text{desired}} = 2^{-4}$, and $CFL(0) = 0.5$

Method	Error		CPU		
	L_1 ($\times 10^{-3}$)	E (%)	Time (%)	Memory (%)	Leaves (%)
$J = 10$					
MR/CT-ϵ	1.2146	0.0340	16	23	15
MR/CTS-ϵ_j	0.7961	0.0288	26	37	25
$J = 11$					
MR/CT-ϵ	2.0135	0.0156	7	11	7
MR/CTS-ϵ_j	0.8470	0.0144	14	22	14

In Table 3, we compare the computational efficiency and the precision of the MR/CT methods. The reference computations are given by the FV scheme on the finest regular grid with constant time step $\Delta t = 2.790179 \times 10^{-4}$, for $J = 10$, and $\Delta t = 1.395089 \times 10^{-4}$, for $J = 11$, corresponding to $CFL(0) = 0.5$. This test problem also shows an almost constant $\approx$1.77 maximum eigenvalue. Concerning CPU time and compression gains with respect to the FV reference scheme, the effectiveness of the MR/CT schemes increases with J. As expected, with respect to the MR/CT-ϵ_j scheme, the MR/CT-ϵ scheme requires about half of the memory, with equivalent gain in CPU time, but with a consequent increase in L^1-error. Nevertheless, the kinetic energy errors are comparable.

Contour plots for the density, velocity, and energy at $t = 0.3$ are presented in Fig. 8 for the MR/CT-ϵ (*left*) and MR/CT-ϵ_j (*right*), for $J = 11$. They show that both MR solutions are similar, but with a better definition of the details for the MR/CT-ϵ_j case. This improved approximation is a consequence of a more refined MR grid close to strong variation regions, as shown in Fig. 9. Density cuts at the line $x = 0.5$ are shown in Fig. 10 (*top, left*). We find a rather good agreement between both FV and MR-CT computations. Zooms around the left, center, and right sides are also shown. In the center, where both schemes have a well refined grid, the results are very accurate. However, the MR/CT-ϵ scheme is not able to resolve the transition

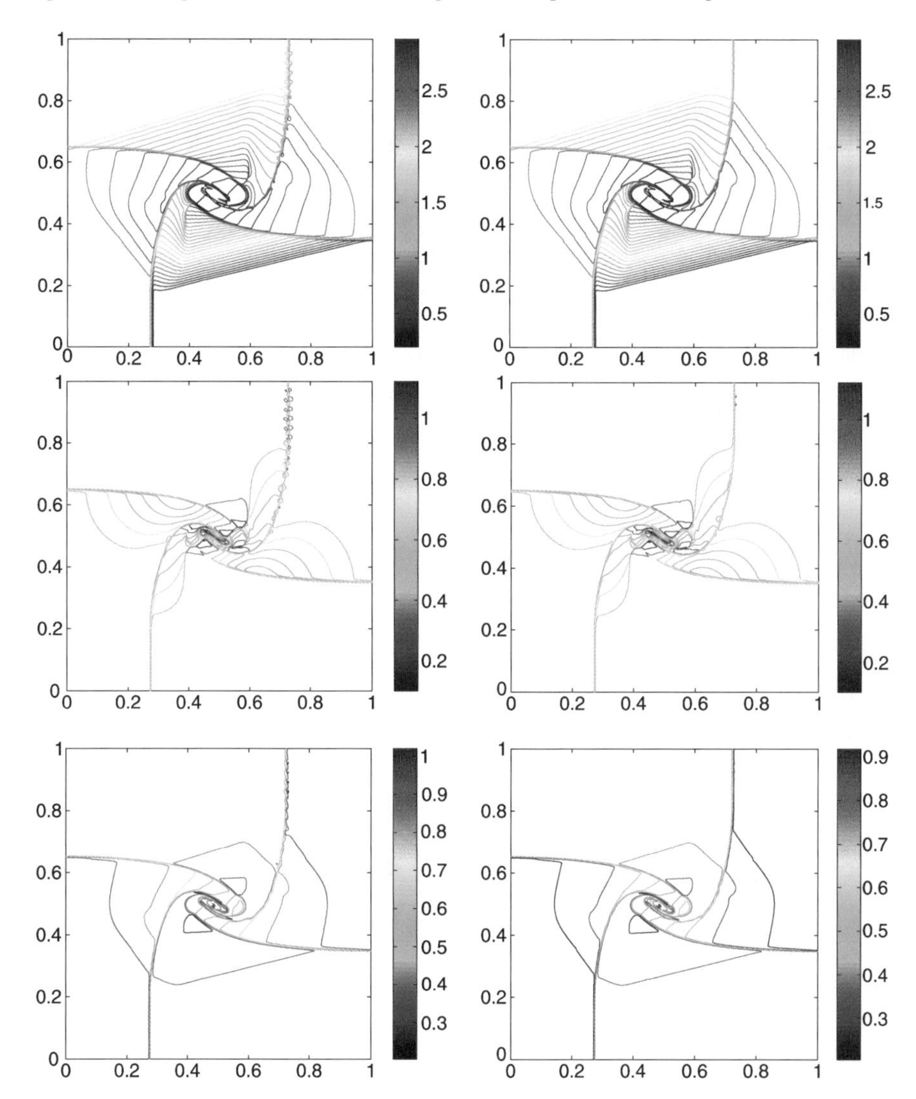

Fig. 8 2D test case: Density ρ (*top*), velocity $|\boldsymbol{v}|$ (*center*), and energy e (*bottom*) profiles at $t = 0.3$, for the MR/CT-ϵ (*left*), and MR/CT-ϵ_j (*right*), with $J = 11$, $\epsilon = 10^{-3}$, and $\delta_{\text{desired}} = 2^{-4}$. Contour lines from 0, with intervals 0.1

regions of the constant states $\rho = 3$ and $\rho = 2$, where the constant ϵ strategy is not sensible enough.

Concerning time adaptivity, the influence of the threshold strategies seems to be insignificant in the choices of the Δt parameters in the MR/CT schemes, as presented in Fig. 11 (*left*). The plots in Fig. 11 (*right*) correspond to the distribution of the percentage of leaves of the MR adaptive grids on each scale level, with respect

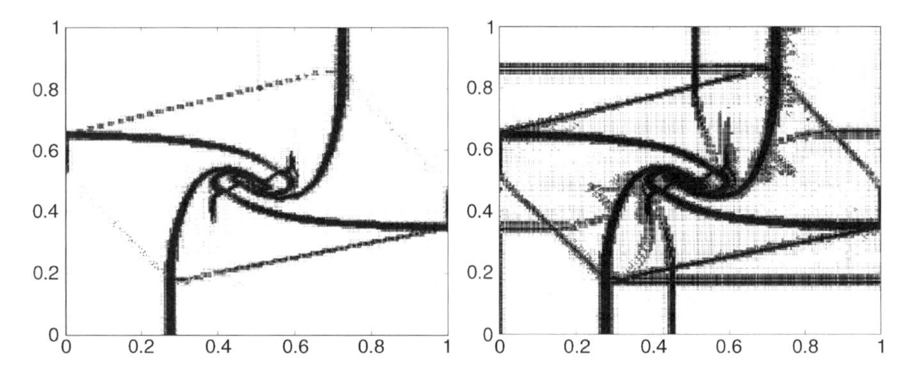

Fig. 9 2D test case: Adaptive grids at $t = 0.3$, for the MR/CT-ϵ (*left*), and the MR/CT-ϵ_j schemes (*right*), with $J = 11$, $\epsilon = 10^{-3}$, and $\delta_{\text{desired}} = 2^{-4}$

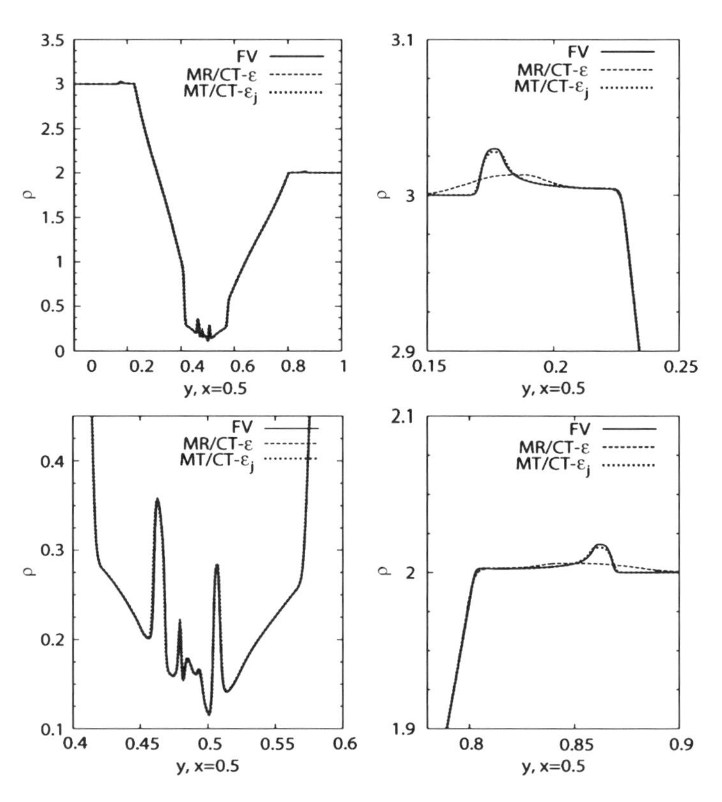

Fig. 10 2D test-case: Density cuts at $x = 0.5$ for $t = 0.3$, $J = 11$ and $\epsilon = 10^{-3}$, $\delta_{\text{desired}} = 2^{-4}$, for the FV and MR/CT methods (*top*, *left*). Corresponding zooms around the regions $[0.15, 0.25]$ (*top*, *right*), $[0.4, 0.6]$ (*bottom*, *left*) and $[0.78, 0.9]$ (*bottom*, *right*)

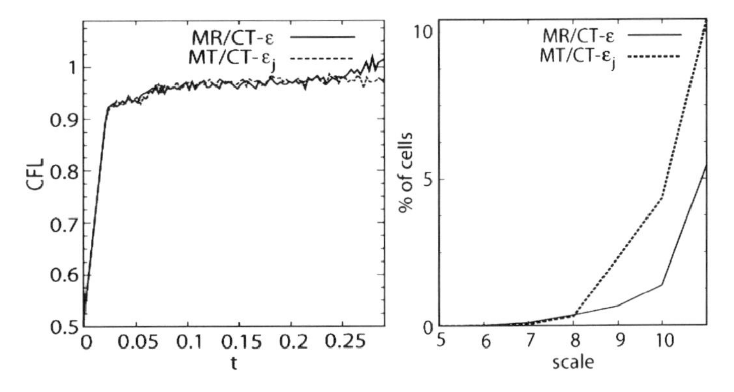

Fig. 11 2D test-case: Evolution of the CFL number (*left*) and the percentage of cells for each scale level (*right*) for the MR/CT methods at $t = 0.3$, with $J = 11$, $\epsilon = 10^{-3}$, and $\delta_{\text{desired}} = 2^{-4}$

to the full uniform grid at the finest resolution. The higher degree of refinement given by the MR/CT-ϵ_j scheme at higher levels is noticeable.

4 Conclusions

The present work on adaptive multiresolution techniques in space and time illustrates the potential of such advanced numerical methods for solving the compressible Euler equations in one and two space dimensions with reduced computational complexity, i.e., reducing both memory and CPU time requirements with respect to computations of the finest regular grid. The accuracy of the adaptive computations is nevertheless guaranteed by suitable thresholding of the wavelet coefficients and the convergence order of the underlying finite volume scheme on the regular grid is maintained. Furthermore, the adaptive time control mechanism maintains the stability of the simulations and the solution satisfies an imposed precision in time. The time step is adapted automatically by the numerical scheme during the time evolution.

In the future work, we plan to perform computations in three space dimensions and to benchmark extensively our multiresolution code against adaptive mesh refinement (AMR) strategies in terms of precision and CPU time. Preliminary results comparing MR with AMR can be found in Deiterding et al. [9].

Acknowledgements M.O. Domingues and S. Gomes acknowledge financial support from Fundação de Amparo à Pesquisa do Estado de São Paulo (FAPESP grant number: 2012/07281-2, 07/52015-0), and from CNPq (grant number 306828/2010-3, 307511/2010-3, 483226/2011-4)—the Brazilian Research Council, Brazil. They are also grateful for the financial support for visiting positions at École Centrale de Marseille (M.O. Domingues and S.M. Gomes) and Université de Provence (S.M. Gomes). K. Schneider thanks Prof. Carlos de Moura for the invitation to the conference "CFL-Condition: 80 years gone by", held in Rio de Janeiro in May, 2010. He also acknowledges financial support from the PEPS program of INSMI–CNRS. The authors are grateful to Dominique Fougère, Varlei E. Menconni, and Michel Pognant for their helpful computational assistance.

References

1. Abgrall, R.: Multiresolution analysis on unstructured meshes: applications to CFD. In: Chetverushkin, B.E.A. (ed.) Experimentation, Modelling and Computation in Flow, Turbulence and Combustion. Wiley, New York (1997)
2. Bendahmane, M., Bürger, R., Ruiz, R., Schneider, K.: Adaptive multiresolution schemes with local time stepping for two-dimensional degenerate reaction-diffusion systems. Appl. Numer. Math. **59**, 1668–1692 (2009)
3. Berger, M.J., Collela, P.: Local adaptive mesh refinement for shock hydrodynamics. J. Comput. Phys. **82**, 67–84 (1989)
4. Bihari, B.L.: Multiresolution schemes for conservation laws with viscosity. J. Comput. Phys. **123**, 207–225 (1996)
5. Bürger, R., Ruiz, R., Schneider, K.: Adaptive multiresolution methods for the simulation of waves in excitable media. J. Sci. Comput. **43**, 262–290 (2010)
6. Chiavassa, G., Donat, R.: Point value multi-scale algorithms for 2D compressible flow. SIAM J. Sci. Comput. **23**(3), 805–823 (2001)
7. Cohen, A., Kaber, S.M., Müller, S., Postel, M.: Fully adaptive multiresolution finite volume schemes for conservation laws. Math. Comput. **72**, 183–225 (2003)
8. Courant, R., Friedrichs, K., Lewy, H.: Über die partiellen Differenzengleichungen der mathematischen Physik. Math. Ann. **100**, 32–74 (1928)
9. Deiterding, R., Domingues, M.O., Gomes, S.M., Roussel, O., Schneider, K.: Adaptive multiresolution or adaptive mesh refinement? A case study for 2D Euler equations. ESAIM Proc. **29**, 28–42 (2009)
10. Domingues, M.O., Gomes, S.M., Roussel, O., Schneider, K.: An adaptive multiresolution scheme with local time stepping for evolutionary PDEs. J. Comput. Phys. **227**(8), 3758–3780 (2008)
11. Domingues, M.O., Gomes, S.M., Roussel, O., Schneider, K.: Space-time adaptive multiresolution methods for hyperbolic conservation laws: Applications to compressible Euler equations. Appl. Numer. Math. **59**, 2303–2321 (2009)
12. Domingues, M.O., Roussel, O., Schneider, K.: An adaptive multiresolution method for parabolic PDEs with time-step control. Int. J. Numer. Methods Eng. **78**, 652–670 (2009)
13. Ferm, L., Löstedt, P.: Space-time adaptive solutions of first order PDEs. J. Sci. Comput. **26**(1), 83–110 (2006)
14. Harten, A.: Adaptive multiresolution schemes for shock computations. J. Comput. Phys. **115**, 319–338 (1994)
15. Harten, A.: Multiresolution algorithms for the numerical solution of hyperbolic conservation laws. Commun. Pure Appl. Math. **48**, 1305–1342 (1995)
16. Kaibara, M., Gomes, S.M.: A fully adaptive multiresolution scheme for shock computations. In: Toro, E.F. (ed.) Godunov Methods: Theory and Applications. Kluwer Academic/Plenum Publishers, New York (2001)
17. Lax, P.D., Liu, X.D.: Solution of two-dimensional Riemann problems of gas dynamics by positive schemes. SIAM J. Sci. Comput. **19**(2), 319–340 (1998)
18. Leveque, R.J.: Finite Volume Methods for Hyperbolic Systems. Cambridge University Press, Cambridge (2002)
19. Liou, M.S.: A sequel to AUSM: AUSM+. J. Comput. Phys. **129**, 364–382 (1996)
20. Müller, S.: Adaptive Multiscale Schemes for Conservation Laws. Lectures Notes in Computational Science and Engineering, vol. 27. Springer, Heidelberg (2003)
21. Müller, S., Stiriba, Y.: Fully adaptive multiscale schemes for conservation laws employing locally varying time stepping. J. Sci. Comput. **30**(3), 493–531 (2007)
22. Roussel, O., Schneider, K.: An adaptive multiresolution method for combustion problems: application to flame ball–vortex interaction. Comput. Fluids **34**(7), 817–831 (2005)
23. Roussel, O., Schneider, K., Tsigulin, A., Bockhorn, H.: A conservative fully adaptive multiresolution algorithm for parabolic PDEs. J. Comput. Phys. **188**, 493–523 (2003)

24. Schneider, K., Vasilyev, O.: Wavelet methods in computational fluid dynamics. Annu. Rev. Fluid Mech. **42**, 473–503 (2010)
25. Schulz-Rinne, C.W., Collis, J.P., Glaz, H.M.: Numerical solution of the Riemann problem for two-dimensional gas dynamics. SIAM J. Sci. Comput. **14**, 1394–1414 (1993)
26. Stoer, J., Bulirsch, R.: Introduction to Numerical Analysis, 2nd edn. Text in Applied Mathematics, vol. 12. Springer, Berlin (1991)
27. Wesseling, P.: Principles of Computational Fluid Dynamics. Springer, Berlin (2001)
28. Zhang, T., Zheng, Y.: Conjecture on the structure of solutions the Riemann problem for two-dimensional gas dynamics systems. SIAM J. Math. Anal. **21**, 593–630 (1990)

A Framework for Late-Time/Stiff Relaxation Asymptotics

Philippe G. LeFloch

Abstract We consider solutions to nonlinear hyperbolic systems of balance laws with stiff relaxation and formally derive a parabolic-type effective system describing the late-time asymptotics of these solutions. We show that many examples from continuous physics fall into our framework, including the Euler equations with (possibly nonlinear) friction. We then turn our attention to the discretization of these stiff problems and introduce a new finite volume scheme which preserves the late-time asymptotic regime. Importantly, our scheme requires only the classical CFL (Courant–Friedrichs–Lewy) condition associated with the hyperbolic system under consideration, rather than the more restrictive, parabolic-type stability condition.

Keywords Hyperbolic system · Late-time · Stiff relaxation · Finite volume method · Asymptotic preserving

1 Introduction

This short presentation is based on the joint work [3] in collaboration with C. Berthon and R. Turpault. We are interested in hyperbolic models arising in continuum physics and, especially, describing complex multi-fluid flows involving several time-scales. The partial differential equations under consideration are nonlinear hyperbolic systems of balance laws with stiff relaxation sources. We investigate here the late-time behavior of entropy solutions.

Precisely, our objective is to derive the relevant effective system—which turns out to be of parabolic type—and to investigate the role of a convex entropy associated with the given system of balance laws. As we show it, many examples from continuous physics fall into our framework. In addition, we investigate here the discretization of such problems, and we propose a new finite volume scheme which preserves the late-time asymptotic regime identified in the first part of this paper.

P.G. LeFloch (✉)
Laboratoire Jacques-Louis Lions, Centre National de la Recherche Scientifique, Université Pierre et Marie Curie (Paris 6), 4 Place Jussieu, 75252 Paris, France
e-mail: contact@philippelefloch.org
url: philippelefloch.org

C.A. de Moura, C.S. Kubrusly (eds.), *The Courant–Friedrichs–Lewy (CFL) Condition*,
DOI 10.1007/978-0-8176-8394-8_8,

An outline of this paper follows. In Sect. 2, we present a formal derivation of the effective equations associated with our problem. In Sect. 3, we demonstrate that many examples of continuous physics are covered by our theory. Finally, in Sect. 4, we are in a position to present the new discretization method and state several properties of interest.

2 Late-Time/Stiff Relaxation Framework

2.1 Hyperbolic Systems of Balance Laws

We consider systems of partial differential equations of the form

$$\varepsilon\, \partial_t U + \partial_x F(U) = -\frac{R(U)}{\varepsilon}, \qquad U = U(t,x) \in \Omega \subset \mathbb{R}^N, \tag{1}$$

where $t > 0$, $x \in \mathbb{R}$ denote the (time and space) independent variables. We make the following standard assumptions. The flux $F : \Omega \to \mathbb{R}^N$ is defined on a convex and open subset Ω. The first-order part of (1) is a hyperbolic system, that is, the matrix-valued field $A(U) := D_U F(U)$ admits real eigenvalues and a basis of eigenvectors.

We are interested in the singular limit problem $\varepsilon \to 0$ in the limit of late-time and stiff relaxation. In fact, two distinct regimes for systems like (1) can be considered. In the hyperbolic-to-hyperbolic regime, one replaces $\varepsilon \partial_t U$ by $\partial_t U$, and establishes that solutions to

$$\partial_t U + \partial_x F(U) = -\frac{R(U)}{\varepsilon}, \qquad U = U(t,x),$$

are driven by an effective system of equations ($\varepsilon \to 0$) of *hyperbolic type*. Such a study was pioneered by Chen, Levermore, and Liu [7]. On the other hand, in the hyperbolic-to-parabolic regime under consideration in the present work, we obtain effective equations of *parabolic type*. Earlier work by Marcati et al. [9] discussed this regime too and established several important convergence theorems.

Our objective here is to introduce a general framework to deal with such problems. We make the following assumptions.

Assumption 1 There exists an $n \times N$ matrix Q with (maximal) rank $n < N$ such that

$$QR(U) = 0, \qquad U \in \Omega.$$

Hence, $QU \in Q\Omega =: \omega$ satisfies

$$\varepsilon \partial_t (QU) + \partial_x \big(QF(U)\big) = 0.$$

Assumption 2 There exists a map $\mathcal{E} : \omega \subset \mathbb{R}^m \to \Omega$ describing the equilibria $u \in \omega$, with

$$R\big(\mathcal{E}(u)\big) = 0, \qquad u = Q\mathcal{E}(u).$$

It is then convenient to introduce the equilibrium submanifold $\mathcal{M} := \{U = \mathcal{E}(u)\}$.

Assumption 3 It is also assumed that

$$QF\big(\mathcal{E}(u)\big) = 0, \quad u \in \omega.$$

To motivate this condition observe that, at least formally, the term $\partial_x(QF(\mathcal{E}(u)))$ must vanish identically, so that $QF(\mathcal{E}(u))$ must be a constant, conveniently normalized to be 0.

Assumption 4 For all $u \in \omega$,

$$\begin{aligned}
&\dim\big(\ker\big(B\big(\mathcal{E}(u)\big)\big)\big) = n,\\
&\ker\big(B\big(\mathcal{E}(u)\big)\big) \cap \operatorname{Im}\big(B\big(\mathcal{E}(u)\big)\big) = \{0\}.
\end{aligned}$$

Hence, the $N \times N$ matrix $B := DR_U$ has "maximal" kernel on the equilibrium manifold.

2.2 Chapman–Engskog-Type Expansion

We proceed by using a Chapman–Engskog expansion in order to derive effective equations satisfied by the local equilibria $u = u(t,x) \in \omega$. So, we write

$$U^\varepsilon = \mathcal{E}(u) + \varepsilon U_1 + \varepsilon^2 U_2 + \cdots, \qquad u := QU^\varepsilon,$$

which should satisfy $\varepsilon \partial_t U^\varepsilon + \partial_x F(U^\varepsilon) = -R(U^\varepsilon)/\varepsilon$. It follows that

$$QU_1 = QU_2 = \cdots = 0.$$

For the flux we find

$$F\big(U^\varepsilon\big) = F\big(\mathcal{E}(u)\big) + \varepsilon\, A\big(\mathcal{E}(u)\big)U_1 + \mathcal{O}\big(\varepsilon^2\big),$$

and for the relaxation

$$\frac{R(U^\varepsilon)}{\varepsilon} = B\big(\mathcal{E}(u)\big)U_1 + \frac{\varepsilon}{2} D_U^2 R\big(\mathcal{E}(u)\big).(U_1, U_1) + \varepsilon B\big(\mathcal{E}(u)\big)U_2 + \mathcal{O}\big(\varepsilon^2\big).$$

In turn, we deduce that

$$\begin{aligned}
&\varepsilon\, \partial_t\big(\mathcal{E}(u)\big) + \partial_x\big(F\big(\mathcal{E}(u)\big)\big) + \varepsilon\, \partial_x\big(A\big(\mathcal{E}(u)\big)\, U_1\big)\\
&\quad = -B\big(\mathcal{E}(u)\big)\, U_1 - \frac{\varepsilon}{2} D_U^2 R\big(\mathcal{E}(u)\big).(U_1, U_1) - \varepsilon B\big(\mathcal{E}(u)\big)\, U_2 + \mathcal{O}\big(\varepsilon^2\big).
\end{aligned}$$

We begin by considering the *zero-order terms* and thus deduce that $U_1 \in \mathbb{R}^N$ satisfies the linear system

$$B(\mathcal{E}(u))U_1 = -\partial_x(F(\mathcal{E}(u))) \in \mathbb{R}^N.$$

We can solve this equation in U_1, provided we recall that $QU_1 = 0$ and observe the following.

Lemma 1 (Technical lemma) *Let C be an $N \times N$ matrix satisfying* $\dim\ker C = n$, *and* $\ker C \cap \operatorname{Im} C = \{0\}$, *and let Q be an $n \times N$ matrix of rank n. Then, for all $J \in \mathbb{R}^N$, the system*

$$CV = J,$$
$$QV = 0,$$

admits a unique solution $V \in \mathbb{R}^m$ if and only if $QJ = 0$.

We can thus conclude with the following.

Proposition 1 (First-order corrector problem) *The first-order corrector U_1 is characterized uniquely by*

$$B(\mathcal{E}(u))U_1 = -\partial_x(F(\mathcal{E}(u))),$$
$$QU_1 = 0.$$

We now turn our attention to the *first-order terms* and we arrive at

$$\partial_t(\mathcal{E}(u)) + \partial_x(A(\mathcal{E}(u))\,U_1) = -\frac{1}{2}D_U^2 R(\mathcal{E}(u)).(U_1, U_1) - B(\mathcal{E}(u))\,U_2.$$

Multiplying by Q and using $Q\mathcal{E}(u) = u$, we find

$$\partial_t u + \partial_x(QA(\mathcal{E}(u))\,U_1) = -\frac{1}{2}QD_U^2 R(\mathcal{E}(u)).(U_1, U_1) - QB(\mathcal{E}(u))U_2.$$

But, by differentiating the identity $QR(U) = 0$, it follows that

$$QD_U^2 R.(U_1, U_1) \equiv 0, \qquad QBU_2 \equiv 0.$$

Theorem 1 (Late time/stiff relaxation effective equations) *One has*

$$\partial_t u = -\partial_x(QA(\mathcal{E}(u))\,U_1) =: \partial_x(\mathcal{M}(u)\partial_x u)$$

for some $n \times n$ matrix $\mathcal{M}(u)$, where U_1 is the unique solution to the first-order corrector problem

$$B(\mathcal{E}(u))U_1 = -A(\mathcal{E}(u))\partial_x(\mathcal{E}(u)),$$
$$QU_1 = 0.$$

2.3 The Role of a Mathematical Entropy

Next, we investigate the consequences of assuming existence of a mathematical entropy $\Phi : \Omega \to \mathbb{R}$, satisfying by definition:

Assumption 5 There exists an entropy-flux $\Psi : \Omega \to \mathbb{R}$ such that

$$D_U \Phi\, A = D_U \Psi \quad \text{in } \Omega.$$

So, all smooth solutions satisfy

$$\varepsilon \partial_t \big(\Phi(U^\varepsilon)\big) + \partial_x \big(\Psi(U^\varepsilon)\big) = -D_U \Phi(U^\varepsilon) \frac{R(U^\varepsilon)}{\varepsilon}$$

and, consequently, the matrix $D_U^2 \Phi\, A$ is symmetric in Ω. In addition, we assume that the map Φ is convex, i.e., the $N \times N$ matrix $D_U^2 \Phi$ is positive definite on $\mathcal{M}$.

Assumption 6 The following compatibility property with the relaxation term holds:

$$D_U \Phi\, R \geq 0 \quad \text{in } \Omega,$$
$$D_U \Phi(U) = \nu(U) Q \in \mathbb{R}^N, \quad \nu(U) \in \mathbb{R}^m.$$

Returning to the effective equations

$$\partial_t u = \partial_x \mathcal{D}, \qquad \mathcal{D} := -QA\big(\mathcal{E}(u)\big)\, U_1$$

and multiplying it by the Hessian of the entropy, we conclude that the term $U_1 \in \mathbb{R}^N$ is now characterized by

$$\mathcal{L}(u) U_1 = -\big(D_U^2 \Phi\big)\big(\mathcal{E}(u)\big) \partial_x \big(F\big(\mathcal{E}(u)\big)\big),$$
$$Q U_1 = 0,$$

where $\mathcal{L}(u) = D_U^2 \Phi(\mathcal{E}(u)) B(\mathcal{E}(u))$.

Then, using the notation $\mathcal{L}(u)^{-1}$ for the generalized inverse with constraint and setting

$$S(u) := QA\big(\mathcal{E}(u)\big),$$

we obtain

$$\mathcal{D} = S \mathcal{L}^{-1} \big(D_U^2 \Phi\big)(\mathcal{E}) \partial_x \big(F(\mathcal{E})\big).$$

Finally, one can check that

$$\big(D_U^2 \Phi\big)(\mathcal{E}) \partial_x \big(F(\mathcal{E})\big) = S^T v,$$

with $v := \partial_x (D_u \Phi(\mathcal{E}))^T$.

Theorem 2 (Entropy structure of the effective system) *When the system of balance laws is endowed with a mathematical entropy, the effective equations take the form*

$$\partial_t u = \partial_x \big(L(u)\, \partial_x \big(D_u \Phi \big(\mathcal{E}(u) \big) \big)^T \big),$$

where

$$\begin{aligned} L(u) &:= S(u)\mathcal{L}(u)^{-1} S(u)^T, \\ S(u) &:= QA\big(\mathcal{E}(u)\big), \\ \mathcal{L}(u) &:= \big(D_U^2 \Phi\big)\big(\mathcal{E}(u)\big) B\big(\mathcal{E}(u)\big), \end{aligned}$$

where, for all b satisfying $Qb = 0$, the unique solution to

$$\mathcal{L}(u) V = b, \qquad QV = 0$$

is denoted by $\mathcal{L}(u)^{-1} b$ (generalized inverse).

Alternatively, the above result can be reformulated in terms of the so-called entropy variable $(D_u \Phi(\mathcal{E}(u)))^T$. Furthermore, an important dissipation property can be deduced from our assumptions, as follows. From the entropy property and the equilibrium property $R(\mathcal{E}(u)) = 0$, we find

$$\begin{aligned} &D_U \Phi R \geq 0 \quad \text{in } \Omega, \\ &(D_U \Phi R)|_{U = \mathcal{E}(u)} = 0 \quad \text{in } \omega. \end{aligned}$$

Thus, the matrix $D_U^2 (D_U \Phi R)|_{U=\mathcal{E}(u)}$ is non-negative definite. It follows that

$$D_U^2 (D_U \Phi R) = D_U^2 \Phi B + \big(D_U^2 \Phi B \big)^T \quad \text{when } U = \mathcal{E}(u),$$

so that

$$D_U^2 \Phi \, B|_{U = \mathcal{E}(u)} \geq 0 \quad \text{in } \omega.$$

The *equilibrium entropy* $\Phi(\mathcal{E}(u))$ has the property that its associated (entropy) flux $u \mapsto \Psi(\mathcal{E}(u))$ is constant on the equilibrium manifold ω. Indeed, for the map $\Psi(\mathcal{E})$, we have

$$\begin{aligned} D_u \big(\Psi(\mathcal{E}) \big) &= D_U \Psi(\mathcal{E}) D_u \mathcal{E} \\ &= D_U \Phi(\mathcal{E}) A(\mathcal{E}) D_u \mathcal{E}. \end{aligned}$$

Observing that $(D_U \Phi)(\mathcal{E}) = D_u(\Phi(\mathcal{E}))Q$, we obtain

$$\begin{aligned} D_u \big(\Psi \big(\mathcal{E}(u) \big) \big) &= D_u \Phi \big(\mathcal{E}(u) \big) QA \big(\mathcal{E}(u) \big) D_u \mathcal{E}(u) \\ &= D_u \big(\Phi \big(\mathcal{E}(u) \big) \big) D_u QF \big(\mathcal{E}(u) \big). \end{aligned}$$

Since $QF(\mathcal{E})=0$, then $D_u QF(\mathcal{E})=0$ and the proof is completed. Therefore, $D_u(\Psi(\mathcal{E}(u)))=0$ for all $u\in\omega$.

Recalling the asymptotic expansion

$$U^\varepsilon = \mathcal{E}(u)+\varepsilon U_1+\cdots,$$

where U_1 is given by the first-order corrector problem, we write

$$\Psi\big(U^\varepsilon\big)=\Psi\big(\mathcal{E}(u)\big)+\varepsilon\, D_U\Psi\big(\mathcal{E}(u)\big)U_1+\mathcal{O}\big(\varepsilon^2\big),$$

and deduce

$$\partial_x\Psi\big(U^\varepsilon\big)=\varepsilon\partial_x D_U\Psi\big(\mathcal{E}(u)\big)U_1+\mathcal{O}\big(\varepsilon^2\big).$$

Similarly, for the relaxation source term, we have

$$D_U\Phi\big(U^\varepsilon\big)R\big(U^\varepsilon\big)=\varepsilon^2 D_U^2\Phi\big(\mathcal{E}(u)\big)D_U R\big(\mathcal{E}(u)\big)U_1+\mathcal{O}\big(\varepsilon^3\big).$$

We thus obtain

$$\begin{aligned}&\partial_t\big(\Phi\big(\mathcal{E}(u)\big)\big)+\partial_x\big(D_U\Psi\big(\mathcal{E}(u)\big)U_1\big)\\&\quad=-U_1^T\big(D_U^2\Phi\big(\mathcal{E}(u)\big)B\big(\mathcal{E}(u)\big)\big)U_1.\end{aligned}$$

But, we have already established

$$X\big(D_U^2\Phi\big)(\mathcal{E})B(\mathcal{E})X\geq 0,\quad X\in\mathbb{R}^N.$$

Proposition 2 (Monotonicity of the entropy) *The entropy is non-increasing in the sense that*

$$\partial_t\big(\Phi\big(\mathcal{E}(u)\big)\big)+\partial_x\big(D_U\Psi\big(\mathcal{E}(u)\big)U_1\big)\leq 0.$$

In the notation given earlier, one thus have

$$\partial_t\big(\Phi\big(\mathcal{E}(u)\big)\big)=\partial_x\big(D_u\big(\Phi\big(\mathcal{E}(u)\big)\big)L(u)\partial_x\big(D_u\big(\Phi\big(\mathcal{E}(u)\big)\big)^T\big).$$

Remark 1 The above framework can be extended to handle certain *nonlinear diffusion regime*, corresponding to the scaling

$$\varepsilon\,\partial_t U+\partial_x F(U)=-\frac{R(U)}{\varepsilon^q}.$$

The parameter $q\geq 1$ introduces an *additional scale*, and is indeed necessary for certain problems where the relaxation is nonlinear. The relaxation term is supposed to be such that

$$R\big(\mathcal{E}(u)+\varepsilon U\big)=\varepsilon^q R\big(\mathcal{E}(u)+M(\varepsilon)U\big),\quad U\in\Omega,\ u\in\omega,$$

for some matrix $M(\varepsilon)$. In that regime, the effective equations turn out to be of *nonlinear parabolic* type.

3 Examples from Continuum Physics

3.1 Euler Equations with Friction Term

The simplest example of interest is provided by the Euler equations of compressible fluids with friction:

$$\begin{aligned} &\varepsilon \partial_t \rho + \partial_x(\rho v) = 0, \\ &\varepsilon\, \partial_t(\rho v) + \partial_x\big(\rho v^2 + p(\rho)\big) = -\frac{\rho v}{\varepsilon}, \end{aligned} \tag{2}$$

in which the density $\rho \geq 0$ and the velocity component v represent the main unknowns while the pressure $p : \mathbb{R}^+ \to \mathbb{R}^+$ is prescribed and satisfy the hyperbolicity condition $p'(\rho) > 0$ for all $\rho > 0$. Then, the first-order homogeneous system is strictly hyperbolic and (2) fits into our late-time/stiff relaxation framework provided we set

$$U = \begin{pmatrix} \rho \\ \rho v \end{pmatrix}, \qquad F(U) = \begin{pmatrix} \rho v \\ \rho v^2 + p(\rho) \end{pmatrix}, \qquad R(U) = \begin{pmatrix} 0 \\ \rho v \end{pmatrix},$$

and $Q = (1\ 0)$. In this case, the local equilibria $u = \rho$ are scalar-valued, with

$$\mathcal{E}(u) = \begin{pmatrix} \rho \\ 0 \end{pmatrix},$$

and we do have $QF(\mathcal{E}(u)) = 0$.

The diffusive regime for the Euler equations with friction is analyzed as follows. First, according to the general theory, equilibrium solutions satisfy

$$\partial_t \rho = -\partial_x\big(QA\big(\mathcal{E}(u)\big)U_1\big),$$

where

$$D_U F\big(\mathcal{E}(u)\big) = \begin{pmatrix} 0 & 1 \\ p'(\rho) & 0 \end{pmatrix}.$$

Here, U_1 is the unique solution to

$$\begin{aligned} &B\big(\mathcal{E}(u)\big)U_1 = -\partial_x\big(F\big(\mathcal{E}(u)\big)\big), \\ &QU_1 = 0 \end{aligned}$$

with

$$B\big(\mathcal{E}(u)\big) = \begin{pmatrix} 0 & 0 \\ 0 & 1 \end{pmatrix}, \qquad \partial_x\big(F\big(\mathcal{E}(u)\big)\big) = \begin{pmatrix} 0 \\ \partial_x(p(\rho)) \end{pmatrix}.$$

This leads us to the effective diffusion equation for the Euler equations with friction

$$\partial_t \rho = \partial_x^2\big(p(\rho)\big), \tag{3}$$

which is a nonlinear parabolic equation, at least away from vacuum, since $p'(\rho) > 0$ by assumption. Interestingly, at vacuum, this equation may be degenerate since $p'(\rho)$ typically vanishes at $\rho = 0$. For instance, in the case of polytropic gases $p(\rho) = \kappa\rho^{\gamma}$ with $\kappa > 0$ and $\gamma \in (1, \gamma)$ we obtain

$$\partial_t \rho = \kappa\gamma\, \partial_x\big(\rho^{\gamma-1}\partial\rho\big). \tag{4}$$

In addition, by defining the internal energy $e(\rho) > 0$ by

$$e'(\rho) = \frac{p(\rho)}{\rho^2},$$

we easily check that all smooth solutions to (2) satisfy

$$\varepsilon\, \partial_t\left(\rho\frac{v^2}{2} + \rho e(\rho)\right) + \partial_x\left(\rho\frac{v^3}{2} + \big(\rho e(\rho) + p(\rho)\big)v\right) = -\frac{\rho v^2}{\varepsilon}, \tag{5}$$

so that the function

$$\Phi(U) = \rho\frac{v^2}{2} + \rho e(\rho)$$

is convex entropy compatible with the relaxation. All the conditions of the general framework are therefore satisfied by the Euler equations with friction.

3.2 *M*1 *Model of Radiative Transfer*

Our next model of interest arises in the theory of radiative transfer, i.e.,

$$\begin{aligned}
\varepsilon\partial_t e + \partial_x f &= \frac{\tau^4 - e}{\varepsilon},\\
\varepsilon\partial_t f + \partial_x\big(\chi(f/e)e\big) &= -\frac{f}{\varepsilon},\\
\varepsilon\partial_t \tau &= \frac{e - \tau^4}{\varepsilon},
\end{aligned} \tag{6}$$

where the radiative energy $e > 0$ and the radiative flux f are the main unknowns, restricted by the condition $|f/e| \le 1$, while $\tau > 0$ denotes the temperature. The function $\chi : [-1, 1] \to \mathbb{R}^+$ is called the Eddington factor and, typically,

$$\chi(\xi) = \frac{3 + 4\xi^2}{5 + 2\sqrt{4 - 3\xi^2}}.$$

This system fits into our late-time/stiff relaxation framework if we introduce

$$U = \begin{pmatrix} e \\ f \\ \tau \end{pmatrix}, \qquad F(U) = \begin{pmatrix} f \\ \chi(\frac{f}{e})e \\ 0 \end{pmatrix}, \qquad R(U) = \begin{pmatrix} e - \tau^4 \\ f \\ \tau^4 - e \end{pmatrix}.$$

Now, the equilibria are given by $u = \tau + \tau^4$, with

$$\mathcal{E}(u) = \begin{pmatrix} \tau^4 \\ 0 \\ \tau \end{pmatrix}, \qquad Q := (1 \quad 0 \quad 1).$$

We do have $QF(\mathcal{E}(u)) = 0$ and the assumptions in Sect. 2 are satisfied.

We determine the diffusive regime for the $M1$ model from the expression

$$(D_U F)\big(\mathcal{E}(u)\big) = \begin{pmatrix} 0 & 1 & 0 \\ \chi(0) & \chi'(0) & 0 \\ 0 & 0 & 0 \end{pmatrix} = \begin{pmatrix} 0 & 1 & 0 \\ \frac{1}{3} & 0 & 0 \\ 0 & 0 & 0 \end{pmatrix},$$

where U_1 is the solution to the linear problem

$$\begin{pmatrix} 1 & 0 & -4\tau^3 \\ 0 & 1 & 0 \\ -1 & 0 & 4\tau^3 \end{pmatrix} U_1 = \begin{pmatrix} 0 \\ \partial_x(\tau^4/3) \\ 0 \end{pmatrix},$$
$$(1 \quad 0 \quad 1) U_1 = 0.$$

Therefore, we have

$$U_1 = \begin{pmatrix} 0 \\ \frac{4}{3}\tau^3 \partial_x \tau \\ 0 \end{pmatrix},$$

and the effective diffusion equation for the $M1$ system reads

$$\partial_t \big(\tau + \tau^4\big) = \partial_x \left(\frac{4}{3}\tau^3 \partial_x \tau\right). \tag{7}$$

Again, an entropy can be associated to this model.

3.3 Coupled Euler/$M1$ Model

By combining the previous two examples, we arrive at a more involved model:

$$\begin{aligned}
&\varepsilon \partial_t \rho + \partial_x(\rho v) = 0, \\
&\varepsilon \partial_t \rho v + \partial_x\big(\rho v^2 + p(\rho)\big) = -\frac{\kappa}{\varepsilon}\rho v + \frac{\sigma}{\varepsilon} f, \\
&\varepsilon \partial_t e + \partial_x f = 0, \\
&\varepsilon \partial_t f + \partial_x \left(\chi\left(\frac{f}{e}\right) e\right) = -\frac{\sigma}{\varepsilon} f,
\end{aligned} \tag{8}$$

in which the same notation as before is used and κ and σ are positive constants. In the applications, a typical choice for the pressure is

$$p(\rho) = C_p \rho^\eta, \quad C_p \ll 1,\ \eta > 1.$$

To fit this model within the late-time/stiff relaxation framework, we need to set

$$U = \begin{pmatrix} \rho \\ \rho v \\ e \\ f \end{pmatrix}, \qquad F(U) = \begin{pmatrix} \rho v \\ \rho v^2 + p(\rho) \\ f \\ \chi(\frac{f}{e})e \end{pmatrix}, \qquad R(U) = \begin{pmatrix} 0 \\ \kappa\rho v - \sigma f \\ 0 \\ \sigma f \end{pmatrix}.$$

The local equilibria are given by

$$\mathcal{E}(u) = \begin{pmatrix} \rho \\ 0 \\ e \\ 0 \end{pmatrix}, \quad u = QU = \begin{pmatrix} \rho \\ e \end{pmatrix}, \ Q = \begin{pmatrix} 1 & 0 & 0 & 0 \\ 0 & 0 & 1 & 0 \end{pmatrix},$$

and once again, one has $QF(\mathcal{E}(u)) = 0$.

We can then compute

$$D_U F\big(\mathcal{E}(u)\big) = \begin{pmatrix} 0 & 1 & 0 & 0 \\ p'(\rho) & 0 & 0 & 0 \\ 0 & 0 & 0 & 1 \\ 0 & 0 & \frac{1}{3} & 0 \end{pmatrix}, \qquad U_1 = \begin{pmatrix} 0 \\ \frac{1}{\kappa}(-\partial_x p(\rho) - \frac{1}{3}\partial_x e) \\ 0 \\ -\frac{1}{3\sigma}\partial_x e \end{pmatrix},$$

and we arrive at the effective diffusion system for the coupled Euler/$M1$ model:

$$\begin{aligned} &\partial_t \rho - \frac{1}{\kappa}\partial_x^2 p(\rho) - \frac{1}{3\kappa}\partial_x^2 e = 0, \\ &\partial_t e - \frac{1}{3\sigma}\partial_x^2 e = 0. \end{aligned} \tag{9}$$

The second equation is a standard heat equation, and its solution serves as a source-term in the first equation.

3.4 Shallow Water with Nonlinear Friction

Our final example requires a more general theory of nonlinear relaxation mentioned in Remark 1, and it reads

$$\begin{aligned} &\varepsilon\partial_t h + \partial_x(hv) = 0, \\ &\varepsilon\partial_t(hv) + \partial_x\big(h v^2 + p(h)\big) = -\frac{\kappa^2(h)}{\varepsilon^2}\, g\, hv|hv|, \end{aligned} \tag{10}$$

where h denotes the fluid height and v the fluid velocity. The pressure is taken to be $p(h) = g\,h^2/2$, and $g > 0$ is called the gravity constant. The friction coefficient $\kappa : \mathbb{R}^+ \to \mathbb{R}^+$ is a positive function, and a standard choice is $\kappa(h) = \frac{\kappa_0}{h}$ with $\kappa_0 > 0$.

The nonlinear version of the late-time/stiff relaxation framework applies if we set

$$U = \begin{pmatrix} h \\ hv \end{pmatrix}, \qquad F(U) = \begin{pmatrix} hv \\ hv^2 + p(h) \end{pmatrix}, \qquad R(U) = \begin{pmatrix} 0 \\ \kappa^2(h) ghv|hv| \end{pmatrix}.$$

The scalar equilibria $u = h$ are associated with

$$\mathcal{E}(u) = \begin{pmatrix} h \\ 0 \end{pmatrix}, \qquad Q = (1 \quad 0).$$

Here, the relaxation is nonlinear and satisfies

$$R\big(\mathcal{E}(u) + \varepsilon U\big) = \varepsilon^2 R\big(\mathcal{E}(U) + M(\varepsilon)U\big),$$

with

$$M(\varepsilon) := \begin{pmatrix} \varepsilon & 0 \\ 0 & 1 \end{pmatrix}$$

in turn, we may derive a *nonlinear* effective equation for the Euler equations with nonlinear friction, that is,

$$\partial_t h = \partial_x \left(\frac{\sqrt{h}}{\kappa(h)} \frac{\partial_x h}{\sqrt{|\partial_x h|}} \right), \tag{11}$$

which is a nonlinear parabolic equation.

In addition, by introducing the internal energy $e(h) := gh/2$, we observe that all smooth solutions to (10) satisfy the entropy inequality

$$\varepsilon \partial_t \left(h\frac{v^2}{2} + g\frac{h^2}{2} \right) + \partial_x \left(h\frac{v^2}{2} + gh^2 \right) v = -\frac{\kappa^2(h)}{\varepsilon^2} ghv^2 |hv|. \tag{12}$$

The entropy

$$\Phi(U) := h\frac{v^2}{2} + g\frac{h^2}{2}$$

satisfies the compatibility properties relevant to the nonlinear late-time/stiff relaxation theory, with in particular

$$R\big(\mathcal{E}(u) + M(0)\bar{U}_1\big) = \begin{pmatrix} 0 \\ \partial_x p(h) \end{pmatrix},$$

where $\bar{U}_1 = (0 \ \beta)$. We obtain here $R(\mathcal{E}(u) + M(0)\bar{U}_1) = c(u)\bar{U}_1$ with

$$c(u) = g\kappa(h)\sqrt{h|\partial_x h|} \geq 0.$$

4 Asymptotic-Preserving Finite Volume Schemes

4.1 General Strategy

We are now going to design finite volume schemes that are consistent with the asymptotic regime $\varepsilon \to 0$ determined in the previous section and, indeed, to recover an effective diffusion equation that is independent of the mesh-size. The discretization of hyperbolic-to-hyperbolic regimes was investigated first by Ji and Xin [11]. Here we propose a framework to cover hyperbolic-to-parabolic regimes. For earlier work on this latter regime, see [1, 2, 4–6].

Step 1. Our construction is based on a standard finite volume scheme for the homogeneous system

$$\partial_t U + \partial_x F(U) = 0,$$

and we will begin by describing such a scheme.

Step 2. We will then modify the above scheme and include a matrix-valued free parameter, allowing us to approximate the non-homogeneous system

$$\partial_t U + \partial_x F(U) = -\gamma R(U),$$

for a fixed coefficient $\gamma > 0$.

Step 3. Finally, we will perform an asymptotic analysis after replacing the discretization parameter Δt by $\varepsilon \Delta t$, and γ by $1/\varepsilon$. Our goal then will be to determine the free parameters to ensure the asymptotic-preserving property.

Let us briefly define the so-called HLL discretization of the homogeneous system, as proposed by Harten, Lax, and van Leer [10]. For simplicity, we present here the solver based on a single constant intermediate state. The mesh is assumed to be the uniform mesh made of cells of length Δx:

$$[x_{i-1/2}, x_{i+1/2}], \quad x_{i+1/2} = x_i + \frac{\Delta x}{2}$$

for all $i = \dots, -1, 0, 1, \dots$. The time discretization is based on a parameter Δt, restricted by the famous CFL condition (Courant, Friedrichs, Lewy, cf. [8]) with

$$t^{m+1} = t^m + \Delta t.$$

Starting from some initial data (lying in the convex set Ω),

$$U^0(x) = \frac{1}{\Delta x} \int_{x_{i-1/2}}^{x_{i+1/2}} U(x,0)\, dx, \quad x \in [x_{i-1/2}, x_{i+1/2}),$$

we construct piecewise constant approximations at each time t^m:

$$U^m(x) = U_i^m, \quad x \in [x_{i-1/2}, x_{i+1/2}),\ i \in \mathbb{Z}.$$

Following Harten, Lax, and van Leer [10], at each cell interface we use the *approximate Riemann solver*:

$$\widetilde{U}_{\mathcal{R}}\left(\frac{x}{t}; U_L, U_R\right) = \begin{cases} U_L, & \frac{x}{t} < -b, \\ \widetilde{U}^\star, & -b < \frac{x}{t} < b, \\ U_R, & \frac{x}{t} > b, \end{cases}$$

where $b > 0$ is sufficiently large. The "numerical cone" (and numerical diffusion) is thus determined by the parameter $b > 0$ and here, for simplicity in the presentation, we have assumed a single constant b but, more generally, one can introduce two speeds $b_{i+1/2}^- < b_{i+1/2}^+$ at each interface.

We introduce the intermediate state

$$\widetilde{U}^\star = \frac{1}{2}(U_L + U_R) - \frac{1}{2b}\big(F(U_R) - F(U_L)\big),$$

and assume the CFL condition

$$b\frac{\Delta t}{\Delta x} \leq \frac{1}{2},$$

so that the underlying approximate Riemann solutions are non-interacting. Our global approximate solutions

$$\widetilde{U}_{\Delta x}^m\big(x, t^m + t\big), \quad t \in [0, \Delta t), \ x \in \mathbb{R}$$

are then obtained as follows.

The approximations at the next time level t^{m+1} are determined from

$$\widetilde{U}_i^{m+1} = \frac{1}{\Delta x}\int_{x_{i-1/2}}^{x_{i+1/2}} \widetilde{U}_{\Delta x}^m\big(x, t^m + \Delta t\big)\, dx.$$

Then, recalling $\widetilde{U}_{i+1/2}^\star = \frac{1}{2}(U_i^m + U_{i+1}^m) - \frac{1}{2b}(F(U_{i+1}^m) - F(U_i^m))$, and integrating out the expression given by the Riemann solutions, we arrive at the *scheme for the homogeneous system*

$$\widetilde{U}_i^{m+1} = U_i^m - \frac{\Delta t}{\Delta x}\big(F_{i+1/2}^{\mathrm{HLL}} - F_{i-1/2}^{\mathrm{HLL}}\big),$$

where

$$F_{i+1/2}^{\mathrm{HLL}} = \frac{1}{2}\big(F\big(U_i^m\big) + F\big(U_{i+1}^m\big)\big) - \frac{b}{2}\big(U_{i+1}^m - U_i^m\big).$$

(More generally, one could take into account two speeds $b_{i+1/2}^- < b_{i+1/2}^+$.)

We observe that the above scheme enjoys invariant domains. The intermediate states $\widetilde{U}_{i+1/2}^\star$ can be written in the form of a convex combination

$$\widetilde{U}_{i+1/2}^\star = \frac{1}{2}\left(U_i^m + \frac{1}{b}F\big(U_i^m\big)\right) + \frac{1}{2}\left(U_{i+1}^m - \frac{1}{b}F\big(U_{i+1}^m\big)\right) \in \Omega,$$

provided b is large enough. An alternative decomposition is given by

$$\widetilde{U}^{\star}_{i+1/2} = \frac{1}{2}\Big(I + \frac{1}{b}\overline{A}(U_i^m, U_{i+1}^m)\Big) U_i^m + \frac{1}{2}\Big(I - \frac{1}{b}\overline{A}(U_i^m, U_{i+1}^m)\Big) U_{i+1}^m,$$

where $\overline{A}$ is an "average" of $D_U F$. By induction, it follows that $\widetilde{U}_i^m$ in Ω for all m, i.

4.2 Discretization of the Relaxation Term

We start from the following *modified Riemann solver:*

$$U_{\mathcal{R}}\Big(\frac{x}{t}; U_L, U_R\Big) = \begin{cases} U_L, & \frac{x}{t} < -b, \\ U^{\star L}, & -b < \frac{x}{t} < 0, \\ U^{\star R}, & 0 < \frac{x}{t} < b, \\ U_R, & \frac{x}{t} > b, \end{cases}$$

with the following states at the interface:

$$U^{\star L} = \underline{\alpha}\widetilde{U}^{\star} + (I - \underline{\alpha})\big(U_L - \bar{R}(U_L)\big),$$
$$U^{\star R} = \underline{\alpha}\widetilde{U}^{\star} + (I - \underline{\alpha})\big(U_R - \bar{R}(U_R)\big).$$

We have here introduced certain $N \times N$-matrix and N-vector defined by

$$\underline{\alpha} = \Big(I + \frac{\gamma \Delta x}{2b}(I + \underline{\sigma})\Big)^{-1}, \qquad \bar{R}(U) = (I + \underline{\sigma})^{-1} R(U).$$

The term $\underline{\sigma}$ is a parameter matrix to be chosen so that (all inverse matrices are well-defined and) the correct asymptotic regime is recovered at the discrete level.

At each interface $x_{i+1/2}$, we use the modified Riemann solver

$$U_{\mathcal{R}}\Big(\frac{x - x_{i+1/2}}{t - t^m}; U_i^m, U_{i+1}^m\Big)$$

and we superimpose non-interacting Riemann solutions

$$U_{\Delta x}^m\big(x, t^m + t\big), \quad t \in [0, \Delta t), \; x \in \mathbb{R}.$$

The new approximate solution at the next time t^{m+1} is

$$U_i^{m+1} = \int_{x_{i-1/2}}^{x_{i+1/2}} U_{\Delta x}^m\big(x, t^m + \Delta t\big)\, dx.$$

By integrating out the Riemann solutions, we arrive at the discretized balance law

$$\frac{1}{\Delta t}\big(U_i^{m+1} - U_i^m\big) + \frac{1}{\Delta x}\big(\underline{\alpha}_{i+1/2} F^{\text{HLL}}_{i+1/2} - \underline{\alpha}_{i-1/2} F^{\text{HLL}}_{i-1/2}\big)$$

$$= \frac{1}{\Delta x}(\underline{\alpha}_{i+1/2} - \underline{\alpha}_{i-1/2}) F\big(U_i^m\big) - \frac{b}{\Delta x}(I - \underline{\alpha}_{i-1/2}) \bar{R}_{i-1/2}\big(U_i^m\big)$$
$$- \frac{b}{\Delta x}(I - \underline{\alpha}_{i+1/2}) \bar{R}_{i+1/2}\big(U_i^m\big). \tag{13}$$

Observing that the discretized source can be rewritten as

$$\frac{b}{\Delta x}(I - \underline{\alpha}_{i+1/2}) \bar{R}_{i+1/2}\big(U_i^m\big) = \frac{b}{\Delta x} \underline{\alpha}_{i+1/2}\big(\underline{\alpha}_{i+1/2}^{-1} - I\big) \bar{R}_{i+1/2}\big(U_i^m\big)$$
$$= \frac{\gamma}{2} \underline{\alpha}_{i+1/2} R\big(U_i^m\big)$$

and, similarly,

$$\frac{b}{\Delta x}(I - \underline{\alpha}_{i-1/2}) \bar{R}_{i-1/2}\big(U_i^m\big) = \frac{\gamma}{2} \underline{\alpha}_{i-1/2} R\big(U_i^m\big),$$

we conclude that the proposed *finite volume schemes for late-time/stiff-relaxation problems* take the form

$$\frac{1}{\Delta t}\big(U_i^{m+1} - U_i^m\big) + \frac{1}{\Delta x}\big(\underline{\alpha}_{i+1/2} F_{i+1/2}^{\mathrm{HLL}} - \underline{\alpha}_{i-1/2} F_{i-1/2}^{\mathrm{HLL}}\big)$$
$$= \frac{1}{\Delta x}(\underline{\alpha}_{i+1/2} - \underline{\alpha}_{i-1/2}) F\big(U_i^m\big) - \frac{\gamma}{2}(\underline{\alpha}_{i+1/2} + \underline{\alpha}_{i-1/2}) R\big(U_i^m\big). \tag{14}$$

Theorem 3 (Properties of the finite volume scheme) *Provided*

$$\underline{\sigma}_{i+1/2} - \underline{\sigma}_{i-1/2} = \mathcal{O}(\Delta x)$$

and the matrix-valued map $\underline{\sigma}$ *is sufficiently smooth, the modified finite volume scheme is* consistent *with the hyperbolic system with relaxation. The following* invariant domain *property holds*: *If all the states at the interfaces*

$$U_{i+1/2}^{\star L} = \underline{\alpha}_{i+1/2} \widetilde{U}_{i+1/2}^{\star} + (I - \underline{\alpha}_{i+1/2})\big(U_i^m - \bar{R}\big(U_i^m\big)\big),$$
$$U_{i+1/2}^{\star R} = \underline{\alpha}_{i+1/2} \widetilde{U}_{i+1/2}^{\star} + (I - \underline{\alpha}_{i+1/2})\big(U_{i+1}^m - \bar{R}\big(U_{i+1}^m\big)\big)$$

belong to Ω, *then all the states* U_i^m *belong to* Ω.

4.3 Discrete Late-Time Asymptotic Regime

As explained earlier, we now replace Δt by $\Delta t/\varepsilon$ and γ by $1/\varepsilon$, and consider

$$\frac{\varepsilon}{\Delta t}\big(U_i^{m+1} - U_i^m\big) + \frac{1}{\Delta x}\big(\underline{\alpha}_{i+1/2} F_{i+1/2}^{\mathrm{HLL}} - \underline{\alpha}_{i-1/2} F_{i-1/2}^{\mathrm{HLL}}\big)$$
$$= \frac{1}{\Delta x}(\underline{\alpha}_{i+1/2} - \underline{\alpha}_{i-1/2}) F\big(U_i^m\big) - \frac{1}{2\varepsilon}(\underline{\alpha}_{i+1/2} + \underline{\alpha}_{i-1/2}) R\big(U_i^m\big),$$

where

$$\underline{\alpha}_{i+1/2} = \left(I + \frac{\Delta x}{2\varepsilon b}(I + \underline{\sigma}_{i+1/2}) \right)^{-1}.$$

Plugging in an expansion near an equilibrium state

$$U_i^m = \mathcal{E}(u_i^m) + \varepsilon (U_1)_i^m + \mathcal{O}(\varepsilon^2),$$

we find

$$F_{i+1/2}^{\mathrm{HLL}} = \frac{1}{2} F(\mathcal{E}(u_i^m)) + \frac{1}{2} F(\mathcal{E}(u_{i+1}^m)) - \frac{b}{2}(\mathcal{E}(u_{i+1}^m) - \mathcal{E}(u_i^m)) + \mathcal{O}(\varepsilon),$$

$$\frac{1}{\varepsilon} R(U_i^m) = B(\mathcal{E}(u_i^m))(U_1)_i^m + \mathcal{O}(\varepsilon),$$

$$\underline{\alpha}_{i+1/2} = \frac{2b\varepsilon}{\Delta x}(I + \underline{\sigma}_{i+1/2})^{-1} + \mathcal{O}(1).$$

The *first-order* terms for the discrete scheme lead us to

$$\begin{aligned} &\frac{1}{\Delta t}(\mathcal{E}(u_i^{m+1}) - \mathcal{E}(u_i^m)) \\ &\quad = -\frac{2b}{\Delta x^2}\big((I + \underline{\sigma}_{i+1/2})^{-1} F_{i+1/2}^{\mathrm{HLL}}|_{\mathcal{E}(u)} - (I + \underline{\sigma}_{i-1/2})^{-1} F_{i-1/2}^{\mathrm{HLL}}|_{\mathcal{E}(u)}\big) \\ &\qquad + \frac{2b}{\Delta x^2}\big((I + \underline{\sigma}_{i+1/2})^{-1} - (I + \underline{\sigma}_{i-1/2})^{-1}\big) F(\mathcal{E}(u_i^m)) \\ &\qquad - \frac{b}{\Delta x}\big((I + \underline{\sigma}_{i+1/2})^{-1} + (I + \underline{\sigma}_{i-1/2})^{-1}\big) B(\mathcal{E}(u_i^m))(U_1)_i^m. \end{aligned}$$

At this juncture, we assume the existence of an $n \times n$ matrix $\mathcal{M}_{i+1/2}$ such that

$$Q(I + \underline{\sigma}_{i+1/2})^{-1} = \frac{1}{b^2} \mathcal{M}_{i+1/2} Q.$$

We then multiply the equation above by the $n \times N$ matrix Q and obtain

$$\frac{1}{\Delta t}(u_i^{m+1} - u_i^m) = -\frac{2}{b \Delta x^2}\big(\mathcal{M}_{i+1/2} Q F_{i+1/2}^{\mathrm{HLL}}|_{\mathcal{E}(u)} - \mathcal{M}_{i-1/2} Q F_{i-1/2}^{\mathrm{HLL}}|_{\mathcal{E}(u)}\big),$$

with

$$\begin{aligned} Q F_{i+1/2}^{\mathrm{HLL}}|_{\mathcal{E}(u)} &= \frac{Q}{2} F(\mathcal{E}(u_i^m)) + \frac{Q}{2} F(\mathcal{E}(u_{i+1}^m)) - \frac{b}{2} Q(\mathcal{E}(u_{i+1}^m) - \mathcal{E}(u_i^m)) \\ &= -\frac{b}{2}(u_{i+1}^m - u_i^m). \end{aligned}$$

The discrete asymptotic system is thus

$$\frac{1}{\Delta t}(u_i^{m+1} - u_i^m) = \frac{1}{\Delta x^2}\big(\mathcal{M}_{i+1/2}(u_{i+1}^m - u_i^m) + \mathcal{M}_{i-1/2}(u_{i-1}^m - u_i^m)\big). \quad (15)$$

Recall that for some $n \times n$ matrix $\mathcal{M}(u)$ the effective equation reads

$$\partial_t u = \partial_x \big(\mathcal{M}(u) \partial_x u\big).$$

Theorem 4 (Discrete late-time asymptotic-preserving property) *Assume the following conditions on the matrix-valued coefficients*:

- *The matrices*

$$I + \underline{\sigma}_{i+1/2}, \qquad \left(1 + \frac{\Delta x}{2\varepsilon b}\right) I + \underline{\sigma}_{i+1/2}$$

are invertible for $\varepsilon \in [0, 1]$.

- *There exists an* $n \times n$ *matrix* $\mathcal{M}_{i+1/2}$ *ensuring the commutation condition*

$$Q(I + \underline{\sigma}_{i+1/2})^{-1} = \frac{1}{b^2} \mathcal{M}_{i+1/2} Q.$$

- *The discrete form of* $\mathcal{M}(u)$ *at each cell interface* $x_{i+1/2}$ *satisfies*

$$\mathcal{M}_{i+1/2} = \mathcal{M}(u) + \mathcal{O}(\Delta x).$$

Then, the effective system associated with the discrete scheme coincides with the one of the late-time/stiff relaxation framework.

We refer the reader to [3] for numerical experiments with this scheme, which turns out to efficiently compute the late-time behavior of solutions. It is observed therein that asymptotic solutions may have large gradients but are actually regular. We also emphasize that we rely here on the CFL stability condition based on the homogeneous hyperbolic system, i.e., a restriction on $\Delta t/\Delta x$ only is imposed. In typical tests, about 10000 time-steps were used to reach the late-time behavior and, for simplicity, the initial data were taken in the image of Q. A *reference* solution, needed for a comparison, was obtained by solving the parabolic equation, under a (stronger) restriction on $\Delta t/(\Delta x)^2$.

The proposed theoretical framework for late-time/stiff relaxation problems thus led us to the development of a good strategy to design asymptotic-preserving schemes involving matrix-valued parameter. The convergence analysis ($\varepsilon \to 0$) and the numerical analysis ($\Delta x \to 0$) for the problems under consideration are important and challenging open problems. It will also very interesting to apply our technique to, for instance, plasma mixtures in a multi-dimensional setting.

Acknowledgements The author was partially supported by the Agence Nationale de la Recherche (ANR) through the grant 06-2-134423, and by the Centre National de la Recherche Scientifique (CNRS).

References

1. Berthon, C., Turpault, R.: Asymptotic preserving HLL schemes. Numer. Methods Partial Differ. Equ. doi:10.1002/num.20586

2. Berthon, C., Charrier, P., Dubroca, B.: An HLLC scheme to solve the M1 model of radiative transfer in two space dimensions. J. Sci. Comput. **31**, 347–389 (2007)
3. Berthon, C., LeFloch, P.G., Turpault, R.: Late-time relaxation limits of nonlinear hyperbolic systems. A general framework. Math. Comput. (2012)
4. Bouchut, F., Ounaissa, H., Perthame, B.: Upwinding of the source term at interfaces for Euler equations with high friction. J. Comput. Math. Appl. **53**, 361–375 (2007)
5. Buet, C., Cordier, S.: An asymptotic preserving scheme for hydrodynamics radiative transfer models: numerics for radiative transfer. Numer. Math. **108**, 199–221 (2007)
6. Buet, C., Després, B.: Asymptotic preserving and positive schemes for radiation hydrodynamics. J. Comput. Phys. **215**, 717–740 (2006)
7. Chen, G.Q., Levermore, C.D., Liu, T.P.: Hyperbolic conservation laws with stiff relaxation terms and entropy. Commun. Pure Appl. Math. **47**, 787–830 (1995)
8. Courant, R., Friedrichs, K., Lewy, H.: Über die partiellen Differenzengleichungen der mathematischen Physik. Math. Ann. **100**, 32–74 (1928)
9. Donatelli, D., Marcati, P.: Convergence of singular limits for multi-D semilinear hyperbolic systems to parabolic systems. Trans. Am. Math. Soc. **356**, 2093–2121 (2004)
10. Harten, A., Lax, P.D., van Leer, B.: On upstream differencing and Godunov-type schemes for hyperbolic conservation laws. SIAM Rev. **25**, 35–61 (1983)
11. Jin, S., Xin, Z.: The relaxation scheme for systems of conservation laws in arbitrary space dimension. Commun. Pure Appl. Math. **45**, 235–276 (1995)
12. Marcati, P.: Approximate solutions to conservation laws via convective parabolic equations. Commun. Partial Differ. Equ. **13**, 321–344 (1988)
13. Marcati, P., Milani, A.: The one-dimensional Darcy's law as the limit of a compressible Euler flow. J. Differ. Equ. **84**, 129–146 (1990)
14. Marcati, P., Rubino, B.: Hyperbolic to parabolic relaxation theory for quasilinear first order systems. J. Differ. Equ. **162**, 359–399 (2000)

Is the CFL Condition Sufficient? Some Remarks

Kai Schneider, Dmitry Kolomenskiy, and Erwan Deriaz

Abstract We present some remarks about the CFL condition for explicit time discretization methods of Adams–Bashforth and Runge–Kutta type and show that for convection-dominated problems stability conditions of the type $\Delta t \leq C \Delta x^{\alpha}$ are found for high order space discretizations, where the exponent α depends on the order of the time scheme. For example, for second order Adams–Bashforth and Runge–Kutta schemes we find $\alpha = 4/3$.

Keywords Explicit time discretization · Stability · CFL condition · Runge–Kutta · Adams–Bashforts · Computational fluid dynamics · Convection dominated problems

1 Introduction

This discussion paper presents some reflections about the stability of time discretization schemes for convection-dominated problems, presented by the first author at the conference "CFL-condition, 80 years gone by", held in Rio de Janeiro in May 2010. In Computational Fluid Dynamics, explicit schemes are typically used for the nonlinear convection term. Thus for stability reasons, the celebrated Courant–Friedrichs–Lewy (CFL) condition [3] has to be satisfied, which states that the time step should be proportional to the space step, with a constant depending on the magnitude of the velocity.

The aim of the paper is to revisit the time-stability issue for some higher order time schemes. We present several numerical experiments using either one-step methods of Runge–Kutta type or multi-step methods of Adams–Bashforth type applied to the one-dimensional Burgers equations and to the two-dimensional Euler/Navier–Stokes equations. The numerical results, using a spectral discretization in space, illustrate that for stability the classical CFL condition is not sufficient and that the time step is limited by non-integer powers (larger than one) of the spatial grid size.

K. Schneider (✉) · D. Kolomenskiy · E. Deriaz
M2P2–CNRS, Aix-Marseille Université, 38 rue Joliot-Curie, 13451 Marseille cedex 20, France
e-mail: kschneid@cmi.univ-mrs.fr

C.A. de Moura, C.S. Kubrusly (eds.), *The Courant–Friedrichs–Lewy (CFL) Condition*,
DOI 10.1007/978-0-8176-8394-8_9, © Springer Science+Business Media New York 2013

The remainder of the manuscript is organized as follows. In Sect. 2, explicit one-step and multi-step time schemes are recalled, together with their stability domains. Section 3 presents some numerical examples for the inviscid Burgers equation in one space dimension and for the two-dimensional incompressible Navier–Stokes equation. Finally, some conclusions are drawn.

2 Stability of Time Schemes

We consider the general form of an evolutionary partial differential equation

$$\partial_t u = H(u) \tag{1}$$

where $H(u)$ contains all the spatial derivatives. The above equation is completed with suitable initial and boundary conditions. In fluid mechanics, one typically encounters equations where H is the sum of a nonlinear term with first order derivatives and a linear term with second order derivatives. For simplicity, we consider here a convection–diffusion equation, i.e., $H(u) = -a\partial_x u + \nu\partial_{xx} u$, where a is a constant convection velocity and $\nu \geq 0$ is the viscosity. For time discretization of (1), we use here explicit schemes, either one-step methods of Runge–Kutta type, or multi-step methods of Adams–Bashforth type. The time step is denoted by Δt, and u^n is an approximation of $u(x,t)$ at time $t^n = n\Delta t$ for $n = 0, 1, \ldots$.

In the following, we will recall some results on the stability of one-step and multi-step methods; details can be found, e.g., [8].

2.1 Runge–Kutta Schemes

Explicit Runge–Kutta schemes are one-step schemes and use two time levels, t^n and t^{n+1}. However, they imply s intermediate stages to increase the order. Typically, they have the approximation order $(\Delta t)^k$ where $k = s$. They can be written in the general form:

$$K_1 = H\big(u^n\big), \tag{2}$$

$$K_i = H\left(u^n + \Delta t \sum_{j=1}^{i-1} a_{i,j} K_j\right) \quad \text{for } i = 2, \ldots, s, \tag{3}$$

$$u^{n+1} = u^n + \Delta t \sum_{j=1}^{s} b_j K_j. \tag{4}$$

For $s = 1$, we recover the explicit Euler method with $b_0 = 1$. For the second order Runge–Kutta method (RK2), we have $a_{1,1} = 1/2$, $b_0 = 0$, and $b_1 = 1$. The coefficients of RK3 are given by $a_{1,1} = 1/2$, $a_{2,1} = -1$, $a_{2,2} = 2$ and $b_1 = 1/6$, $b_2 = 2/3$, $b_3 = 1/6$. For the classical RK4 scheme, we have $a_{1,1} = 1/2$, $a_{2,1} = 0$, $a_{2,2} = 1/2$, $a_{3,1} = a_{3,2} = 0$, $a_{3,3} = 1$ and $b_1 = 1/6$, $b_2 = 1/3$, $b_3 = 1/3$, $b_4 = 1/6$. For more details, we refer, e.g., to the textbook of Deuflhard [5].

Table 1 Coefficients of the Adams–Bashforth schemes

Scheme	Order	b_0	b_1	b_2	b_3
AB2	2	$\frac{3}{2}$	$-\frac{1}{2}$		
AB3	3	$\frac{23}{12}$	$-\frac{16}{12}$	$\frac{5}{12}$	
AB4	4	$\frac{55}{24}$	$-\frac{59}{24}$	$\frac{37}{24}$	$-\frac{9}{24}$

2.2 Adams–Bashforth Schemes

Adams–Bashforth schemes of order $(\Delta t)^k$ approximate the time derivative of (1) using a finite difference with two time levels, while the term $H(u)$ is approximated by evaluations using k time levels, and thus they belong to the family of multi-step methods. Their general form is given by

$$\frac{u^{n+1} - u^n}{\Delta t} = \sum_{j=0}^{k-1} b_j H\big(u^{n-j}\big). \tag{5}$$

The coefficients for some Adams–Bashforth schemes can be found in Table 1.

2.3 Stability

To investigate the stability of the above time schemes, the Fourier method, also called von Neumann stability analysis [2], is typically used; for details again, we refer, e.g., to [8]. Thus we are looking for a solution u of (1) in terms of a truncated Fourier series, $u(x,t) \approx \sum_{|k|\le K} \widehat{u}_k(t)e^{ikx}$ and we obtain the following ordinary differential equation for its Fourier coefficients $\widehat{u}_k(t)$,

$$d_t\widehat{u}_k = \lambda_k \widehat{u}_k \quad \text{for } |k| \le K \tag{6}$$

with the complex numbers $\lambda_k = -iak - \nu k^2$, denoted in the following for ease of notation by λ.

Applying Runge–Kutta methods to (6), we obtain

$$v^{n+1} = gv^n \tag{7}$$

where $g(z)$ with $z = \lambda\Delta t$ is the Taylor expansion of $e^{\lambda\Delta t}$. For absolute stability, the amplification factor g has to satisfy $g(z) \le 1$. The stability domains $\mathcal{S}$ in the complex plane are determined by solving $|g(z)| = e^{i\theta}$ for $0 \le \theta < 2\pi$. For RK1 (explicit Euler), RK2, RK3, and RK4, the corresponding domains are shown in Fig. 1 (left).

Applying Adams–Bashforth methods to (6) yields the following finite difference equation:

$$u^{n+1} - u^n - \lambda\Delta t \sum_{j=0}^{k-1} b_j u^{n-j} = 0 \tag{8}$$

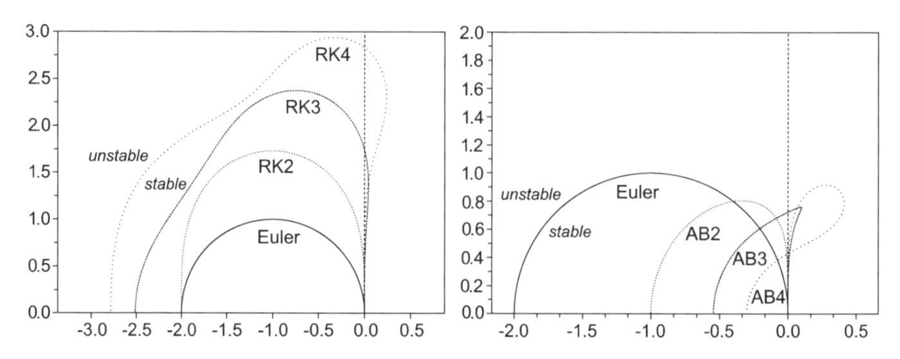

Fig. 1 Stability domains $|g(z)| \leq 1$ of Runge–Kutta methods (*left*) and $|f(r)| \leq 1$ of Adams–Bashforth methods (*right*). The domains have to be completed by symmetry with respect to the horizontal axis

which has solutions of the form r^n. Here r is the root of the characteristic equation

$$f(r) = r^k - r^{k-1} - \lambda \Delta t \sum_{j=0}^{k-1} b_j r^{k-1-j} = 0 \tag{9}$$

which has k solutions. For absolute stability (a-stable schemes), all roots have to satisfy the condition $|r| \leq 1$. The boundary of the stability domain $\mathcal{S}$ in the complex plane is then determined by $f(r) = 0$ with $|r| = 1$. A time step Δt is stable if the point $z = \lambda \Delta t$ belongs to the stability domain $\mathcal{S}$. In Fig. 1 (right), the stability domains of AB2, AB3, and AB4 are shown.

From the stability domains, we can draw the following well known conclusions: For Adams–Bashforth schemes, we observe that the stability domain decreases for increasing order, while for Runge–Kutta methods the stability domain becomes larger for increasing order. For pure convection, i.e., $\lambda = -iak$ being purely imaginary, we can observe that the explicit Euler scheme is unstable and that both second order schemes, RK2 and AB2, are also unstable.

2.4 A Refined CFL Condition and Stability

Relaxing the absolute stability condition, but requiring the error to be bounded at time $T = n\Delta t$, we obtain the necessary and sufficient stability condition for the amplification factor $|g(z)| \leq 1 + C\Delta t$ where C is a constant. For the error $\varepsilon(t_n) = u(x, t_n) - u_n(x)$, we get the following L^2-estimate:

$$\left\|\varepsilon(t_n)\right\|_{L^2} = \left\|g(z)^n \varepsilon(t_0)\right\|_{L^2} \leq C'(1 + \Delta t)^n \left\|\varepsilon(t_0)\right\|_{L^2} \leq C' e^{CT} \left\|\varepsilon(t_0)\right\|_{L^2}. \tag{10}$$

The absolute stability is recovered for the case $C = 0$.

For the explicit Euler method (RK1), we have $g(z) = 1 + z$ and applying this scheme to a transport equation $\partial_t u + a\partial u = 0$, i.e., $\lambda = -iak$, we get $|1 - iak\Delta t|^2 \leq$

$1 + 2C\Delta t$; and with $k \propto 1/\Delta x$ we thus obtain a time step constraint of the type $\Delta t \leq 2C(\frac{\Delta x}{a})^2$, as presented in [6].

In the following, we summarize the time step stability conditions for the transport equation, denoting $\rho = \frac{\Delta t}{\Delta x}$. For details, we refer to [4, 7]:

- 1st order explicit Euler explicit scheme

$$\|\varepsilon_{n+1}\|_{L^2} \leq \left(1 + \frac{\rho^2}{2}a^2\right)\|\varepsilon_n\|_{L^2} \quad \text{and the CFL:} \quad \Delta t \leq 2C\left(\frac{\Delta x}{a}\right)^2;$$

- 2nd order Runge–Kutta scheme

$$\|\varepsilon_{n+1}\|_{L^2} \leq \left(1 + \frac{\rho^4}{8}a^4\right)\|\varepsilon_n\|_{L^2} \quad \text{hence the CFL:} \quad \Delta t \leq 2C^{\frac{1}{3}}\left(\frac{\Delta x}{a}\right)^{\frac{4}{3}};$$

- 2nd order Adams–Bashforth scheme

$$\|\varepsilon_{n+1}\|_{L^2} \leq \left(1 + \frac{\rho^4}{4}a^4\right)\|\varepsilon_n\|_{L^2} \quad \text{inducing the CFL:} \quad \Delta t \leq 2^{\frac{2}{3}}C^{\frac{1}{3}}\left(\frac{\Delta x}{a}\right)^{\frac{4}{3}}.$$

We recall that the classical CFL condition [3], which is a necessary condition, yields $\Delta t \leq C\frac{\Delta x}{a}$.

3 Numerical Results

In the following, we present a series of numerical results for the inviscid Burgers equation in one space dimension and for the incompressible Euler/Navier–Stokes equations in two dimensions to illustrate the time stability of different explicit schemes.

3.1 Inviscid Burgers Equation

We consider the inviscid Burgers equation,

$$\partial_t u + u\partial_x u = 0 \quad \text{for } x \in \mathbb{T}(-1, 1), \ t \in [0, 10/\pi] \tag{11}$$

with the initial condition $u_0(x) = 10 - 0.1\sin(\pi x)$ and completed with periodic boundary conditions. For the numerical solution, we use a classical Fourier pseudo-spectral method for space discretization, which is fully de-aliased, see, e.g., [1]. For time integration, we apply either Runge–Kutta schemes of order 1 up to 4 or Adams–Bashforth schemes of order 2, 3, and 4.

The criterion to check the stability of the numerical simulations is based on the total variation of the solution, which remains constant for the exact solution. The divergence criterion we apply thus reads $\|u_N(\cdot, t)\|_{TV} > K\|u_0(\cdot)\|_{TV}$ with $K = 5$. For each scheme we perform a series of computations for an increasing number

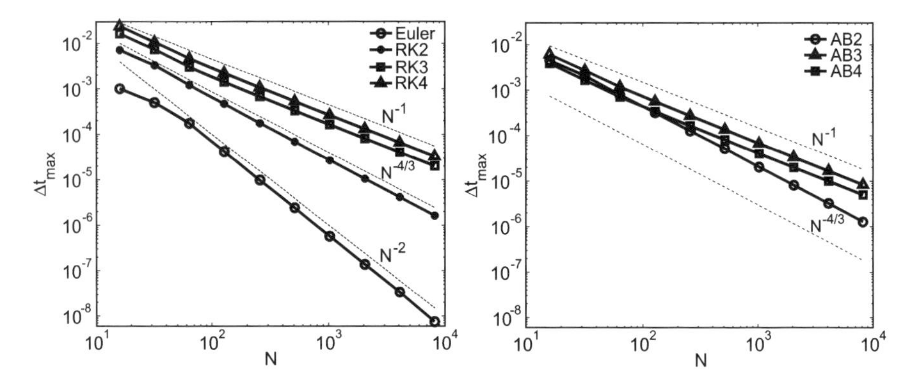

Fig. 2 Time stability for the inviscid Burgers equation. Maximum Δt as a function of the number of grid points. *Left*: Runge–Kutta methods. *Right*: Adams–Bashforth methods

of grid points, $N = 2^4, \ldots, 2^{13}$. In Fig. 2, we plot the maximal time step Δt as a function of the number of grid points for Runge–Kutta (left) and Adams–Bashforth schemes (right) in double logarithmic representation. For sufficiently large N, we indeed observe in all cases straight lines which correspond to the power law behaviors with exponents given in Sect. 2. For example, for the explicit Euler scheme we find that $\Delta t \leq C(\Delta x)^2$, while for the second order Runge–Kutta and Adams–Bashforth schemes we observe $\Delta t \leq C(\Delta x)^{4/3}$.

3.2 Incompressible Euler and Navier–Stokes Equations

Now we consider the two-dimensional incompressible Euler and Navier–Stokes equations written in the vorticity-stream function formulation:

$$\partial_t \omega + \mathbf{u} \cdot \nabla \omega - \nu \nabla^2 \omega = 0, \qquad \nabla^2 \Psi = \omega, \qquad \mathbf{u} = \nabla^\perp \Psi. \tag{12}$$

The above equations are completed with periodic boundary conditions. As the initial condition we take the 'three vortex' initial condition (cf. Fig. 3, left) which corresponds to two vortices with positive vorticity and one vortex with negative vorticity, see, e.g., [10].

The numerical scheme consists of a Fourier pseudo-spectral method in space. In time the AB2 scheme is used for the nonlinear advection term $\mathbf{u} \cdot \nabla \omega$, while exact integration is applied for the linear viscous term $\nu \nabla^2 \omega$. Details on the numerical scheme can be found in [9].

We performed a series of computations up to $t = 10$ for three different values of viscosity, $\nu = 10^{-2}, 10^{-3}$, and $\nu = 0$, the latter corresponding to the inviscid Euler equation. The numerical resolution has been varied from $N = 64^2$ to 2048^2 grid points. In Fig. 3, we show the evolution of the vorticity field for three time instants which illustrate the merging of the two positive vortices. Note that even in the case of the Euler equations the vorticity field remains smooth in the considered time

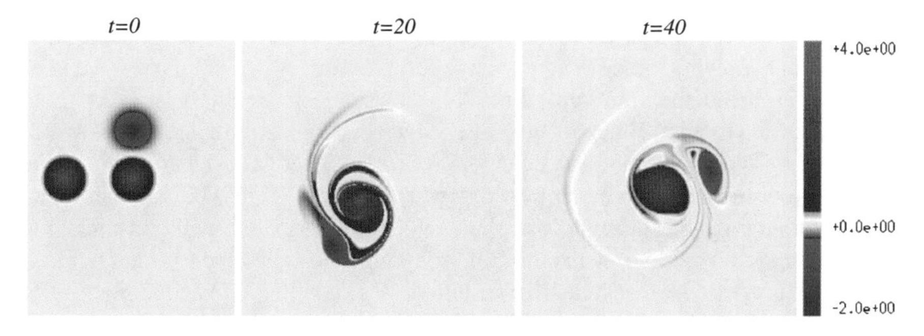

Fig. 3 Interaction of three vorticities. Vorticity field at $t = 0, 20$, and 40

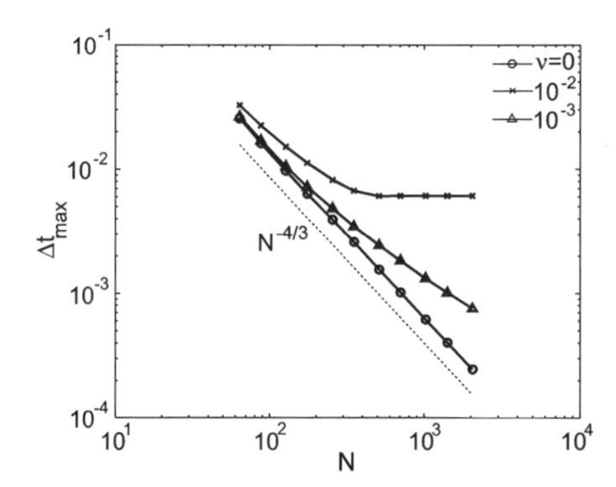

Fig. 4 Time stability of Euler/Navier–Stokes equations using AB2 scheme for different values of ν. Maximum Δt as a function of the number of grid points

interval. The maximum stable time step Δt is determined by considering again as stability criterion the total variation of the stream function ψ, i.e., the largest $\Delta t_{\max}$ is determined such that $\|\psi_N(\cdot,t)\|_{TV} \leq K\|\psi_0(\cdot)\|_{TV}$ with $K = 5$. In Fig. 4, we plot the maximum time step size Δt as a function of the number of grid points for three different values of $\nu = 10^{-2}, 10^{-3}$ and $\nu = 0$. For the inviscid case, we observe a power law with slope $-4/3$, i.e., $\Delta t \leq C(\Delta x)^{4/3}$. In the viscous cases, we observe a slower decay for increasing N (corresponding to decreasing Δx) which leads even to a saturation for sufficiently large values of the viscosity, i.e., the maximum time step size becomes independent of the spatial grid size. This can be explained by remarking that the viscosity damps the small scales and thus the small grid size is not necessary, which is reflected in vanishing (or very small amplitude) large wavenumber Fourier modes.

4 Conclusions

Our numerical experiments give some evidence that the CFL condition is necessary but not sufficient to guarantee the time stability of explicit time schemes given that high order space discretizations are used. For the inviscid Burgers equation, we showed that applying either a second order Runge–Kutta or a second order Adams–Bashforth scheme, the time step Δt has to be chosen to be smaller than $(\Delta x)^{4/3}$. In the case of the explicit Euler scheme, it even necessitates $\Delta t \leq C(\Delta x)^2$ for stability. For the Euler/Navier–Stokes equation, we have numerical evidence that similar results as for the Burgers equation hold, confirmed here for the AB2 scheme. These observations give thus a possible explanation why for spectral methods applied to convection-dominated problems the CFL constant, also called the Courant number, has to be decreased for increasing resolution as a modified CFL condition has to be satisfied. Further details on the theoretical justification of the above results and implications for CFD codes can be found in [4] and [7], respectively. Finally, we are asking the question if these observations are well known?

Acknowledgements K.S. is grateful to Carlos de Moura for the invitation to the conference "CFL-Condition, 80 years gone by", held in Rio de Janeiro in May 2010.

References

1. Canuto, C., Hussaini, M.T., Quarteroni, A., Zang, T.A.: Spectral Methods in Fluid Dynamics. Springer, New York (1988)
2. Charney, J.G., Fjörtoft, R., von Neumann, J.: Numerical integration of the barotropic vorticity equation. Tellus **2**, 237–254 (1950)
3. Courant, R., Friedrichs, K., Lewy, H.: Über die partiellen Differenzengleichungen der mathematischen Physik. Math. Ann. **100**(1), 32–74 (1928)
4. Deriaz, E.: Stability of explicit numerical schemes for smooth convection-dominated problems. Preprint (2011)
5. Deuflhard, P., Bornemann, F.: Scientific Computing with Ordinary Differential Equations. Texts in Applied Mathematics, vol. 42. Springer, Berlin (2002)
6. Gottlieb, D., Tadmor, E.: The CFL condition for spectral approximations to hyperbolic initial-boundary value problems. Math. Comput. **56**(194), 565–588 (1991)
7. Kolomenskiy, D., Schneider, K., Deriaz, E.: In preparation (2012)
8. Peyret, R.: Spectral Methods of Incompressible Viscous Flows. Springer, Berlin (2001)
9. Schneider, K.: Numerical simulation of the transient flow behaviour in chemical reactors using a penalization method. Comput. Fluids **34**, 1223–1238 (2005)
10. Schneider, K., Kevlahan, N., Farge, M.: Comparison of an adaptive wavelet method and nonlinearly filtered pseudo-spectral methods for two-dimensional turbulence. Theor. Comput. Fluid Dyn. **9**(3/4), 191–206 (1997)

Fast Chaotic Artificial Time Integration

Uri Ascher and Kees van den Doel

Abstract Gradient descent methods for large positive definite linear and nonlinear algebraic systems arise when integrating a PDE to steady state and when regularizing inverse problems. However, these methods may converge very slowly when utilizing a constant step size, or when employing an exact line search at each step, with the iteration count growing proportionally to the condition number. Faster gradient descent methods must occasionally resort to significantly larger step sizes, which in turn yields a strongly nonmonotone decrease pattern in the residual vector norm.

In fact, such faster gradient descent methods generate chaotic dynamical systems for the normalized residual vectors. Very little is required to generate chaos here: simply damping steepest descent by a constant factor close to 1 will do. The fastest practical methods of this family in general appear to be the chaotic, two-step ones. Despite their erratic behavior, these methods may also be used as smoothers, or regularization operators. Our results also highlight the need for better theory for these methods.

Keywords Gradient descent · Artificial time integration · Dynamical system · Stability · Chaos · Regularization

1 Introduction

The famous Courant–Friedrichs–Lewy (CFL) condition really provides a consistency, or compatibility bound, rather than a stability one. It states a bound on the allowed time step in terms of the spatial discretization step for explicit difference methods applied to a hyperbolic partial differential equation (PDE). Its immense popularity relates to the fact that this bound typically coincides with that of the stability condition for simple explicit methods in case of hyperbolic PDEs in one space

U. Ascher (✉) · K. van den Doel
Department of Computer Science, University of British Columbia, Vancouver, Canada
e-mail: ascher@cs.ubc.ca

K. van den Doel
e-mail: kvdoel@cs.ubc.ca

C.A. de Moura, C.S. Kubrusly (eds.), *The Courant–Friedrichs–Lewy (CFL) Condition*,
DOI 10.1007/978-0-8176-8394-8_10,

variable, and it does relate to the essential limitation on time-stepping using explicit methods when applied to time dependent PDEs in general. Thus, practitioners over the years have often come to identify the CFL condition with an essential stability restriction in a wide context.

However, occasionally the complete picture is more delicate, and this is indicated already in the following basic example.

Example 1 For the simple initial value advection equation

$$\frac{\partial u}{\partial t} + a\frac{\partial u}{\partial x} = 0, \quad t \geq 0, \ -\infty < x < \infty, \tag{1a}$$

$$u(0,x) = u_0(x), \tag{1b}$$

where a is a known constant, the exact solution is $u(t,x) = u_0(x - at)$.

For an explicit discretization consider the well-known upwind method (e.g., [2]), where we fix the spatial step size but allow the time step to vary. For $a = -1$ we have for all j integer

$$v_j^{k+1} = v_j^k + \frac{\alpha_k}{\Delta x}\left(v_{j+1}^k - v_j^k\right), \quad k = 0, 1, \ldots,$$

where v_j^k approximates $u(t_k, x_j)$, $x_j = j\Delta x$ and $t_k = \sum_{i=0}^{k-1} \alpha_i$. We set $v_j^0 = u_0(x_j)$, $\forall j$.

Next, consider the initial value function

$$u_0(x) = \begin{cases} 1 & x \leq 0, \\ 0 & x > 0. \end{cases}$$

The stability condition for this method is

$$\alpha_k \leq \Delta x, \quad \forall k.$$

If this bound is violated, e.g., $\alpha_0 > \Delta x$, then $v_{-1}^1 = u_0(-\Delta x) + \frac{\alpha_0}{\Delta x}(u_0(0) - u_0(-\Delta x)) = 1$, whereas the exact solution is $u(\alpha_0, -\Delta x) = u_0(\alpha_0 - \Delta x) = 0$. This inconsistency demonstrates a violation of the CFL condition which coincides with the stability condition for the upwind method. There is no general way to recover from such an inconsistency error by taking smaller time steps α_k later on.

In contrast, consider the simple initial value heat equation

$$\frac{\partial u}{\partial t} = \frac{\partial^2 u}{\partial x^2}, \quad t \geq 0, \ -\infty < x < \infty, \tag{2a}$$

$$u(0,x) = u_0(x), \tag{2b}$$

and the explicit discretization

$$v_j^{k+1} = v_j^k + \frac{\alpha_k}{\Delta x^2}\left(v_{j+1}^k - 2v_j^k + v_{j-1}^k\right), \quad k = 0, 1, \ldots.$$

The stability condition is well-known to be

$$\alpha_k \leq \frac{1}{2}\Delta x^2, \quad \forall k,$$

and there is no consistency condition such as the CFL.

In this latter case, there is no immediate a priori reason to believe that violating stability by taking $\alpha_0 > \frac{1}{2}\Delta x^2$ cannot be recovered from by using smaller time steps later on: in principle, those error modes that get excited at $k = 0$ may be subsequently calmed down.

Let us generalize (2a)–(2b) for later purposes by considering its analogue on a general bounded domain in several space dimensions and equipped with a source function and garden-variety boundary conditions. Upon applying a finite difference or finite element semi-discretization in space, we are then led to consider a problem of the form

$$\frac{d\mathbf{v}}{dt} = \mathbf{b} - A\mathbf{v}, \quad t \geq 0, \tag{3}$$

where A is a symmetric positive definite $m \times m$ matrix that is potentially large and sparse, $\mathbf{v}$ is the solution mesh function reshaped as an (unknown) vector, and $\mathbf{b}$ is a likewise reshaped vector of known inhomogeneities.

One may next wonder if it makes sense to violate the stability restriction, given that this is indeed possible. The answer depends on the purpose of the computation. If what we want is a pointwise accurate solution (trajectory) then respecting the stability restriction at each and every time step, at least approximately, is wise. Indeed, typical mesh selection algorithms achieve this, being "greedy" in nature [3, 10]. However, there are occasions where the goal is different. Such is the case in many geometric integration applications [11]. Other occasions, on which we concentrate here, are when we wish to integrate the time-dependent problem to steady state or when the purpose of integration is smoothing, or regularization of an ill-posed problem [7, 18]. Such applications correspond to continuation, or homotopy methods, where one is not interested in the accurate reconstruction of the homotopy path but only in its end.

Thus, consider a forward Euler discretization of (3), with the purpose of integrating the ODE system to steady state (so in particular, higher order discretization methods are not necessarily more attractive than this simplest time integration scheme). The method is written as

$$\mathbf{v}_{k+1} = \mathbf{v}_k + \alpha_k \mathbf{r}_k, \qquad \mathbf{r}_k = \mathbf{b} - A\mathbf{v}_k, \quad k = 0, 1, \ldots. \tag{4}$$

Its absolute stability condition requires

$$\alpha_k \leq \frac{2}{\lambda_1}, \tag{5}$$

where $\lambda_1 \geq \lambda_2 \geq \cdots \geq \lambda_m > 0$ are the eigenvalues of A.[1] This restriction must be strictly obeyed if a uniform step size $\alpha_k = \alpha$ is to be employed, but we next consider varying the step size.

Note that (4) can also be interpreted as a *gradient descent* (GD) method for the problem of minimizing the convex quadratic function

$$f(\mathbf{v}) = \frac{1}{2}\mathbf{v}^T A\mathbf{v} - \mathbf{b}^T\mathbf{v}, \tag{6}$$

which is equivalent to solving the steady state equations

$$A\mathbf{v} = \mathbf{b}. \tag{7}$$

The best uniform step size for (4) is

$$\alpha_k = \alpha^* = \frac{2}{\lambda_1 + \lambda_m}. \tag{8}$$

The *steepest descent* step size, obtained by exact line search for (6) at each step, or iteration, is

$$\alpha_k = \alpha_k^{\mathrm{SD}} = \frac{\mathbf{r}_k^T\mathbf{r}_k}{\mathbf{r}_k^T A\mathbf{r}_k}. \tag{9}$$

These are both slow methods requiring $O(\kappa)$ iterations to reduce the residual by a fixed amount, where $\kappa = \kappa(A)$ is the condition number. As it turns out, the steepest descent step sizes never grow much larger than what (5) allows [1, 5].

To obtain a faster explicit method integrating the PDE to steady state, or a faster GD method, a methodology must be found that automatically and significantly increases the step size α_k every few steps in such a way that the resulting overall method still converges. This may look like a lot to ask for at first, but as it turns out there are many methods of this sort, all being variants of steepest descent that are often much faster when the latter method is slow [6, 8, 9, 14, 15]. All of these methods yield chaotic dynamical systems in terms of the normalized residual vectors that their iteration sequences generate [17]. Furthermore, they retain some of the smoothing properties that steepest descent (SD) and conjugate gradient (CG) methods possess [12].

2 Faster Gradient Descent Methods

Many efforts have been devoted in the two decades that have passed since the pioneering paper of [6] to the design, analysis, extension, and application of faster gradient descent methods for function minimization. In the context of the iteration (4), we mention here the following variants of steepest descent.

[1] Let us assume throughout for simplicity that $\lambda_1 > \lambda_2$ and $\lambda_{m-1} > \lambda_m$.

Table 1 Iteration counts for the model Poisson problem using gradient descent with different step size choices for initial vectors (a) $\mathbf{v}_0 = \mathbf{0}$ and (b) $\mathbf{v}_0 = 10^{-3} \cdot \mathbf{1}$

m	$\mathbf{x}_0$	CG	SD	α^*	HLSD	SD(0.9)
49	(a)	10	341	348	69	97
	(b)	10	341	348	62	99
225	(a)	33	1414	1417	179	291
	(b)	32	1414	1419	151	258
961	(a)	71	5721	5689	279	497
	(b)	70	5721	5698	417	609

1. The half-lagged steepest descent (HLSD) method [15] simply updates the step size α only every second step, reading

$$\alpha_{2j} = \alpha_{2j+1} = \alpha_{2j}^{\mathrm{SD}}, \quad j = 0, 1, 2, \ldots. \tag{10}$$

2. The relaxed steepest descent method, denoted SD(ω), calls for calculating α_k^{SD} at each iterate k and setting

$$\alpha_k = \omega \alpha_k^{\mathrm{SD}}, \tag{11}$$

with $0 < \omega < 1$ a fixed constant [14]. Below we choose $\omega = 0.9$.

Example 2 Let us describe the *model Poisson problem*. The PDE

$$-\Delta u = q, \quad 0 < x, y < 1, \tag{12}$$

with $q(x, y)$ known and subject to homogeneous Dirichlet boundary conditions, is discretized using the standard 5-point difference scheme. Utilizing a uniform mesh width $\Delta x = 1/(J+1)$, and denoting by $\mathbf{b}$ the reshaped mesh function of $q(i\Delta x, j\Delta x)$, $1 \le i, j \le J$, and also letting $\mathbf{v}$ be likewise composed of solution mesh values, we have a problem in the form (7) with $m = J^2$ unknowns.

Table 1 records iteration counts required by different methods to bring the relative residual norm $\|\mathbf{r}_k\|/\|\mathbf{r}_0\|$ below a tolerance of 10^{-12}, for a right hand side $\mathbf{b} = \mathbf{1}$ of all ones.[2]

We use two starting guesses, equally smooth and differing from each other by 10^{-3} in the maximum norm.

The results in Table 1 exhibit the traits of gradient descent methods observed also in [5, 17]. Thus, we note:

1. Steepest descent is essentially as good for this example as the best constant step size, α^*, and both can be unacceptably slow.

[2]The vector ℓ_2-norm is utilized here and elsewhere, unless otherwise specified.

2. The iteration counts for the faster gradient descent methods HLSD and SD(0.9) are much better than those for steepest descent and increase much slower than $\kappa = O(m)$, behaving more like those of CG in trend.
3. The two-step method HLSD is better than the one-step method, although (not shown) it is neither consistently better nor worse than other two-step methods such as the method of [6] which we call LSD.
4. The progress of the iteration counts as a function of $\kappa = O(m)$ is less consistent for HLSD and SD(0.9) than that of CG. The latter observation relates directly to the fact that the quantities $\|\mathbf{r}_k\|$ oscillate wildly as a function of k; see Fig. 2 of [5], where the behavior of the resulting step size sequence is also depicted.
5. Finally, we observe the sensitivity for HLSD and SD(0.9) of the total number of iterations required to achieve a fixed accuracy to small changes in the initial vector $\mathbf{v}_0$. This is in marked contrast to the behavior of the CG iteration, or the SD iteration, and it suggests a chaotic behavior of the iterative process for the faster gradient descent methods.

There are many more experiments of this sort reported in [5, 17] that support the above observations.

3 Chaos

The gradient descent family of methods (4) is completely characterized by the residual evolution

$$\mathbf{r}_{k+1} = (1 - \alpha_k A)\mathbf{r}_k, \quad k = 0, 1, \ldots. \tag{13}$$

Furthermore, if we write $A = U \Lambda U^T$ with U orthogonal and Λ the diagonal matrix of eigenvalues of A, and let $\hat{\mathbf{r}} = U^T \mathbf{r}$, then (13) becomes

$$\hat{\mathbf{r}}_{k+1} = (1 - \alpha_k \Lambda)\hat{\mathbf{r}}_k, \tag{14}$$

with $\|\hat{\mathbf{r}_k}\| = \|\mathbf{r}_k\|$. Thus, (14) has precisely the same convergence behavior as (13), and hence we may consider (for analysis purposes) a diagonal A without loss of generality. Note that now, if $\alpha_k = \lambda_i^{-1}$ for some i, $1 \le i \le m$, then the ith component of the next residual vanishes: $r_i^{(k+1)} = 0$. If m is large, though, then even if we knew the eigenvalues we would not want to use them in this way in practice.

Below it is convenient to omit the iteration index k where no confusion can possibly arise. An alternative notation to (13), for instance, is

$$\mathbf{r} \leftarrow (1 - \alpha A)\mathbf{r}.$$

To study the behavior of these residual vectors, we associate with $\mathbf{r}$ the Akaike "probability" $\mathbf{p}$, see [1], which is the component-wise square of the normalized residual, given by

$$p_i = (r_i)^2 / \|\mathbf{r}\|^2, \quad i = 1, \ldots, m. \tag{15}$$

In [17], we considered in detail the Jacobian of the transformation $\mathbf{p}_{k+1} \leftarrow \mathbf{p}_k$. Using these over a long sequence of iterations, we calculated the Lyapunov exponent μ for various gradient descent methods. Recall that for a stable system, $\mu \le 0$, whereas for a chaotic system, $\mu > 0$ and nearby orbits separate exponentially, cf. [16].

The summary of those experiments over many examples is that for all the faster methods we have $\mu > 0$, whereas SD is marginally stable: $\mu = 0$. The over-relaxed SD(ω) variant with $\omega > 1$ has $\mu < 0$ and is stable and slow. The faster SD(ω) methods are obtained for $\omega < 1$.

Having established that there is chaos in the faster methods, the question is what characterizes them (other than the fact that they converge). Ironically, as it turns out, the more special methods that are variants of steepest descent are the slower ones. To get the feeling for this, it is better to consider the under-relaxed steepest descent methods SD(ω), say $0.5 \le \omega < 1$. These are not quite as fast as LSD or HLSD, but they are much faster than SD, as we have seen, and this in itself may be considered surprising. Note that these are simple, memoryless, one-step methods. There are no random parameters and no switches in the step size selection strategy. Standard arguments (e.g., [13]) imply that the methods yield monotonic decrease in $f(\mathbf{v})$ of (6) and converge Q-linearly, which is not known to occur for the two-step methods. Finally, note that although this method with $\omega < 1$ takes at each iteration a fraction of the SD step size, its average step size is much larger than that of SD! Here then is one of the simplest and cleanest instances of both a chaotic system and the peril of greed (in numerical algorithms at least).

Recall that for a diagonal $A = \operatorname{diag}(\lambda_i)$ we can write

$$\gamma_k^{\mathrm{SD}} = 1/\alpha_k^{\mathrm{SD}} = \frac{\sum_i \lambda_i r_i^2}{\sum_i r_i^2} = \sum_i \lambda_i p_i, \quad i = 1, \dots, m, \tag{16a}$$

$$\alpha_k = \omega \alpha_k^{\mathrm{SD}}, \tag{16b}$$

$$r_i^{(k+1)} = (1 - \alpha_k \lambda_i) r_i^{(k)}, \quad i = 1, \dots, m. \tag{16c}$$

We make the following observations (see [5, 17] for the full details):

- If p_i are roughly equal then $\lambda_1 p_1$ (or the first few) dominate, so $\alpha_k^{\mathrm{SD}} \sim 1/\lambda_1$ is within the forward Euler stability bound, as is α_k.
- The corresponding step is effectively a *smoother*: it reduces high frequency (i.e., large-eigenvalue) residuals much more than low frequency ones. This effect may be repeated over a few steps.
- The *usual* case is that the high frequency residuals become so small in magnitude due to repeated smoothing that other frequencies temporarily dominate. This in turn means that a much larger step size α_k^{SD} (and hence α_k) is obtained. The ensuing step is unstable, increasing the high frequency probabilities p_i; see (16a)–(16c), (15). Once these components are large, their $\lambda_i p_i$ dominate again and the next few step sizes will be small and smoothing. This closes a chaotic cycle.

– The *unusual* case is that of steepest descent! With SD the high frequency $\lambda_1 p_1$ always dominates [1]. In the limit p_1 and p_m alternate (maintaining roughly the same order of magnitude) while $p_i \approx 0$ for $1 < i < m$. Thus, chaos (and good news) are both avoided.

4 Preconditioning and Regularization

Considered over a wide range of problems (3) and initial guesses, the LSD and HLSD variants have been found in our experiments to be as good as any other practical gradient descent method and better than most. Of course, in practical situations one would use preconditioning, but this does not change the relative merit of the methods.

Furthermore, as our Table 1 indicates, the CG method is really better than any of the gradient descent methods in the usual circumstances of (7) considered hitherto. What the gradient descent methods offer is the attraction of simplicity, the direct connection to artificial time integration (which is a very popular approach in practice [4]), and robustness when the gradients are computed only approximately [17].

In particular, when these methods are used as a combination of smoothers and solvers, which is often the case in computational inverse problems, the faster gradient descent methods often perform as well as any other method, including CG. In [5, 12], we examined this question experimentally for several applications, including image deblurring, image denoising, and DC-resistivity recovery of piecewise constant functions. These are really nonlinear problems, and freezing them at each iteration in order to determine the step size in the manner described here does more damage to CG than to the faster GD methods. Rather than repeating the details of these papers, let us only mention that in such circumstances often the SD method does not look so awful either because its slowness develops only when many such iterations are performed in an uninterrupted succession.

Acknowledgements The first author thanks IMPA, Rio de Janeiro, for support and hospitality during several months in 2011 when this work was completed.

U. Ascher supported in part by NSERC Discovery Grant 84306.

References

1. Akaike, H.: On a successive transformation of probability distribution and its application to the analysis of the optimum gradient method. Ann. Inst. Stat. Math. Tokyo **11**, 1–16 (1959)
2. Ascher, U.: Numerical Methods for Evolutionary Differential Equations. SIAM, Philadelphia (2008)
3. Ascher, U., Mattheij, R., Russell, R.: Numerical Solution of Boundary Value Problems for Ordinary Differential Equations. SIAM, Philadelphia (1995)
4. Ascher, U., Huang, H., van den Doel, K.: Artificial time integration. BIT Numer. Math. **47**, 3–25 (2007)

5. Ascher, U., van den Doel, K., Huang, H., Svaiter, B.: Gradient descent and fast artificial time integration. Modél. Math. Anal. Numér. **43**, 689–708 (2009)
6. Barzilai, J., Borwein, J.: Two point step size gradient methods. IMA J. Numer. Anal. **8**, 141–148 (1988)
7. Engl, H.W., Hanke, M., Neubauer, A.: Regularization of Inverse Problems. Kluwer, Dordrecht (1996)
8. Fletcher, R.: On the Barzilai–Borwein method. In: Qi, L., Teo, K., Yang, X. (eds.) Optimization and Control with Applications. Kluwer Series in Applied Optimization, vol. 96, pp. 235–256 (2005)
9. Friedlander, A., Martinez, J., Molina, B., Raydan, M.: Gradient method with retard and generalizations. SIAM J. Numer. Anal. **36**, 275–289 (1999)
10. Hairer, E., Norsett, S.P., Wanner, G.: Solving Ordinary Differential Equations I: Nonstiff Problems. Springer, Berlin (1993)
11. Hairer, E., Lubich, C., Wanner, G.: Geometric Numerical Integration. Springer, Berlin (2002)
12. Huang, H., Ascher, U.: Faster gradient descent and the efficient recovery of images. Math. Program. (2013, to appear)
13. Nocedal, J., Wright, S.: Numerical Optimization. Springer, New York (1999)
14. Pronzato, L., Wynn, H., Zhigljavsky, A.: Dynamical Search: Applications of Dynamical Systems in Search and Optimization. Chapman & Hall/CRC, Boca Raton (2000)
15. Raydan, M., Svaiter, B.: Relaxed steepest descent and Cauchy–Barzilai–Borwein method. Comput. Optim. Appl. **21**, 155–167 (2002)
16. Shimada, I., Nagashima, T.: A numerical approach to ergodic problems of dissipative dynamical systems. Prog. Theor. Phys. **61**(6), 1605–1616 (1979)
17. van den Doel, K., Ascher, U.: The chaotic nature of faster gradient descent methods. J. Sci. Comput. **48** (2011). doi:10.1007/s10915-011-9521-3
18. Vogel, C.: Computational Methods for Inverse Problem. SIAM, Philadelphia (2002)

Appendix A
Hans Lewy's Recovered String Trio

Lori Courant Lax

Played by Lori Courant Lax, viola, Dorothy Strahl, violin, and Carol Buck, cello

Hans Lewy was a musical prodigy. As a child he gave rave-review violin concerts. He also learned music theory and started to compose. But at 19 or 20 he began to turn to mathematics. He studied with my father, Richard Courant, and often came to the chamber music evenings of my mother, Nina.

On one occasion he brought a string trio he had written, and my mother and her friends tried to play it. They didn't do very well, and Hans, with his typical explosiveness, grabbed the parts, tore them up, and threw them on the floor.

"But I liked it!" said my mother as she gathered up the scraps. She laboriously copied the parts, made a score, numbered the measures—all without the help of whiteout or xerox.

I found the Lewy trio among my mother's papers after her death. Enclosed here is a recording of one movement of it, a romantic, gracious waltz, played by Dorothy Strahl, violin, Carol Buck, cello, and myself on the viola.

L. Courant Lax (✉)
New York, USA
e-mail: loriblax@gmail.com

C.A. de Moura, C.S. Kubrusly (eds.), *The Courant–Friedrichs–Lewy (CFL) Condition*,
DOI 10.1007/978-0-8176-8394-8,

Fig. 1 A piece of Lewy's recovered string trio score

Fig. 2 Musicians Dorothy Strahl (violin) and Lori Courant Lax (viola); and Carol Buck (cello) (Photos from L. Courant Lax files)

Appendix B

Reprint of CFL original paper

C.A. de Moura, C.S. Kubrusly (eds.), *The Courant–Friedrichs–Lewy (CFL) Condition*,
DOI 10.1007/978-0-8176-8394-8, © Springer Science+Business Media New York 2013

Über die partiellen Differenzengleichungen der mathematischen Physik.

Von

R. Courant, K. Friedrichs und H. Lewy in Göttingen.

Ersetzt man bei den klassischen linearen Differentialgleichungsproblemen der mathematischen Physik die Differentialquotienten durch Differenzenquotienten in einem — etwa rechtwinklig angenommenen — Gitter, so gelangt man zu algebraischen Problemen von sehr durchsichtiger Struktur. Die vorliegende Arbeit untersucht nach einer elementaren Diskussion dieser algebraischen Probleme vor allem die Frage, wie sich die Lösungen verhalten, wenn man die Maschen des Gitters gegen Null streben läßt. Dabei beschränken wir uns vielfach auf die einfachsten, aber typischen Fälle, die wir derart behandeln, daß die Anwendbarkeit der Methoden auf allgemeinere Differenzengleichungen und solche mit beliebig vielen unabhängigen Veränderlichen deutlich wird.

Entsprechend den für Differentialgleichungen geläufigen Fragestellungen behandeln wir Randwert- und Eigenwertprobleme für elliptische Differenzengleichungen und das Anfangswertproblem für hyperbolische bzw. parabolische Differenzengleichungen. Wir werden an einigen typischen Beispielen beweisen, daß der Grenzübergang stets möglich ist, nämlich daß die Lösungen der Differenzengleichungen gegen die Lösungen der entsprechenden Differentialgleichungsprobleme konvergieren; ja wir werden sogar erkennen, daß bei elliptischen Gleichungen i. a. die Differenzenquotienten beliebig hoher Ordnung gegen die entsprechenden Differentialquotienten streben. Die Lösbarkeit der Differentialgleichungsprobleme setzen wir nirgends voraus; vielmehr erhalten wir durch den Grenzübergang hierfür einen einfachen Beweis [1]). Während aber beim elliptischen

[1]) Unsere Beweismethode läßt sich ohne Schwierigkeit so erweitern, daß sie bei beliebigen linearen elliptischen Differentialgleichungen das Rand- und Eigenwertproblem und bei beliebigen linearen hyperbolischen Differentialgleichungen das Anfangswertproblem zu lösen gestattet.

Falle einfache und weitgehend von der Wahl des Gitters unabhängige Konvergenzverhältnisse herrschen, werden wir bei dem Anfangswertproblem hyperbolischer Gleichungen erkennen, daß die Konvergenz allgemein nur dann vorhanden ist, wenn die Verhältnisse der Gittermaschen in verschiedenen Richtungen gewissen Ungleichungen genügen, die durch die Lage der Charakteristiken zum Gitter bestimmt werden.

Das typische Beispiel ist für uns im elliptischen Falle das Randwertproblem der Potentialtheorie. Seine Lösung von der Lösung des entsprechenden Differenzengleichungsproblems her ist übrigens in den letzten Jahren mehrfach behandelt worden [2]). Allerdings werden dabei im Gegensatz zu der vorliegenden Arbeit meist spezielle Eigenschaften der Potentialgleichung benutzt, so daß die Anwendbarkeit der Methode auf andere Probleme nicht ohne weiteres zu übersehen ist.

Abgesehen von dem gekennzeichneten Hauptziel der Arbeit werden wir im Anschluß an die elementare algebraische Diskussion des Randwertproblems elliptischer Gleichungen dessen Zusammenhang mit dem aus der Statistik bekannten Probleme der Irrwege erörtern.

I. Der elliptische Fall.

§ 1.

Vorbemerkungen.

1. Definitionen.

Wir betrachten zunächst in der Ebene mit den rechtwinkligen Koordinaten x, y ein quadratisches Punktgitter der Maschenweite $h > 0$, etwa alle Punkte mit den Koordinaten $x = nh$, $y = mh$,

$$m, n = 0, \pm 1, \pm 2, \ldots .$$

---

[2]) J. le Roux, Sur le problème de Dirichlet, Journ. de mathém. pur. et appl. (6) **10** (1914), p. 189. R. G. D. Richardson, A new method in boundary problems for differential equations, Transactions of the Americ. Mathem. Soc. **18** (1917), p. 489 ff. H. B. Philips and N. Wiener, Nets and the Dirichlet Problem, Publ. of the Mass. Institute of Technology (1925).

Leider waren diese Abhandlungen dem ersten der drei Verfasser bei der Abfassung seiner Note „Zur Theorie der partiellen Differenzengleichungen", Gött. Nachr. 23. X. 1925, an welche die vorliegende Arbeit anschließt, entgangen.

Vgl. ferner: L. Lusternik, Über einige Anwendungen der direkten Methoden in der Variationsrechnung, Recueil de la Société Mathém. de Moscou, 1926. G. Bouligand, Sur le problème de Dirichlet, Ann. de la soc. polon. de mathém. **4**, Krakau 1926.

Über die Bedeutung des Differenzenansatzes und über weitere sie verwendende Arbeiten vgl. R. Courant, Über direkte Methoden in der Variationsrechnung, Math. Annalen **97**, S. 711 und die dort angegebene Literatur.

Es sei G ein Gebiet der Ebene, begrenzt von einer stetigen, doppelpunktfreien geschlossenen Kurve. Dann soll das zugehörige — bei genügend kleiner Maschenweite eindeutig bestimmte — Gittergebiet G_h aus allen denjenigen Gitterpunkten bestehen, welche in G liegen und sich von einem festen vorgegebenen Gitterpunkt aus G durch eine zusammenhängende Kette von Gitterpunkten verbinden lassen. Wir nennen zusammenhängende Kette von Gitterpunkten eine Folge solcher Punkte, bei der jeder Punkt einer der vier Nachbarpunkte des folgenden ist. Als Randpunkt von G_h bezeichnen wir einen solchen, dessen vier Nachbarpunkte nicht alle zu G_h gehören. Alle anderen Punkte von G_h nennen wir innere Punkte.

Wir betrachten Funktionen $u, v, \ldots$ des Ortes im Gitter, d. h. Funktionen, welche nur für die Gitterpunkte definiert sind. Wir bezeichnen sie auch mit $u(x, y)$, $v(x, y), \ldots$. Für ihre vorderen und hinteren Differenzenquotienten verwenden wir die folgenden Abkürzungen:

$$\frac{1}{h}(u(x+h, y) - u(x, y)) = u_x, \qquad \frac{1}{h}(u(x, y+h) - u(x, y)) = u_y,$$

$$\frac{1}{h}(u(x, y) - u(x-h, y)) = u_{\bar{x}}, \qquad \frac{1}{h}(u(x, y) - u(x, y-h)) = u_{\bar{y}}.$$

Entsprechend bilden wir Differenzenquotienten höherer Ordnung, z. B.

$$(u_x)_{\bar{x}} = u_{x\bar{x}} = u_{\bar{x}x} = \frac{1}{h^2}(u(x+h, y) - 2u(x, y) + u(x-h, y))$$

usw.

2. Differenzenausdrücke und Greensche Umformungen.

Zu der einfachsten allgemeinen Übersicht über lineare Differenzenausdrücke zweiter Ordnung gelangen wir nach dem Muster der Theorie der partiellen Differentialgleichungen, indem wir aus zwei Funktionen u und v und ihren vorderen Differenzenquotienten einen bilinearen Ausdruck

$$B(u, v) = a u_x v_x + b u_x v_y + c u_y v_x + d u_y v_y + \alpha u_x v + \beta u_y v + \gamma u v_x + \delta u v_y + g u v$$

bilden, wobei

$$a = a(x, y), \ldots, \quad \alpha = \alpha(x, y), \ldots, \quad g = g(x, y)$$

Funktionen im Gitter sind.

Aus dem Bilinearausdruck erster Ordnung leiten wir einen Differenzenausdruck zweiter Ordnung in folgender Weise ab: Wir bilden die Summe

$$h^2 \sum_{G_h}\sum B(u, v)$$

über alle Punkte eines Gebietes G_h im Gitter, wobei in $B(u, v)$ für die Differenzenquotienten zwischen einem Randpunkte und einem nicht zu G_h

gehörigen Punkte Null zu setzen ist. Die Summe formen wir nun durch partielle Summation um (d. h. wir ordnen nach v), und zerspalten sie in eine Summe über die Menge der inneren Punkte G_h' und eine Summe über die Menge der Randpunkte Γ_h. Wir erhalten so:

$$(1) \qquad h^2 \sum_{G_h}\sum B(u, v) = -h^2 \sum_{G_h'}\sum v\, L(u) - h \sum_{\Gamma_h} v\, \mathfrak{R}(u).$$

$L(u)$ ist der für alle inneren Punkte von G_h definierte lineare „Differenzenausdruck zweiter Ordnung“:

$$L(u) = (a u_x)_{\bar{x}} + (b u_x)_{\bar{y}} + (c u_y)_{\bar{x}} + (d u_y)_{\bar{y}}$$
$$- \alpha u_x - \beta u_y + (\gamma u)_{\bar{x}} + (\delta u)_{\bar{y}} - g u.$$

$\mathfrak{R}(u)$ ist für jeden Randpunkt ein linearer Differenzenausdruck, dessen genaue Gestalt wir hier nicht angeben.

Ordnet man $\sum_{G_h}\sum B(u, v)$ nach u, so erhält man

$$(2) \qquad h^2 \sum_{G_h}\sum B(u, v) = -h^2 \sum_{G_h'}\sum u\, M(v) - h \sum_{\Gamma_h} u\, \mathfrak{S}(v).$$

$M(v)$ heißt der zu $L(u)$ adjungierte Differenzenausdruck; er lautet:

$$M(v) = (a v_x)_{\bar{x}} + (b v_y)_{\bar{x}} + (c v_x)_{\bar{y}} + (d v_y)_{\bar{y}}$$
$$+ (\alpha v)_{\bar{x}} + (\beta v)_{\bar{y}} - \gamma v_x - \delta v_y - g v,$$

während $\mathfrak{S}(v)$ ein $\mathfrak{R}(u)$ entsprechender Differenzenausdruck für den Rand ist.

Die Formeln (1), (2) und die aus ihnen folgende Formel

$$(3) \qquad h^2 \sum_{G_h'}\sum (v\, L(u) - u\, M(v)) + h \sum_{\Gamma_h} (v\, \mathfrak{R}(u) - u\, \mathfrak{S}(v)) = 0$$

nennen wir die Greenschen Formeln.

Der einfachste und wichtigste Fall ergibt sich, wenn die Bilinearform symmetrisch ist, d. h. wenn die Gleichungen $b = c$, $\alpha = \gamma$, $\beta = \delta$ bestehen. In diesem Falle stimmt der Ausdruck $L(u)$ mit seinem adjungierten $M(u)$ überein; wir nennen ihn deshalb selbstadjungiert, und er ist schon aus dem quadratischen Ausdruck

$$B(u, u) = a u_x^2 + 2 b u_x u_y + d u_y^2 + 2 \alpha u_x u + 2 \beta u_y u + g u^2$$

ableitbar.

Wir beschränken uns im folgenden meist auf Ausdrücke $L(u)$, die sich selbst adjungiert sind. Der Charakter des Differenzenausdruckes $L(u)$ hängt vor allem von der Natur derjenigen Glieder aus der quadratischen Form $B(u, u)$ ab, die in den ersten Differenzenquotienten quadratisch sind. Wir nennen diesen Teil von $B(u, u)$ die charakteristische Form:

$$P(u, u) = a u_x^2 + 2 b u_x u_y + d u_y^2.$$

Je nachdem nun $P(u, u)$ in den Differenzenquotienten (positiv) definit oder indefinit ist, nennen wir den zugehörigen Differenzenausdruck $L(u)$ elliptisch oder hyperbolisch.

Der Differenzenausdruck

$$\Delta u = u_{x\bar{x}} + u_{y\bar{y}},$$

mit dem wir uns vorzugsweise in den folgenden Paragraphen beschäftigen werden, ist elliptisch. Er entsteht nämlich aus dem quadratischen Ausdruck

$$B(u, u) = u_x^2 + u_y^2 \quad \text{bzw.} \quad u_{\bar{x}}^2 + u_{\bar{y}}^2.$$

Die zugehörigen Greenschen Formeln lauten also:

$$h^2 \sum_{G_h}\sum (u_x^2 + u_y^2) = -h^2 \sum_{G_h'}\sum u\,\Delta u - h \sum_{\Gamma_h} u\,\Re(u),^{3)} \tag{4}$$

$$h^2 \sum_{G_h'}\sum (v\,\Delta u - u\,\Delta v) + h \sum_{\Gamma_h} (v\,\Re(u) - u\,\Re(v)) = 0. \tag{5}$$

Der Differenzenausdruck $\Delta u = u_{x\bar{x}} + u_{y\bar{y}}$ ist offenbar das Analogon des Differentialausdruckes $\frac{\partial^2 u}{\partial x^2} + \frac{\partial^2 u}{\partial y^2}$ für eine Funktion $u(x, y)$ der kontinuierlichen Variablen x und y. Ausführlich geschrieben lautet der Differenzenausdruck

$$\Delta u = \frac{1}{h^2}\{u(x+h, y) + u(x, y+h) + u(x-h, y) + u(x, y-h) - 4u(x, y)\}.$$

Es ist also $\frac{h^2}{4}\Delta u$ der Überschuß des arithmetischen Mittels der Funktionswerte in den vier Nachbarpunkten über den Funktionswert in dem betreffenden Punkt.

Ganz ähnliche Überlegungen führen zu linearen Differenzenausdrücken vierter Ordnung und entsprechenden Greenschen Formeln, wenn wir von bilinearen Differenzenausdrücken ausgehen, welche aus Differenzenquotienten zweiter Ordnung gebildet sind. Wir begnügen uns mit dem Beispiel des Differenzenausdruckes

$$\Delta\Delta u = u_{xx\bar{x}\bar{x}} + 2u_{x\bar{x}y\bar{y}} + u_{yy\bar{y}\bar{y}}.$$

Er entspringt aus dem quadratischen Ausdruck

$$B(u, u) = (u_{x\bar{x}} + u_{y\bar{y}})^2 = (\Delta u)^2,$$

wenn wir die Summe

$$h^2 \sum_{G_h'}\sum \Delta u\,\Delta v$$

---

[3] Der Randausdruck $\Re(u)$ läßt sich hier so beschreiben: Sind $u_0, u_1, \ldots, u_\nu$ die Funktionswerte in dem betreffenden Randpunkte und seinen ν Nachbarpunkten ($\nu \leq 3$), so ist

$$\Re(u) = \frac{1}{h}(u_1 + \ldots + u_\nu - \nu u_0).$$

nach v ordnen, etwa indem wir in der Formel (5) an Stelle von u den Ausdruck Δu setzen. Wir müssen dabei beachten, daß in dem Ausdruck $\Delta\Delta u$ der Funktionswert an einer Stelle mit den Funktionswerten in seinen Nachbarpunkten und deren Nachbarpunkten verknüpft ist und daher nur für solche Punkte des Gebietes G_h definiert ist, die innere Punkte auch von G_h' sind (vgl. 5) und deren Gesamtheit wir mit G_h'' bezeichnen wollen. Wir erhalten dann die Greensche Formel

$$(6) \qquad h^2 \sum_{G_h'}\sum \Delta u \cdot \Delta v = h^2 \sum_{G_h''}\sum v \cdot \Delta\Delta u + h \sum_{\Gamma_h + \Gamma_h'} v \cdot \mathfrak{R}(u),$$

wo $\mathfrak{R}(u)$ ein für jeden Punkt des Randstreifens $\Gamma_h + \Gamma_h'$ definierbarer linearer Differenzenausdruck ist, den wir nicht näher angeben. Γ_h' bedeutet dabei die Menge der Randpunkte von G_h'.

§ 2.

Randwertprobleme und Eigenwertprobleme.

1. Die Theorie des Randwertproblems.

Die Randwertaufgabe für lineare elliptische homogene Differenzengleichungen zweiter Ordnung, welche der klassischen Randwertaufgabe für partielle Differentialgleichungen entspricht, formulieren wir folgendermaßen:

In einem Gittergebiete G_h sei ein selbstadjungierter elliptischer linearer Differenzenausdruck zweiter Ordnung $L(u)$ gegeben. Er möge aus einem quadratischen Ausdruck $B(u, u)$ entspringen, der positiv-definit ist in dem Sinne, daß er nicht verschwinden kann, wenn nicht u_x und u_y selbst verschwinden.

Man bestimme nun in G_h eine solche der Differenzengleichung

$$L(u) = 0$$

genügende Funktion u, welche in den Randpunkten dieses Gittergebietes mit vorgegebenen Werten übereinstimmt.

Unsere Forderung wird dargestellt durch ebenso viele lineare Gleichungen wie es innere Gitterpunkte des Gittergebietes, also zu bestimmende Funktionswerte u gibt[4]). Einige dieser Gleichungen, nämlich soweit sie zu Gitterpunkten gehören, welche mit ihren vier Nachbarn im Innern liegen, sind homogen; andere, bei welchen Randpunkte des Gittergebietes mit eingehen, sind inhomogen. Setzen wir die rechten Seiten dieses in-

---

[4]) Bildet man zu einer beliebigen Differenzengleichung zweiter Ordnung $L(u) = 0$, indem man sie als ein lineares Gleichungssystem auffaßt, das transponierte Gleichungssystem, so wird dieses durch die adjungierte Differenzengleichung $M(v) = 0$ dargestellt. Die oben betrachtete selbstadjungierte Differenzengleichung stellt also ein lineares Gleichungssystem mit symmetrischem Koeffizientenschema dar.

homogenen Gleichungssystems, d. h. die Randwerte von u gleich Null, so folgt aus der Greenschen Formel (1), wenn wir dort $u = v$ setzen, sofort das Verschwinden von $B(u, u)$ und wegen des Definitheitscharakters von $B(u, u)$ das Verschwinden von u_x, u_y und damit auch von u. Die Differenzengleichung hat also die Lösung $u = 0$, wenn die Randwerte verschwinden, oder mit anderen Worten, die Lösung ist durch die Randwerte, wenn überhaupt, eindeutig bestimmt, da die Differenz zweier Lösungen mit denselben Randwerten verschwinden muß. Wenn aber ein lineares Gleichungssystem mit ebenso vielen Unbekannten wie Gleichungen die Eigenschaft besitzt, daß bei verschwindenden rechten Seiten auch die Unbekannten sämtlich verschwinden müssen, so besagt der Fundamentalsatz der Gleichungstheorie, daß bei beliebig vorgegebenen rechten Seiten genau eine Lösung vorhanden sein muß. In unserem Falle folgt somit die Existenz einer Lösung der Randwertaufgabe.

Wir sehen also, daß bei unseren elliptischen Differenzengleichungen die eindeutige Bestimmtheit und die Existenz der Lösung der Randwertaufgabe durch den Fundamentalsatz der Theorie der linearen Gleichungen miteinander zusammenhängen, während in der Theorie der partiellen Differentialgleichungen bekanntlich beide Tatsachen mit ganz verschiedenen Methoden bewiesen werden müssen. Der Grund für diese Schwierigkeit ist darin zu erblicken, daß Differentialgleichungen nicht mehr mit endlich vielen Gleichungen äquivalent sind; und man sich daher nicht mehr auf die Gleichheit der Anzahl von Unbekannten und Gleichungen berufen kann.

Da die Differenzengleichung

$$\Delta u = 0$$

aus dem positiv-definiten quadratischen Ausdruck

$$h^2 \sum_{G_h}\sum (u_x^2 + u_y^2)$$

entspringt, ist also das Randwertproblem dieser Differenzengleichung stets eindeutig lösbar.

Ganz entsprechend wie für die Differenzengleichungen zweiter Ordnung entwickelt sich die Theorie für Differenzengleichungen höherer, z. B. vierter Ordnung, wofür das Beispiel der Differenzengleichung

$$\Delta\Delta u = 0$$

genügen möge. Hier müssen die Werte der Funktion u in dem Randstreifen $\Gamma_h + \Gamma_h'$ vorgegeben werden. Offenbar liefert auch die Differenzengleichung $\Delta\Delta u = 0$ ebensoviel lineare Gleichungen wie unbekannte Funktionswerte in den Punkten von G_h''. Um die eindeutige Lösbarkeit der Randwertaufgabe nachzuweisen, brauchen wir wieder nur zu zeigen, daß eine Lösung, deren Werte im Randstreifen $\Gamma_h + \Gamma'_h$ Null sind, notwendig identisch verschwindet. Zu dem Zweck bemerken wir, daß die Summe über den zugehörigen quadratischen Ausdruck:

$$h^2 \sum_{G''_h}\sum (\Delta u)^2 \tag{7}$$

für eine solche Funktion verschwindet, wie wir sofort erkennen, wenn wir diese Summe nach der Greenschen Formel (6) umformen. Das Verschwinden der Summe (7) zieht aber das Verschwinden von Δu in allen Punkten von G'_h nach sich, und das kann bei verschwindenden Randwerten nach dem oben Bewiesenen nur stattfinden, wenn die Funktion u überall den Wert Null annimmt. Damit ist aber unsere Behauptung bewiesen und die eindeutige Lösbarkeit der Randwertaufgabe des Differenzenausdruckes sichergestellt [5]).

2. Beziehungen zu Minimumproblemen.

Die obige Randwertaufgabe steht in Zusammenhang mit dem folgenden Minimumproblem: Unter allen im Gittergebiet G_h definierten Funktionen $\varphi(x, y)$, welche in den Randpunkten vorgeschriebene Werte annehmen, ist eine solche $\varphi = u(x, y)$ zu suchen, für welche die über das Gittergebiet erstreckte Summe

$$h^2 \sum_{G_h}\sum B(\varphi, \varphi)$$

einen möglichst kleinen Wert annimmt. Dabei setzen wir voraus, daß der quadratische Differenzenausdruck erster Ordnung $B(u, u)$ in dem oben (vgl. S. 36) genannten Sinne positiv-definit ist. Daß sich aus dieser Minimumforderung als Bedingung für die Lösung $\varphi = u(x, y)$ die Differenzengleichung $L(\varphi) = 0$ ergibt, wo $L(\varphi)$ der in der obigen (vgl. S. 35 (1)) Weise aus $B(\varphi, \varphi)$ abgeleitete Differenzenausdruck zweiter Ordnung ist, erkennt man entweder nach den Regeln der Differentialrechnung, indem man die Summe $h^2 \sum_{G_h}\sum B(\varphi, \varphi)$ als Funktion der endlichen vielen Werte von φ in den Gitterpunkten ansieht oder analog dem üblichen Verfahren in der Variationsrechnung.

Beispielsweise ist das Randwertproblem, eine Lösung der Gleichung $\Delta\varphi = 0$ zu finden, die vorgegebene Randwerte annimmt, mit der Aufgabe

---

[5]) Vergleiche für die wirkliche Durchführung der Lösung unserer Randwertprobleme durch iterierende Verfahren u. a. die Abhandlung: Über Randwertaufgaben bei partiellen Differenzengleichungen von R. Courant, Zeitschr. f. angew. Mathematik u. Mechanik **6** (1925), S. 322—325. Im übrigen sei verwiesen auf den Bericht von H. Henky, Zeitschr. f. angew. Math. u. Mech. **2** (1922), S. 58ff.

gleichwertig, die Summe $h^2 \sum_{G_h}\sum (\varphi_x^2 + \varphi_y^2)$ unter allen Funktionen, die die Randwerte annehmen, zum Minimum zu machen.

Ganz Ähnliches gilt für Differenzengleichungen vierter Ordnung, wobei wir uns wiederum auf das Beispiel von $\varDelta\varDelta\varphi = 0$ beschränken. Die zu dieser Differenzengleichung gehörige Randwertaufgabe ist mit dem Problem gleichwertig, die Summe $h^2 \sum_{G_h'}\sum (\varDelta\varphi)^2$ unter allen Funktionen $\varphi(x, y)$ zum Minimum zu machen, deren Werte in dem Randstreifen Γ_h' vorgegeben sind. Außer dieser Summe führen auch noch andere in den zweiten Ableitungen quadratische Ausdrücke durch die Forderung, sie zum Minimum zu machen, auf die Gleichung $\varDelta\varDelta u = 0$, so z. B. die Summe:

$$h^2 \sum_{G_h'}\sum (u_{xx}^2 + 2u_{xy}^2 + u_{yy}^2),$$

in der sämtliche in G_h auftretenden zweiten Differenzenquotienten vorkommen sollen.

Daß die gestellten Minimumprobleme immer eine Lösung besitzen, folgt aus dem Satz, daß eine stetige Funktion von endlichen vielen Veränderlichen (den Funktionswerten von φ in den Gitterpunkten) stets ein Minimum besitzen muß, wenn diese Funktion nach unten beschränkt ist und wenn sie gegen Unendlich strebt, sobald mindestens eine der unabhängigen Veränderlichen es tut[6]).

3. Die Greensche Funktion.

Ähnlich wie die Randwertaufgabe der homogenen Gleichung $L(u) = 0$ kann man auch die Randwertaufgabe der unhomogenen Gleichung $L(u) = -f$ behandeln. Es genügt, bei der unhomogenen Gleichung sich auf den Fall zu beschränken, daß die Randwerte von u überall verschwinden, da wir für andere Randwerte die Lösung durch Addition einer geeigneten Lösung der homogenen Gleichung erhalten. Um das lineare Gleichungssystem, welches durch die Randwertaufgabe von $L(u) = -f$ repräsentiert ist, zu lösen, wählen wir zunächst die Funktion $f(x, y)$ so, daß sie in einem Gitterpunkte mit den Koordinaten $x = \xi$, $y = \eta$ den Wert $-\frac{1}{h^2}$, in allen andern Gitterpunkten den Wert Null annimmt. Ist $K(x, y; \xi, \eta)$ die am Rande verschwindende Lösung der so entstehenden speziellen noch vom Parameterpunkt (ξ, η) abhängigen Differenzengleichung, so wird die zu

[6]) Daß die Voraussetzungen für die Anwendungen dieses Satzes gegeben sind, ist sehr leicht einzusehen.

einer beliebigen Funktion gehörige Lösung durch die Summe

$$u(x, y) = h^2 \sum_{(\xi, \eta) \text{ in } G_h} \sum K(x, y; \xi, \eta) f(\xi, \eta)$$

dargestellt.

Die Funktion $K(x, y; \xi, \eta)$ in ihrer Abhängigkeit von den Punkten (x, y) und (ξ, η) nennen wir die Greensche Funktion des Differentialausdruckes $L(u)$. Bezeichnen wir mit $\overline{K}(x, y; \xi, \eta)$ die Greensche Funktion des adjungierten Ausdruckes $M(v)$, so gilt die Relation

$$K(\bar{\xi}, \bar{\eta}; \xi, \eta) = \overline{K}(\xi, \eta; \bar{\xi}, \bar{\eta}),$$

die man auch unmittelbar aus der Greenschen Formel (5) folgert, wenn man dort $u = K(x, y; \xi, \eta)$ und $v = \overline{K}(x, y; \bar{\xi}, \bar{\eta})$ setzt. Für einen selbstadjungierten Differenzenausdruck ergibt sich aus der obigen Beziehung die Symmetrierelation:

$$K(\bar{\xi}, \bar{\eta}; \xi, \eta) = K(\xi, \eta; \bar{\xi}, \bar{\eta}).$$

4. Eigenwertprobleme.

Selbstadjungierte Differenzenausdrücke $L(u)$ geben Anlaß zu Eigenwertproblemen von folgendem Typ: Es sind die Werte eines Parameters λ — die Eigenwerte — zu suchen, für die die Differenzengleichung

$$L(u) + \lambda u = 0$$

in G_h eine auf dem Rande Γ_h verschwindende Lösung — die Eigenfunktion — besitzt.

Das Eigenwertproblem ist äquivalent dem Hauptachsenproblem der quadratischen Form $B(u, u)$. Es gibt ebenso viele Eigenwerte $\lambda^{(1)}, \ldots, \lambda^{(N)}$ wie innere Gitterpunkte im Gebiet G_h und ebenso viele zugehörige Eigenfunktionen $u^{(1)}, \ldots, u^{(N)}$. Das System der Eigenfunktionen und Eigenwerte und ihre Existenz ergibt sich aus dem Minimumproblem:

Unter allen am Rande verschwindenden Funktionen $\varphi(x, y)$, die den $m - 1$ Orthogonalitätsbedingungen

$$h^2 \sum_{G_h} \sum \varphi u^{(\nu)} = 0 \qquad (\nu = 1, \ldots, m-1)$$

und der Normierungsbedingung

$$h^2 \sum_{G_h} \sum \varphi^2 = 1$$

genügen, ist diejenige $\varphi = u$ gesucht, für die die Summe

$$h^2 \sum_{G_h} \sum B(\varphi, \varphi)$$

den kleinsten Wert annimmt. Der Wert dieses Minimums ist der m-te Eigenwert und die Funktion, für die es angenommen wird, ist die m-te Eigenfunktion [7]).

§ 3.[8])

Zusammenhänge mit dem Problem der Irrwege.

Unser Thema steht in Beziehung zu einer Frage der Wahrscheinlichkeitsrechnung, nämlich dem Problem der Irrwege in einem begrenzten Gebiet [9]). Man stelle sich in einem Gittergebiet G_h die Gitterstrecken als Wege vor, längs deren ein Partikel von einem Gitterpunkt zu einem Nachbarpunkt wandern kann. In diesem Straßennetz möge nun unser Partikel ziellos herumirren, indem es an jeder Straßenecke unter den vier verfügbaren Richtungen eine nach dem Zufall auswählt — alle vier seien gleich wahrscheinlich —. Die Irrfahrt endet, sobald ein Randpunkt von G_h erreicht ist, wo unsere Partikel absorbiert werden mögen.

Wir fragen:

1. Welches ist die Wahrscheinlichkeit $w(P; R)$ daß man bei der Irrfahrt von einem Punkte P ausgehend irgend einmal in dem Randpunkte R ankommt?

2. Welches ist die mathematische Hoffnung $v(P; Q)$, daß man bei einer solchen von P ausgehenden Irrfahrt, ohne den Rand zu treffen, einen Punkt Q von G_h berührt?

---

[7]) Wegen der Orthogonalität $h^2 \sum\sum_{G_h} u^{(\nu)} u^{(\mu)} = 0$ $(\nu \neq \mu)$ der Eigenfunktionen läßt sich jede am Rande verschwindende Funktion $g(x, y)$ des Gitters nach den Eigenfunktionen in der Form

$$y = \sum_{\nu=1}^{N} c^{(\nu)} u^{(\nu)}$$

entwickeln, wo die Koeffizienten $c^{(\nu)}$ durch die Gleichung

$$c^{(\nu)} = \sum\sum_{G_h} g\, u^{(\nu)}$$

bestimmt sind.

Auf diese Weise erhalten wir insbesondere die folgende Darstellung der Greenschen Funktion:

$$K(x, y; \xi, \eta) = -\frac{1}{h^2} \sum_{\nu=1}^{N} \frac{u^{(\nu)}(x, y) \cdot u^{(\nu)}(\xi, \eta)}{\lambda^{(\nu)}}.$$

[8]) Für die Durchführung des Grenzüberganges in § 4 ist § 3 entbehrlich.

[9]) Gerade in der Art wie hier die Grenzen des Gebietes hineinspielen, liegt ein wesentlicher Unterschied der folgenden Betrachtung gegenüber bekannten Überlegungen, die z. B. im Zusammenhange mit der Brownschen Molekularbewegung durchgeführt worden sind.

Diese Wahrscheinlichkeit bzw. mathematische Hoffnung wollen wir durch folgenden Prozeß genauer erklären. Wir denken uns im Punkte P die Einheit irgendeiner Substanzmenge vorhanden. Die Substanz möge sich in unserem Straßennetz mit einer konstanten Geschwindigkeit ausbreiten, etwa in der Zeiteinheit eine Gitterstrecke zurücklegen. In jedem Gitterpunkte soll nach jeder der vier Richtungen genau ein Viertel der dort ankommenden Substanz weiterströmen. Die Substanzmenge, die in einem Randpunkte ankommt, soll dort festgehalten werden. Ist der Ausgangspunkt P ein Randpunkt, so soll die Substanzmenge überhaupt dort bleiben.

Unter der Wahrscheinlichkeit $w(P;R)$ überhaupt bei einer von P ausgehenden Irrfahrt an den Randpunkt R zu gelangen, ohne vorher den Rand berührt zu haben, verstehen wir die Substanzmenge, die sich nach unendlicher Zeit in diesem Randpunkte angesammelt hat.

Unter der Wahrscheinlichkeit $E_n(P;Q)$ in genau n Schritten vom Punkte P zum Punkte Q zu gelangen, ohne den Rand zu berühren, verstehen wir die im Punkte Q nach n Zeiteinheiten befindliche Substanzmenge, falls P und Q innere Punkte sind. Ist P oder Q ein Randpunkt, so setzen wir sie gleich Null.

Die Größe $E_n(P;Q)$ ist gerade die Anzahl der von P nach Q führenden den Rand nicht treffenden Wege von n Schritten, durch 4^n dividiert; es ist also $E_n(P;Q) = E_n(Q;P)$.

Unter der mathematischen Hoffnung $v(P;Q)$ bei einem oben gekennzeichneten Irrwege überhaupt einmal von P aus zum Punkte Q zu gelangen, verstehen wir die unendliche Summe aller dieser Wahrscheinlichkeiten

$$v(P;Q) = \sum_{\nu=0}^{\infty} E_\nu(P;Q),^{10)}$$

also für innere Punkte P und Q die Summe aller Substanzmengen, die in den verschiedenen Zeitmomenten den Punkt Q durchlaufen haben. Es wird also dem Erreichen des Punktes Q der Erwartungswert 1 zugeschrieben. Für Randpunkte ist diese Hoffnung gleich Null.

Bezeichnen wir die im Randpunkte R mit genau n Schritten ankommende Menge mit $F_n(P;R)$, so ist die Wahrscheinlichkeit $w(P;R)$ durch die unendliche Reihe

$$w(P;R) = \sum_{\nu=0}^{\infty} F_\nu(P;R)$$

dargestellt, deren sämtliche Glieder positiv sind, und deren Teilsummen nie größer als Eins sein können, weil die am Rande ankommende Substanz

[10]) Ihre Konvergenz werden wir sogleich beweisen.

nur einen Teil der ursprünglichen Substanzmenge ausmacht. Damit ist aber die Konvergenz dieser Reihe gesichert.

Man kann nun leicht einsehen, daß die Wahrscheinlichkeiten $E_n(P;Q)$, d. h. die nach genau n Schritten in einem Punkte Q anlangende Substanzmenge mit wachsendem n gegen Null strebt. Ist nämlich in irgendeinem Punkte Q, von dem aus ein Randpunkt R in m Schritten zu erreichen sei, $E_n(P;Q) > \alpha > 0$, so wird nach m Schritten in diesem Randpunkt R mindestens die Substanzmenge $\frac{\alpha}{4^m}$ ankommen; da aber wegen der Konvergenz der Summe $\sum_{\nu=0}^{\infty} F_\nu(P;R)$ die an den Randpunkt R ankommende Substanzmenge mit der Zeit gegen Null strebt, so müssen auch die Größen $E_n(P;Q)$ selber mit wachsendem n gegen Null streben; d. h. die Wahrscheinlichkeit bei einem unendlich langen Wege im Innern zu bleiben, ist Null.

Hieraus ergibt sich, daß die gesamte Substanzmenge schließlich an den Rand ankommen muß; mit anderen Worten, daß die über alle Randpunkte R erstreckte Summe

$$\sum_R w(P;R) = 1$$

ist.

Wir haben noch die Konvergenz der unendlichen Reihe für die mathematische Hoffnung $v(P;Q)$

$$v(P;Q) = \sum_{\nu=0}^{\infty} E_\nu(P;Q)$$

zu beweisen.

Zu dem Zweck bemerken wir, daß die Größen $E_n(P;Q)$ der folgenden Relation genügen

$$E_{n+1}(P;Q) = \frac{1}{4}\{E_n(P;Q_1) + E_n(P;Q_2) + E_n(P;Q_3) + E_n(P;Q_4)\} \qquad [n \geqq 1],$$

wo Q_1 bis Q_4 die vier Nachbarpunkte des inneren Punktes Q sind. D. h. die nach $n+1$ Schritten im Punkte Q ankommende Substanzmenge besteht aus dem vierten Teil der nach n Schritten in den vier Nachbarpunkten von Q ankommenden Substanzmenge. Ist einer der Nachbarpunkte von Q z. B. $Q_1 = R$ Randpunkt, so kommt die Tatsache, daß zum Punkte Q von diesem Randpunkte aus keine Substanzmenge weiter fließt, dadurch zum Ausdruck, daß wir $E_n(P;R)$ gleich Null gesetzt haben. Ferner ist für einen inneren Punkt $E_0(P;P) = 1$ und sonst $E_0(P;Q) = 0$.

Aus diesen Relationen ergeben sich für die Teilsummen

$$v_n(P;Q) = \sum_{\nu=0}^{n} E_\nu(P;Q)$$

die Gleichungen

$$v_{n+1}(P;Q) = \frac{1}{4}\{v_n(P;Q_1) + v_n(P;Q_2) + v_n(P;Q_3) + v_n(P;Q_4)\},$$

wenn P nicht mit Q zusammenfällt; andernfalls ist

$$v_{n+1}(P;P) = 1 + \frac{1}{4}\{v_n(P;P_1) + v_n(P;P_2) + v_n(P;P_3) + v_n(P;P_4)\},$$

d. h. die Hoffnung, von einem Punkte zu sich selbst zurückzukommen, setzt sich zusammen aus der Hoffnung, auf einem nicht verschwindenden Wege den Punkt P wieder zu erreichen, nämlich $\frac{1}{4}\{v_n(P;P_1) + v_n(P;P_2) + v_n(P;P_3) + v_n(P;P_4)\}$ und aus der Hoffnung Eins, die ausdrückt, daß ursprünglich die gesamte Substanz in diesem Punkte vorhanden war.

Es genügen also die Größen $v_n(P;Q)$ der folgenden Differenzengleichung[11])

$$\Delta v_n(P;Q) = \frac{4}{h^2} E_n(P;Q), \qquad \text{wenn } P \neq Q \text{ ist},$$

$$\Delta v_n(P;Q) = \frac{4}{h^2}(E_n(P;Q) - 1), \qquad \text{wenn } P = Q \text{ ist}.$$

$v_n(P;Q)$ ist gleich Null, wenn Q ein Randpunkt ist.

Die Lösung dieser Randwertaufgabe ist, wie schon früher auseinandergesetzt, für irgendwelche rechten Seiten eindeutig bestimmt (vgl. S. 38); sie hängt stetig von den rechten Seiten ab. Da nun die Größen $E_n(P;Q)$ gegen Null streben, so konvergieren die Lösungen $v_n(P;Q)$ gegen die Lösungen $v(P;Q)$ der Differenzengleichung

$$\Delta v(P;Q) = 0, \qquad \text{wenn } P \neq Q \text{ ist},$$

$$\Delta v(P;Q) = -\frac{4}{h^2}, \qquad \text{wenn } P = Q \text{ ist},$$

mit den Randwerten $v(P;R) = 0$.

---

[11]) Dabei bezieht sich die Δ-Operation auf den variablen Punkt Q.

Diese Gleichung läßt sich als eine Gleichung vom Wärmeleitungstypus auffassen. Betrachtet man nämlich die Funktion $v_n(P;Q)$ anstatt als Funktion des Index n unserer oben zugrunde gelegten Vorstellung gemäß als Funktion der Zeit t, die zu n proportional ist, indem man $t = n\tau$ und $v_n(P;Q) = v(P;Q;t) = v(t)$ setzt, so können wir die obigen Gleichungen in der folgenden Form schreiben;

$$\Delta v(t) = \frac{4\tau}{h^2} \cdot \frac{v(t+\tau) - v(t)}{\tau} \quad \text{für } P \neq Q,$$

$$\Delta v(t) = \frac{4\tau}{h^2}\left(\frac{v(t+\tau) - v(t)}{\tau} - 1\right) \quad \text{für } P = Q.$$

Über den Grenzübergang von einer ähnlichen Differenzengleichung zu einer parabolischen Differentialgleichung vgl. Teil II, § 6, S. 67.

Wir sehen also, daß die mathematische Hoffnung $v(P;Q)$ existiert und nichts anderes ist als die zur Differenzengleichung $\Delta u = 0$ zugehörige Greensche Funktion $K(P;Q)$ noch mit dem Faktor 4 versehen. Die Symmetrie der Greenschen Funktion $K(P;Q) = K(Q;P)$ ist eine unmittelbare Folge der Symmetrie der Größen $E_n(P;Q)$, mit deren Hilfe sie definiert wurde.

Die Wahrscheinlichkeit $w(P;R)$ genügt hinsichtlich P der Relation

$$w(P;R) = \frac{1}{4}\{w(P_1;R) + w(P_2;R) + w(P_3;R) + w(P_4;R)\},$$

also der Differenzengleichung

$$\Delta w = 0.$$

Sind nämlich P_1, P_2, P_3, P_4 die vier Nachbarpunkte des inneren Punktes P, so muß jeder Weg von P nach R über einen dieser vier Wege führen, und jede der vier Wegrichtungen ist gleich wahrscheinlich. Ferner ist die Wahrscheinlichkeit, von einem Randpunkt R zu einem andern R' zu gelangen, $w(R,R') = 0$, außer wenn die beiden Punkte R und R' zusammenfallen, wo $w(R,R) = 1$ gilt. Es ist also $w(P;R)$ die Lösung der Randwertaufgabe $\Delta w = 0$, wobei im Randpunkte R der Randwert 1 in allen anderen Punkten der Randwert 0 vorgeschrieben ist. Die Lösung der Randwertaufgabe bei beliebig vorgegebenen Randwerten $u(R)$ hat dann einfach die Gestalt $u(P) = \sum_R w(P;R)\,u(R)$, wobei über alle Randpunkte R zu summieren ist[12]). Setzen wir hierin für u die Funktion $u \equiv 1$ ein, so erhalten wir wieder die Relation $1 = \sum_R w(P;R)$.

Die hier gegebene Auffassung der Greenschen Funktion als Hoffnung läßt unmittelbar weitere Eigenschaften erkennen. Wir erwähnen nur die Tatsache, daß die Greensche Funktion wächst, wenn man von dem Gebiete G zu einem in G als Teilgebiet enthaltenen Teilgebiete $\overline{G}$ übergeht; es wächst dann nämlich für jedes n die Anzahl der möglichen Gitterwege, von einem Punkte P zu einem anderen Q zu gelangen, ohne den Rand zu berühren.

Natürlich herrschen für mehr als zwei unabhängige Veränderliche entsprechende Beziehungen. Wir begnügen uns mit dem Hinweis, daß auch andere elliptische Differenzengleichungen eine ähnliche Wahrscheinlichkeitsauffassung zulassen.

Führt man den Grenzübergang zu verschwindender Maschenweite durch, was sich mit den Methoden des folgenden Paragraphen einfach ausführen

---

[12]) Man erkennt übrigens leicht, daß die Wahrscheinlichkeit $w(P;R)$, an den Rand zu gelangen, der von der Greenschen Funktion $K(P;Q)$ hinsichtlich Q gebildete Randausdruck $\mathfrak{R}(K(P,Q))$ ist, indem man in der Greenschen Formel (5) $u(x,y)$ mit $w(P,Q)$, $v(x,y)$ mit $v(P,Q)$ identifiziert.

läßt, so geht die Greensche Funktion im Gitter bis auf einen Zahlenfaktor in die Greensche Funktion der Potentialgleichung über; eine ähnliche Beziehung besteht zwischen dem Ausdruck $\frac{w(P;R)}{h}$ und der normalen Ableitung der Greenschen Funktion am Rande des Gebietes. Auf diese Weise ließe sich z. B. die Greensche Funktion der Potentialgleichung als die spezifische mathematische Hoffnung deuten, von einem Punkte zu einem anderen zu gelangen[13]), ohne den Rand zu berühren.

Nach dem Grenzübergang vom Gitter zum Kontinuum ist der Einfluß der bei den Irrwegen vorgeschriebenen Gitterrichtungen verschwunden. Dieser Tatsache Rechnung zu tragen, indem man den Grenzübergang mit einem allgemeineren Irrfahrtenproblem ohne Richtungsbeschränkung vornimmt, ist eine prinzipiell interessante Aufgabe, welche jedoch über den Rahmen dieser Abhandlung hinaus führt und auf die wir bei anderer Gelegenheit zurückzukommen hoffen.

§ 4.

Grenzübergang zur Lösung der Differentialgleichung.

1. Die Randwertaufgabe der Potentialtheorie.

Bei der Durchführung des Grenzüberganges von der Lösung der Differenzengleichungsprobleme zu der Lösung der entsprechenden Differentialgleichungen wollen wir hinsichtlich des Randes und der Randwerte auf die größtmögliche Allgemeinheit in der Formulierung verzichten, um das für unsere Methoden Charakteristische klarer hervortreten zu lassen[14]). Wir setzen demgemäß voraus, daß in der Ebene ein einfach zusammenhängendes Gebiet G vorgegeben ist, dessen Berandung aus endlich vielen mit stetiger Tangente versehenen Kurvenbögen gebildet wird. In einem G im Innern enthaltenden Gebiete sei eine stetige und mit stetigen partiellen Ableitungen erster und zweiter Ordnung versehene Funktion $f(x, y)$ gegeben. Für das zu der Maschenweite h und zum Gebiete G gehörige Gittergebiet G sei die Randwertaufgabe der Differenzengleichung $\Delta u = 0$ mit denjenigen Randwerten, welche von der Funktion $f(x, y)$ in den Randpunkten von G_h angenommen werden, gelöst; die Lösung heiße $u_h(x, y)$. Wir wollen beweisen, daß die Gitterfunktion u_h mit verschwindender Maschenweite h gegen die Lösung u der Randwertaufgabe der partiellen

[13]) Dabei ist dem Erreichen eines Flächenstücks als Erwartungswert sein Flächeninhalt zugeschrieben.

[14]) Es sei jedoch bemerkt, daß die Ausdehnung unserer Methoden auf allgemeinere Ränder und Randwerte keinerlei prinzipielle Schwierigkeiten bereitet.

Differentialgleichung $\frac{\partial^2 u}{\partial x^2} + \frac{\partial^2 u}{\partial y^2} = 0$ für das Gebiet G konvergiert, wobei die Randwerte für das Gebiet G wiederum durch diejenigen Werte geliefert werden, welche die Funktion $f(x, y)$ auf dem Rande von G annimmt. Weiter werden wir zeigen, daß für jedes ganz im Innern von G liegende Gebiet die Differenzenquotienten beliebiger Ordnung von u_h gleichmäßig gegen die entsprechenden partiellen Differentialquotienten der Grenzfunktion $u(x, y)$ streben.

Bei der Durchführung des Konvergenzbeweises ist es bequem, die Forderung, daß $u(x, y)$ die Randwerte annimmt, durch die folgende schwächere Forderung zu ersetzen: Ist S_r derjenige Randstreifen des Gebietes G, dessen Punkte vom Rande eine Entfernung kleiner als r besitzen, so strebt das Integral

$$\frac{1}{r} \iint_{S_r} (u - f)^2 \, dx \, dy$$

mit abnehmendem r gegen Null[15]).

Unser Konvergenzbeweis beruht auf der Tatsache, daß für jedes ganz im Innern des Gebietes G liegende Teilgebiet G^* die Funktion $u_h(x, y)$ und jeder Differenzenquotient bei abnehmendem h beschränkt bleibt und „gleichartig stetig" ist in folgendem Sinne: Es gibt für jede dieser Funktionen $w_h(x, y)$ eine nur von dem Gebiete und nicht von h abhängige Größe $\delta(\varepsilon)$ derart, daß

$$|w_h(P) - w_h(P_1)| < \varepsilon$$

ist, sobald die beiden Gitterpunkte P und P_1 des Gittergebietes G_h in dem gegebenen Teilgebiet liegen und voneinander einen kleineren Abstand als $\delta(\varepsilon)$ besitzen.

---

[15]) Daß tatsächlich unsere schwächere Randwertforderung zur *eindeutigen* Kennzeichnung der Lösung genügt, folgt aus dem leicht zu beweisenden Satze: Wenn für eine im Innern von G der Differentialgleichung $\frac{\partial^2 u}{\partial x^2} + \frac{\partial^2 u}{\partial y^2} = 0$ genügende Funktion die obige Form der Randbedingung mit $f(x, y) = 0$ erfüllt ist und $\iint_G \left(\left(\frac{\partial u}{\partial x} \right)^2 + \left(\frac{\partial u}{\partial y} \right)^2 \right) dx \, dy$ existiert, so ist $u(x, y)$ identisch Null. (Vgl. Courant, „Über die Lösungen der Diff.-Gl. der Physik", Math. Annalen **85**, insbesondere S. 296 ff.)

Im Falle von zwei unabhängigen Veränderlichen läßt sich aus unserer schwächeren Forderung die tatsächliche Annahme der Randwerte folgern; im Falle von mehr Variablen darf man das Entsprechende schon deswegen nicht allgemein erwarten, weil es dort bekanntlich Ausnahmepunkte am Rande geben kann, in denen die Randwerte nicht mehr angenommen zu werden brauchen, während jedoch für die schwächere Forderung stets eine Lösung existiert.

Haben wir einmal die behauptete gleichartige Stetigkeit bewiesen, so können wir bekanntlich eine Teilfolge unserer Funktionen u_h so auswählen, daß sie mit ihren Differenzenquotienten jeder Ordnung in jedem Teilgebiet G^* gleichmäßig gegen eine Grenzfunktion $u(x, y)$ bzw. deren Differentialquotienten strebt. Die Grenzfunktion besitzt dementsprechend Ableitungen beliebig hoher Ordnung in jedem inneren Teilgebiet G^* von G und genügt dort der partiellen Differentialgleichung $\frac{\partial^2 u}{\partial x^2} + \frac{\partial^2 u}{\partial y^2} = 0$. Wenn wir dann noch zeigen, daß sie die Randbedingung befriedigt, so erkennen wir in ihr die Lösung unseres Randwertproblems für das Gebiet G. Da diese Lösung eindeutig bestimmt ist, so zeigt sich nachträglich, daß nicht nur eine Teilfolge der Funktionen u_h, sondern diese Funktionenfolge selbst die ausgesprochene Konvergenzeigenschaft besitzt.

Die gleichartige Stetigkeit unserer Größen wird sich durch den Nachweis folgender Tatsachen ergeben:

1. Bei abnehmendem h bleiben die über das Gittergebiet G_h erstreckten Summen

$$h^2 \sum_{G_h}\sum u^2 \quad \text{und} \quad h^2 \sum_{G_h}\sum (u_x^2 + u_y^2)$$

beschränkt[16]).

2. Genügt $w = w_h$ in einem Gitterpunkt G_h der Differenzengleichung $\Delta w = 0$ und bleibt bei abnehmendem h die Summe

$$h^2 \sum_{G_h^*}\sum w^2,$$

erstreckt über ein zu einem Teilgebiet G^* von G gehöriges Gittergebiet G_h^*, beschränkt, so bleibt für jedes feste ganz im Innern von G^* liegende Teilgebiet G^{**} auch die über das zugehörige Gittergebiet G_h^{**} erstreckte Summe

$$h^2 \sum_{G_h^{**}}\sum (w_x^2 + w_y^2)$$

bei abnehmendem h beschränkt.

Zusammen mit 1. folgt hieraus, da sämtliche Differenzenquotienten w der Funktion u_h wieder der Differenzengleichung $\Delta w = 0$ genügen, daß jede der Summen

$$h^2 \sum_{G_h^*}\sum w^2$$

beschränkt ist.

3. Aus der Beschränktheit dieser Summen folgt schließlich die Beschränktheit und gleichartige Stetigkeit aller Differenzenquotienten selbst.

---

[16]) Hier und gelegentlich im folgenden lassen wir bei Gitterfunktionen den Index h fort.

2. Beweis der Hilfssätze.

Der Beweis der Tatsache 1 folgt daraus, daß die Funktionswerte u_h selbst beschränkt sind. Denn der größte und der kleinste Wert der Funktion wird am Rande angenommen[17]), strebt also gegen vorgegebene endliche Werte. Die Beschränktheit der Summe $h^2 \sum\sum_{G_h} (u_x^2 + u_y^2)$ ist eine unmittelbare Folge der im § 2, 2. formulierten Minimumeigenschaft unserer Gitterfunktion, wonach sicherlich

$$h^2 \sum\sum_{G_h} (u_x^2 + u_y^2) \leqq h^2 \sum\sum_{G_h} (f_x^2 + f_y^2)$$

gilt. Die Summe rechts strebt aber mit abnehmender Maschenweite gegen das Integral $\iint_G \left[\left(\frac{\partial f}{\partial x}\right)^2 + \left(\frac{\partial f}{\partial y}\right)^2\right] dx\,dy$, welches nach unseren Voraussetzungen existiert.

Um den unter 2. formulierten Hilfssatz zu beweisen, betrachten wir die Quadratsumme

$$h^2 \sum\sum_{Q_1} (w_x^2 + w_{\bar{x}}^2 + w_y^2 + w_{\bar{y}}^2),$$

wobei die Summation sich auf alle inneren Punkte eines Quadrates Q_1 bezieht (vgl. Fig. 1). Die Funktionswerte auf den äußeren Seiten S_1 des Quadrates Q_1 bezeichnen wir mit w_1, die auf der zweiten Randreihe S_0 mit w_0. Dann liefert die Greensche Formel

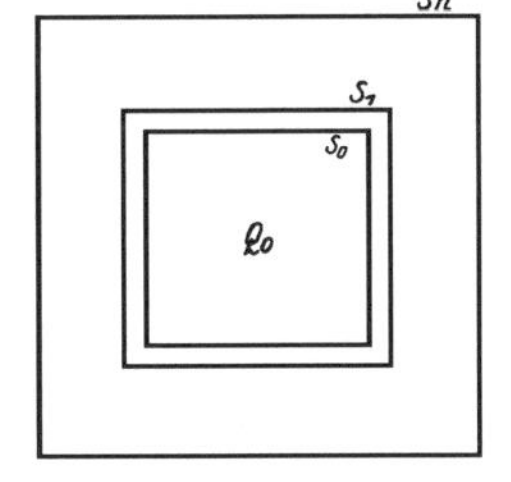

Fig. 1.

$$(8) \qquad h^2 \sum\sum_{Q_1} (w_x^2 + w_{\bar{x}}^2 + w_y^2 + w_{\bar{y}}^2)$$
$$= \sum_S (w_1^2 - w_0^2) \leqq \sum_{S_1} w^2 - \sum_{S_0} w^2,$$

wobei die Summation rechts über die beiden äußeren Randreihen S_1 und S_0 zu erstrecken ist, und wo w_1 und w_0 sich auf benachbarte Punkte beziehen. Wir betrachten nun eine Reihe von konzentrischen Quadraten $Q_0, Q_1, Q_2, \ldots, Q_N$ mit den Rändern $S_0, S_1, \ldots, S_N$, von denen jedes aus dem vorangehenden dadurch entsteht, daß der Kranz der nächsten Nachbarpunkte hinzukommen wird (vgl. Fig. 1). Auf jedes dieser Quadrate wenden wir die Abschätzung (8) an und beachten, daß stets

$$2\,h^2 \sum\sum_{Q_0} (w_x^2 + w_y^2) \leqq h^2 \sum\sum_{Q_k} (w_x^2 + w_{\bar{x}}^2 + w_y^2 + w_{\bar{y}}^2)$$

[17]) Ausdrücklich bemerken wir im Hinblick auf die Übertragung der Methode auf andere Differentialgleichungen, daß wir uns von dieser Eigenschaft unabhängig machen können. Dazu brauchen wir nur die Ungleichung (15) heranzuziehen oder die Schlußweise der Alternative anzuwenden (vgl. S. 55).

für $k \geqq 1$ ist. Addieren wir der Reihe nach die n Ungleichungen

$$2h^2 \sum_{Q_0}\sum (w_x^2 + w_y^2) \leqq \sum_{S_{k+1}} w^2 - \sum_{S_k} w^2 \qquad (0 \leqq k < n),$$

so erhalten wir

$$2nh^2 \sum_{Q_0}\sum (w_x^2 + w_y^2) \leqq \sum_{S_n} w^2 - \sum_{S_0} w^2 \leqq \sum_{S_n} w^2.$$

Diese Ungleichung summieren wir von $n = 1$ bis $n = N$. So ergibt sich

$$N^2 h^2 \sum_{Q_0}\sum (w_x^2 + w_y^2) \leqq \sum\sum w^2,$$

wobei wir die Summe rechts nur vergrößern, wenn wir sie über das ganze Quadrat Q_N erstrecken.

Lassen wir nun bei Verkleinerung der Maschenweite die Quadrate Q_0 und Q_N gegen zwei feste im Innern von G liegende konzentrische Quadrate mit dem Abstande a streben, so konvergiert Nh gegen a, und wir finden, daß unabhängig von der Maschenweite

$$h^2 \sum_{Q_0}\sum (w_x^2 + w_y^2) \leqq \frac{1}{a^2} h^2 \sum_{Q_N}\sum w^2 \tag{9}$$

bleibt.

Diese Ungleichung gilt — bei hinreichend kleiner Maschenweite — natürlich nicht nur für zwei Quadrate Q_0 und Q_N, sondern mit einer anderen Konstanten a für irgend zwei Teilgebiete von G, von denen das eine ganz im Innern des anderen liegt. Damit ist die Behauptung von 2. bewiesen[18]).

Um nun drittens nachzuweisen, daß in jedem inneren Teilgebiet die Funktion u_h und ihre sämtlichen Differenzenquotienten w_h die Verfeinerung der Maschenweite beschränkt und gleichartig stetig bleiben, betrachten wir ein Rechteck R mit den Eckpunkten P_0, Q_0, P, Q (vgl. Fig. 2), dessen Seiten P_0Q_0 und PQ der x-Achse parallel sind und die Länge a haben.

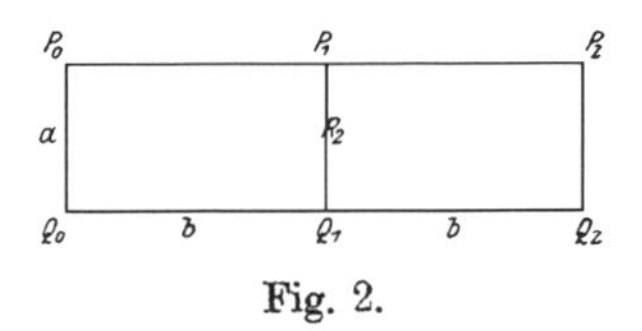

Fig. 2.

Wir gehen aus von der Darstellung

$$w(Q_0) - w(P_0) = h \sum_{PQ} w_x + h^2 \sum_R\sum w_{xy}$$

[18]) Wenn wir nicht annehmen, daß $\Delta w = 0$ ist, so erhalten wir an Stelle der Ungleichung (9)

$$h^2 \sum_{G^{**}}\sum (w_x^2 + w_y^2) \leqq c_1 h^2 \sum_{G^*}\sum w^2 + c_2 h^2 \sum_{G^*}\sum (\Delta w)^2 \tag{10}$$

bei geeigneten von h unabhängigen Konstanten c_1, c_2, wobei G^{**} ganz im Innern des Gebietes G^* liegt, das seinerseits im Innern von G enthalten ist.

und der aus ihr folgenden Ungleichung

$$(11) \qquad |w(Q_0) - w(P_0)| \leqq h \sum_{PQ} |w_x| + h^2 \sum_R \sum |w_{xy}|.$$

Wir lassen nun die Rechtecksseite PQ zwischen einer Anfangslage $P_1 Q_1$ im Abstande b von $P_0 Q_0$ und einer Endlage $P_2 Q_2$ im Abstande $2b$ von $P_0 Q_0$ laufen und summieren die $\frac{b}{h} + 1$ zugehörigen Ungleichungen (11). Wir erhalten so die Abschätzung

$$|w(P_0) - w(Q_0)| \leqq \frac{1}{b+h} h^2 \sum_{R_2} \sum |w_x| + h^2 \sum_{R_2} \sum |w_{xy}|,$$

indem wir die Summationsgebiete auf das ganze Rechteck $R_2 = P_0 Q_0 P_2 Q_2$ ausdehnen. Nach der Schwarzschen Ungleichung folgt daraus:

$$(12) \quad |w(P_0) - w(Q_0)| \leqq \frac{1}{b} \sqrt{2ab} \sqrt{h^2 \sum_{R_2} \sum w_x^2} + \sqrt{2ab} \sqrt{h^2 \sum_{R_2} \sum w_{xy}^2}.$$

Da die hier auftretenden mit h^2 multiplizierten Summen nach Annahme beschränkt bleiben, so folgt, daß die Differenz $|w(P_0) - w(Q_0)|$ zugleich mit ihrem Abstande a gegen Null strebt und zwar unabhängig von der Maschenweite, da wir für jedes Teilgebiet G^* von G die Größe b festhalten können. Damit ist die gleichartige Stetigkeit von $w = w_h$ in der x-Richtung bewiesen. Entsprechend ergibt sie sich für die y-Richtung und damit für jedes innere Teilgebiet G^* von G. Die Beschränktheit der Funktion w_h in G^* folgt schließlich aus ihrer gleichartigen Stetigkeit und der Beschränktheit von $h^2 \sum_{G^*} \sum w_h^2$.

Mit diesem Nachweis ist die Existenz einer Teilfolge von Funktionen u gesichert, welche gegen eine Grenzfunktion $u(x, y)$ konvergiert und zwar mit sämtlichen Differenzenquotienten in dem oben gekennzeichneten Sinne gleichmäßig für jedes innere Teilgebiet von G. Diese Grenzfunktion $u(x, y)$ besitzt also in G überall stetige partielle Differentialquotienten beliebiger Ordnung und genügt der partiellen Differentialgleichung des Potentials

$$\frac{\partial^2 u}{\partial x^2} + \frac{\partial^2 u}{\partial y^2} = 0.$$

3. Die Randbedingung.

Um zu beweisen, daß die Lösung die oben formulierte Randbedingung erfüllt, zeigen wir zunächst, daß für jede Gitterfunktion v die Ungleichung

$$(13) \qquad h^2 \sum_{S_{r,h}} \sum v^2 \leqq A r^2 h^2 \sum_{S_{r,h}} \sum (v_x^2 + v_y^2) + B r h \sum_{\Gamma_h} v^2$$

besteht, wo $S_{r,h}$ derjenige Teil des Gittergebietes G ist, der innerhalb des Randstreifens S_r liegt. Dieser Randstreifen S_r war (vgl. S. 48) aus allen Punkten von G gebildet, deren Abstand vom Rande kleiner als r ist; er wird außer von Γ noch von einer Kurve Γ_r begrenzt. Ferner bedeuten A und B nur vom Gebiet und nicht von der Funktion v oder von der Maschenweite h abhängige Konstanten.

Um die obige Ungleichung nachzuweisen, zerlegen wir den Rand Γ von G in eine endliche Anzahl von Stücken, für welche der Winkel der Tangente entweder mit der x-Achse oder mit der y-Achse oberhalb einer positiven Schranke (etwa 30^0) bleibt. Es sei z. B. γ ein solches zur

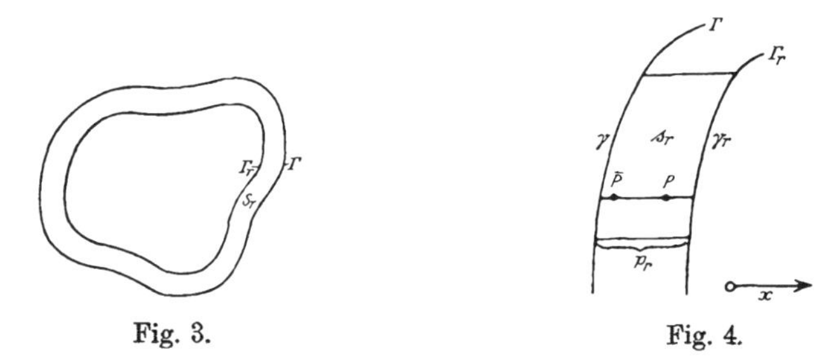

Fig. 3. Fig. 4.

x-Achse hinreichend steil geneigtes Stück von Γ (vgl. Fig. 4). Die Parallelen zur x-Achse durch die Endpunkte des Stückes γ schneiden aus der Näherungskurve Γ_r ein Stück γ_r aus und begrenzen zusammen mit γ und γ_r ein Stück s_r des Randstreifens S_r. Der in dem Streifen s_r enthaltene Teil des Gittergebietes G_h heiße $s_{r,h}$ und der zugehörige Teil des Randes Γ_h heiße γ_h.

Wir denken uns durch einen Gitterpunkt P_h von $s_{r,h}$ die Parallele zur x-Achse gezogen. Sie trifft den Rand γ_h in einem Punkte $\overline{P}_h$. Dasjenige Stück dieser Parallelen, das in $s_{r,h}$ liegt, bezeichnen wir mit $p_{r,h}$. Seine Länge ist sicher kleiner als cr, da r der größte senkrechte Abstand eines Punktes aus S_r von Γ ist. Dabei hängt die Konstante c nur von dem kleinsten Neigungswinkel einer Tangente von γ mit der x-Achse ab.

Nun besteht zwischen dem Wert von v im Punkte P_h und ihrem Werte in $\overline{P}_h$ die Beziehung

$$v(P_h) = v(\overline{P}_h) \pm h \sum_{P_h \overline{P}_h} v_x,$$

woraus sich durch Quadrieren und Anwendung der Schwarzschen Ungleichung

$$v(P_h)^2 \leqq 2\,v(\overline{P}_h)^2 + 2\,c\,r \cdot h \sum_{p_{r,h}} v_x^2$$

ergibt. Summieren wir hinsichtlich P_h in der x-Richtung, so erhalten wir

$$h \sum_{p_r} v^2 \leqq 2cr\, v(\bar{P}_h)^2 + 2c^2 r^2 h \sum_{p_r} v_x^2.$$

Summieren wir noch einmal in der y-Richtung, so entsteht die Relation

$$(14) \qquad h \sum_{S_r}\sum v^2 \leqq 2cr \sum_{\Gamma_h} v(\bar{P}_h) + 2c^2 r^2 h \sum_{S_r}\sum v_x^2,$$

die wir nur noch für die anderen Stücke γ von Γ entsprechend aufzustellen und dann zu addieren haben, um leicht die gewünschte Ungleichung (13) zu erhalten [19]).

Wir setzen nun

$$v_h = u_h - f_h,$$

sodaß v_h am Rande Γ_h verschwindet. Da dann $h^2 \sum_{G_h}\sum (v_x^2 + v_y^2)$ bei abnehmendem h beschränkt bleibt, so erhalten wir aus (13)

$$(16) \qquad \frac{h^2}{r} \sum_{S_{r,h}}\sum v^2 \leqq \varkappa r,$$

wo $\varkappa$ eine nicht von der Funktion v oder der Maschenweite abhängige Konstante ist. Erstrecken wir die Summe links nicht über den ganzen Randstreifen $S_{r,h}$, sondern nur über die Differenz von zwei solchen; $S_{r,h} - S_{\varrho,h}$, so bleibt die Ungleichung (16) mit der selben Konstanten $\varkappa$ gültig, und wir können den Grenzübergang zu verschwindender Maschenweite vollziehen. Aus der Ungleichung (16) entsteht dann

$$\frac{1}{r} \iint_{S_r - S_\varrho} v^2\, dx\, dy \leqq \varkappa r, \qquad v = u - f.$$

Lassen wir nun den kleineren Randstreifen S_ϱ dem Rande zustreben, so erhalten wir die Ungleichung

$$\frac{1}{r} \iint_{S_r} v^2\, dx\, dy = \frac{1}{r} \iint_{S_r} (u-f)^2\, dx\, dy \leqq \varkappa r,$$

die gerade ausdrückt, daß die Grenzfunktion u die von uns geforderte Randbedingung erfüllt.

---

[19]) Durch dieselbe Betrachtungsweise, die zum Nachweis der Ungleichung (13) führt, läßt sich auch die Ungleichung

$$(15) \qquad h^2 \sum_{G_h}\sum v^2 \leqq c_1 h \sum_{\Gamma_h} v^2 + c_2 h^2 \sum_{G_h}\sum (v_x^2 + v_y^2)$$

ableiten, in der die Konstanten c_1, c_2 nur vom Gebiet G, aber nicht von der Mascheneinteilung abhängen.

4. Anwendbarkeit der Methode auf andere Probleme.

Unsere Methode stützt sich wesentlich auf die in dem obigen Hilfssatz ausgesprochene Ungleichheitsbeziehung (10)[20], weil aus ihr die beiden letzten auf S. 49 genannten Hauptpunkte des Beweises folgen; sie macht keinerlei Gebrauch von speziellen Grundlösungen oder sonstigen speziellen Eigenschaften unserer Differenzenausdrücke und läßt sich daher unmittelbar sowohl auf den Fall von beliebig vielen unabhängigen Variablen als auf das Eigenwertproblem der Differentialgleichung $\frac{\partial^2 u}{\partial x^2} + \frac{\partial^2 u}{\partial y^2} + \lambda u = 0$ übertragen und liefert dabei hinsichtlich der Konvergenzverhältnisse genau dieselben Resultate wie oben[21]. Auch eine Übertragung auf lineare Differentialgleichungen anderer Art, insbesondere solche mit nicht konstanten Koeffizienten, erfordert nur einige naheliegende Modifikationen. Der wesentliche *Unterschied* besteht immer nur im Nachweis der Beschränktheit von $h^2 \sum\sum u_h^2$, die allerdings nicht bei einem beliebigen solchen linearen Probleme vorliegt. Aber im Falle der Unbeschränktheit dieser Summe läßt sich zeigen, daß das allgemeine Randwertproblem der betreffenden Differentialgleichung auch wirklich keine Lösung besitzt, daß aber dafür in diesem Falle nicht verschwindende Lösungen des zugehörigen *homogenen* Problems, d. h. Eigenfunktionen, existieren[22].

5. Das Randwertproblem von $\Delta\Delta u = 0$.

Um zu zeigen, daß sich die Methode auch auf den Fall von Differentialgleichungen höherer Ordnung übertragen läßt, behandeln wir im folgenden kurz das Randwertproblem der Differentialgleichung

$$\frac{\partial^4 u}{\partial x^4} + 2\frac{\partial^4 u}{\partial x^2 \partial y^2} + \frac{\partial^4 u}{\partial y^4} = 0.$$

Wir suchen eine Lösung dieser partiellen Differentialgleichung in unserem Gebiete G, für welche die Funktionswerte und ihre ersten Ableitungen am Rande vorgegeben sind, und zwar durch diejenigen Werte, welche von einer vorgegebenen Funktion $f(x, y)$ am Rande definiert werden.

---

[20]) Hinsichtlich der Anwendung entsprechender Integralungleichungen vgl. K. Friedrichs, Die Rand- und Eigenwertprobleme aus der Theorie der elastischen Platten, Math. Annalen **98**, S. 222.

[21]) Es ist dann zugleich bewiesen, daß jede Lösung eines solchen Differentialgleichungsproblems Ableitungen jeder Ordnung besitzt.

[22]) Vgl. Courant-Hilbert, Methoden der mathematischen Physik, **1**, Kap. III, § 3, wo mit Hilfe einer entsprechenden Alternative die Theorie der Integralgleichungen behandelt wird. Vgl. auch die demnächst erscheinende Göttinger Dissertation von W. v. Koppenfels.

Dabei setzen wir wie oben (S. 47) voraus, daß $f(x, y)$ in einem das Gebiet G enthaltenden Gebiete der Ebene mit den ersten und zweiten Ableitungen stetig ist.

Wir ersetzen unser Differentialgleichungsproblem durch die Aufgabe, die Differenzengleichung $\Delta\Delta u = 0$ für das Gittergebiet G zu lösen, wobei in den Punkten des Randstreifens $\Gamma_h + \Gamma_h'$ die Funktion u dieselben Werte wie die vorgegebene Funktion $f(x, y)$ annehmen soll. Nach § 2 wissen wir, daß diese Randwertaufgabe für G_h auf eine und nur eine Weise lösbar ist. Wir werden zeigen, daß bei Verfeinerung der Maschenweite h diese Lösung in jedem inneren Teilgebiet von G mit allen Differenzenquotienten gegen die Lösung unserer Differentialgleichung bzw. gegen die entsprechenden Differentialquotienten konvergiert.

Zu diesem Zwecke bemerken wir erstens, daß für die Lösung $u = u_h$ unseres Differenzenproblems die Summe

$$h^2 \sum_{G_h'}\sum (u_{xx}^2 + 2u_{xy}^2 + u_{yy}^2)$$

bei abnehmender Maschenweite beschränkt bleibt. Wegen der Minimumeigenschaft der Lösung unseres Differenzenproblems (vgl. S. 39) ist nämlich diese Summe nicht größer als die entsprechende Summe

$$h^2 \sum_{G_h'}\sum (f_{xx}^2 + 2f_{xy}^2 + f_{yy}^2)$$

und diese konvergiert bei Verfeinerung der Maschenweite gegen das Integral

$$\iint_G \left(\frac{\partial^2 f}{\partial x^2} + 2\frac{\partial^2 f}{\partial x\,\partial y} + \frac{\partial^2 f}{\partial y^2}\right) dx\,dy,$$

welches nach unseren Voraussetzungen existiert.

Aus der Beschränktheit der Summe

$$h^2 \sum_{G_h'}\sum (u_{xx}^2 + 2u_{xy}^2 + u_{yy}^2)$$

folgt unmittelbar die Beschränktheit von $h^2 \sum_{G_h'}\sum (\Delta u)^2$, weiterhin auch die von

$$h^2 \sum_{G_h}\sum (u_x^2 + u_y^2) \quad \text{und} \quad h^2 \sum_{G_h}\sum u^2.$$

Es besteht nämlich für beliebige w die Ungleichung

$$(15) \qquad h^2 \sum_{G_h}\sum w^2 \leq ch^2 \sum_{G_h}\sum (w_x^2 + w_y^2) + ch \sum_{\Gamma_h} w^2$$

(vgl. (15), S. 54). Indem man in dieser Ungleichung die Funktion w durch die ersten Differenzenquotienten von w ersetzt und auf diejenigen Teilgebiete von G_h anwendet, für welche diese Differenzenquotienten definiert sind, ergibt sich die weitere Ungleichung

$$h^2 \sum_{G_h}\sum (w_x^2 + w_y^2) \leqq c h^2 \sum_{G_h'}\sum (w_{xx}^2 + 2 w_{xy}^2 + w_{yy}^2) + c h \sum_{\Gamma_h + \Gamma_h'} (w_x^2 + w_y^2),$$

wo die Konstanten c wieder nicht von der Funktion und der Maschenweite abhängen. Wir wenden diese Ungleichungen auf $w = u_h$ an und beachten dabei die Beschränktheit der Summen über $\Gamma_h + \Gamma_h'$ auf der rechten Seite — diese Randsummen konvergieren ja definitionsgemäß gegen die entsprechenden mit $f(x, y)$ gebildeten Integrale —. Somit folgt aus der Beschränktheit von

$$h^2 \sum_{G_h'}\sum (u_{xx}^2 + 2 u_{xy}^2 + u_{yy}^2)$$

die Beschränktheit von

$$h^2 \sum_{G_h}\sum (u_x^2 + u_y^2) \quad \text{und} \quad h^2 \sum_{G_h}\sum u^2.$$

Drittens setzen wir in der Ungleichung

$$(10) \qquad h^2 \sum_{G^{**}}\sum (w_x^2 + w_y^2) \leqq c h^2 \sum_{G^*}\sum w^2 + c h^2 \sum_{G^*}\sum (\Delta w)^2$$

(vgl. S. 51), wo G^* ein G^{**} im Innern enthaltendes Teilgebiet von G ist, für w nacheinander die Ausdrücke $\Delta u, \Delta u_x, \Delta u_y, \Delta u_{xx}, \ldots$ ein, die ja alle der Gleichung $\Delta w = 0$ genügen. Es folgt dann sukzessive, daß für alle inneren Teilgebiete G^* von G die Summen

$$h^2 \sum_{G^*}\sum (w_x^2 + w_y^2),$$

d. h.

$$h^2 \sum_{G^*}\sum (\Delta u_x^2 + \Delta u_y^2), \qquad h^2 \sum_{G^*}\sum (\Delta u_{xx}^2 + \Delta u_{xy}^2), \ldots$$

zugleich mit den schon als beschränkt bekannten Summen

$$h^2 \sum_{G_h}\sum u^2, \qquad h^2 \sum_{G_h}\sum (u_x^2 + u_y^2)$$

und

$$h^2 \sum_{G_h}\sum (\Delta u)^2$$

beschränkt bleiben.

Schließlich setzen wir für w in die Ungleichung (10) der Reihe nach die Funktionen $u_{xx}, u_{xy}, u_{yy}, u_{xxx}, \ldots$ ein, für die nach dem eben Bewiesenen

$$h^2 \sum_{G_h^*}\sum (\Delta w)^2, \quad \text{d. h.} \quad h^2 \sum_{G_h^*}\sum (\Delta u_{xx})^2, \ldots$$

beschränkt bleibt. Wir erkennen dann, daß für alle Teilgebiete auch die Summen

$$h^2 \sum_{G_h^*}\sum (u_{xxx}^2 + u_{xxy}^2), \qquad h^2 \sum_{G_h^*}\sum (u_{xyx}^2 + u_{xyy}^2), \ldots$$

beschränkt bleiben.

Aus dieser Tatsache können wir nunmehr wie auf S. 51 ff. schließen, daß sich aus unserer Folge von Gitterfunktionen eine Teilfolge auswählen läßt, die in jedem inneren Teilgebiet von G mit sämtlichen Differenzenquotienten gleichmäßig gegen eine im Innern von G stetige Grenzfunktion bzw. gegen deren Differentialquotienten konvergiert.

Wir haben noch zu zeigen, daß diese Grenzfunktion, die offenbar der Differentialgleichung $\Delta\Delta u = 0$ genügt, auch noch die vorgeschriebenen Randbedingungen erfüllt. Dabei begnügen wir uns analog wie oben damit, diese Randbedingungen in der Form

$$\iint_{S_r} (u-f)^2\,dx\,dy \leqq c r^2, \qquad \iint_{S_r} \left[\left(\frac{\partial u}{\partial x} - \frac{\partial f}{\partial x}\right)^2 + \left(\frac{\partial u}{\partial y} - \frac{\partial f}{\partial y}\right)^2\right] dx\,dy \leqq c r^2$$

auszusprechen [23]). Daß die Grenzfunktion diese Bedingungen erfüllt, ergibt sich aber, indem wir das Schlußverfahren von S. 53 wörtlich auf die Funktion u und ihre ersten Differenzenquotienten anwenden.

Wegen der eindeutigen Bestimmtheit der Lösung unserer Randwertaufgabe erkennt man jetzt nachträglich, daß nicht nur eine ausgewählte Teilfolge, sondern die Funktionenfolge u selbst die angegebenen Konvergenzeigenschaften besitzt.

II. Der hyperbolische Fall.

§ 1.

Die Gleichung der schwingenden Saite.

Im zweiten Teil dieser Arbeit beschäftigen wir uns mit Anfangswertproblemen von hyperbolischen linearen Differentialgleichungen und werden beweisen, daß unter gewissen Voraussetzungen die Lösungen entsprechender Differenzengleichungen bei Verfeinerung der Maschenweite des zugrunde gelegten Gitters gegen die Lösung der Differentialgleichung konvergieren.

Wir können die hier auftretenden Verhältnisse am einfachsten an dem naheliegenden Beispiel der Schwingungsgleichung

$$\frac{\partial^2 u}{\partial t^2} - \frac{\partial^2 u}{\partial x^2} = 0 \tag{1}$$

darlegen. Dabei beschränken wir uns auf dasjenige Anfangswertproblem, in dem auf der Geraden $t = 0$ die Werte der Lösung u und ihrer Ableitungen gegeben sind.

---

[23]) Daß die Randwerte für Funktion und Ableitungen tatsächlich angenommen werden, läßt sich unschwer zeigen. Vgl. die entsprechenden Betrachtungen bei K. Friedrichs loc. cit.

Um die entsprechende Differenzengleichung anzugeben, legen wir in der x, t-Ebene ein quadratisches achsenparalleles Gitter der Maschenweite h. Wir ersetzen die Differentialgleichung (1) durch die Differenzengleichung

$$u_{t\bar{t}} - u_{x\bar{x}} = 0$$

in den Bezeichnungen von S. 34. Greifen wir einen Gitterpunkt P_0 heraus, so verbindet die zugehörige Differenzengleichung den Wert der Funktion u in diesem Punkte mit den Werten in den vier Nachbarpunkten. Kennzeichnen wir wieder die vier Nachbarwerte durch die vier Indizes 1, 2, 3, 4 (s. Fig. 5), so nimmt die Differenzengleichung die einfache Gestalt

$$(2) \qquad u_1 + u_3 - u_2 - u_4 = 0$$

an. Hierbei geht der Wert der Funktion u im Punkte P selbst nicht in die Gleichung ein.

Wir denken uns das Gitter in zwei verschiedene Teilgitter zerlegt, wie in der Fig. 5 durch Kreise und Kreuze angedeutet ist. Die Differenzen-

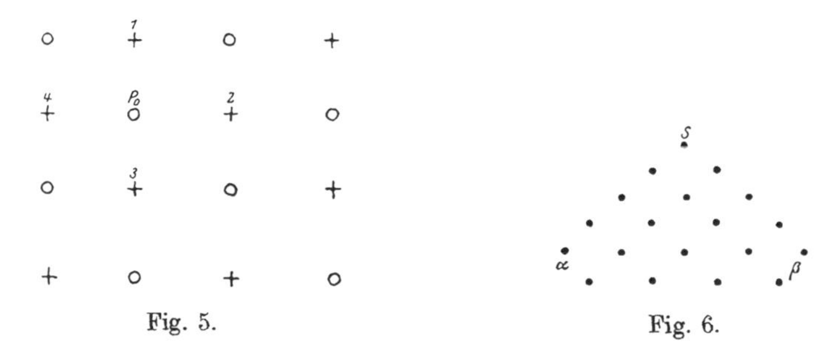

Fig. 5. Fig. 6.

gleichung verbindet dann nur die Werte der Funktion in jedem der Teilgitter untereinander. Wir wollen uns daher auf eines der beiden Teilgitter beschränken. Als Anfangsbedingungen haben wir hier die Werte der Funktion u auf den beiden Gitterreihen $t = 0$ und $t = h$ vorzugeben. Wir geben zunächst die Lösung dieses Anfangswertproblems explizite an; d. h. wir drücken den Wert der Lösung in irgendeinem Punkte S durch die vorgegebenen Werte auf den beiden Anfangsreihen aus. Man erkennt sofort, daß der Wert in einem Punkte der Reihe $t = 2h$ eindeutig bestimmt ist lediglich durch die mit ihm verknüpften drei Werte auf den beiden ersten Reihen. Der Wert in einem Punkte auf der vierten Reihe ist eindeutig bestimmt durch die Werte der Lösung in gewissen drei Punkten der zweiten und dritten Reihe, also auch durch gewisse Werte auf den beiden ersten Reihen. Allgemein wird zu einem Punkte S ein gewisser Abhängigkeitsbereich auf den beiden ersten Reihen gehören; man erhält ihn, wenn man durch den Punkt S die Linien $x + t =$ konst. und $x - t =$ konst. zieht, bis sie die zweite Reihe in den Punkten α und β treffen (vgl. Fig. 6).

Das Dreieck $S\alpha\beta$ nennen wir dann das Bestimmtheitsdreieck, weil in ihm sämtliche u-Werte sich nicht ändern, sobald sie auf den ersten beiden Reihen festgehalten werden. Die Seitenlinien des Dreiecks nennen wir Bestimmtheitslinien.

Bezeichnet man nun die Differenzen von u in Richtung der Bestimmtheitslinien durch u' und $u^{\backprime}$ oder genauer

$$u_1' = u_1 - u_4, \qquad u_1^{\backprime} = u_1 - u_2,$$
$$u_2' = u_2 - u_3, \qquad u_4^{\backprime} = u_4 - u_3,$$

so nimmt die Differenzengleichung etwa die Form

$$u_1' = u_2'$$

an. D. h. auf einer Bestimmtheitslinie sind die Differenzen nach der anderen Bestimmtheitsrichtung konstant, also gleich einer der *vorgegebenen* Differenzen zwischen zwei Punkten der ersten beiden Reihen. Andererseits ist die Differenz $u_S - u_\alpha$ eine Summe über die Differenzen u' längs der Bestimmtheitslinie $S\alpha$, so daß wir unter Benutzung der eben gemachten Bemerkung als Schlußformel (in leicht verständlicher Bezeichnung)

$$u_S = u_\alpha + \sum_{\alpha_1}^{\beta_1} u' \tag{3}$$

erhalten.

Wir lassen nun die Maschenweite h gegen Null streben, wobei die vorgegebenen Werte auf der zweiten oder ersten Reihe gegen eine zweimal stetig differenzierbare Funktion $f(x)$ und die Differenzenquotienten $\frac{u'}{h\sqrt{2}}$ gegen eine stetig differenzierbare Funktion $g(x)$ gleichmäßig konvergieren mögen. Offenbar geht dabei die rechte Seite von (3) gleichmäßig in den Ausdruck

$$f(x-t) + \frac{1}{\sqrt{2}} \int_{x-t}^{x+t} g(\xi)\, d\xi \tag{4}$$

über, wenn S gegen den Punkt (t, x) konvergiert. Dies ist der bekannte Ausdruck der Lösung der Schwingungsgleichung (1) mit den Anfangswerten $u(x, 0) = f(x)$ und $\frac{\partial u}{\partial t}(x, 0) = f'(x) + \sqrt{2}\, g(x)$. Damit ist gezeigt, daß die Lösungen unseres Differenzengleichungsproblems bei abnehmender Maschenweite gegen die Lösung des Differentialgleichungsproblems konvergieren, wenn wir die Anfangswerte (in der oben angegebenen Weise) konvergieren lassen.

§ 2.

Über den Einfluß der Wahl des Gitters. Die Abhängigkeitsgebiete bei Differenzen- und Differentialgleichung.

Die im § 1 betrachteten Verhältnisse legen folgende Überlegungen nahe. Ebenso wie für die Lösung einer linearen hyperbolischen Differentialgleichung im Punkte S nur ein gewisser Teil der Anfangswerte maßgebend ist, nämlich das von den Charakteristiken durch S ausgeschnittene „Abhängigkeitsgebiet“, besitzt auch die Lösung einer Differenzengleichung im Punkte S ein gewisses Abhängigkeitsgebiet, das man erhält, indem man die Bestimmtheitslinien vom Punkte S aus zieht. In § 1 fielen nun die Richtungen der Bestimmtheitslinien der Differenzengleichung mit den charakteristischen Richtungen der Differentialgleichung zusammen, wodurch auch die Abhängigkeitsgebiete in der Grenze übereinstimmten. Diese Tatsache hing aber wesentlich von der Orientierung des Gitters in der (x, t)-Ebene ab und beruhte ferner darauf, daß wir das Gitter quadratisch gewählt hatten. Wir legen jetzt allgemeiner ein rechteckiges achsenparalleles Gitter zugrunde, dessen Maschenweite in der t-Richtung (Zeitmasche) gleich h und diejenige der x-Richtung (Raummasche) gleich $\varkappa h$ mit konstantem $\varkappa$ ist. Das Abhängigkeitsgebiet der Differenzengleichung $u_{t\bar{t}} - u_{x\bar{x}} = 0$ für dieses Gitter wird ganz im Innern des Abhängigkeitsgebietes der Differentialgleichung $\frac{\partial^2 u}{\partial t^2} - \frac{\partial^2 u}{\partial x^2} = 0$ liegen oder wird es selbst in seinem Innern enthalten, je nachdem ob $\varkappa < 1$ oder $\varkappa > 1$ ist.

Hieraus ergibt sich eine merkwürdige Tatsache: Läßt man im Falle $\varkappa < 1$ die Maschenweite h gegen Null abnehmen, so kann die Lösung der Differenzengleichung im allgemeinen *nicht* gegen die Lösung der Differentialgleichung konvergieren. Ändert man nämlich etwa bei der Schwingungsgleichung (1) die Anfangswerte der Lösung der Differentialgleichung in der Umgebung der Endpunkte α und β des Abhängigkeitsgebietes (vgl. Fig. 7), so zeigt die Formel (4), daß sich auch die Lösung selbst im Punkte (x, t) ändert. Für die Lösungen der Differenzengleichungen im Punkte S sind aber die Vorgaben in den Punkten α und β irrelevant, da diese außerhalb des Abhängigkeitsgebietes der Differenzengleichungen liegen. — Daß im Falle $\varkappa > 1$ Konvergenz statthat, werden wir im § 3 beweisen. Vgl. hierzu Fig. 9, S. 62.

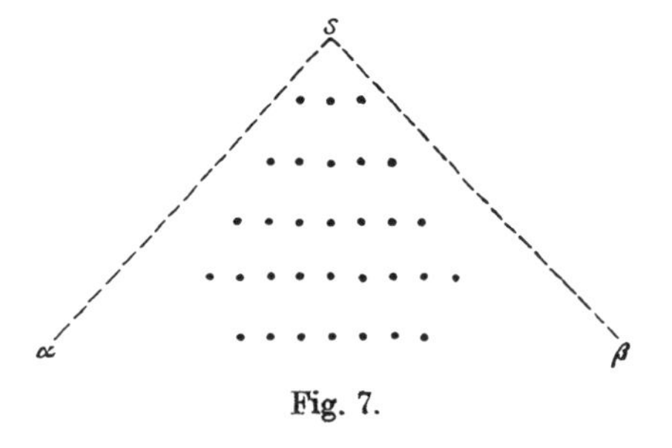

Fig. 7.

Betrachtet man dagegen z. B. die Differentialgleichung

$$(5) \qquad 2\frac{\partial^2 u}{\partial t^2} - \frac{\partial^2 u}{\partial x^2} - \frac{\partial^2 u}{\partial y^2} = 0$$

in den beiden räumlichen Variablen x, y und der zeitlichen t, und ersetzt sie durch entsprechende Differenzengleichungen in geradlinigen Gittern, so ist es im Gegensatz zum Falle von nur zwei unabhängigen Variablen unmöglich, die Mascheneinteilung so zu wählen, daß die Abhängigkeitsgebiete der Differenzen- und Differentialgleichung zusammenfallen; denn das Abhängigkeitsgebiet der Differenzengleichungen wird ein Viereck, während das der Differentialgleichung ein Kreis ist. Wir werden später (vgl. § 4) die Mascheneinteilung so wählen, daß das Bestimmtheitsgebiet der Differenzengleichung das Bestimmtheitsgebiet der Differentialgleichung im Innern enthält, und zeigen, daß wieder Konvergenz stattfindet.

Überhaupt wird ein wesentliches Ergebnis dieses Teils sein, daß man bei jeder linearen hyperbolischen homogenen Differentialgleichung zweiter Ordnung das Gitter so wählen kann, daß die Lösung der Differenzengleichung gegen die Lösung der Differentialgleichung konvergiert, wenn man die Maschenweiten gegen Null streben läßt (vgl. hierzu §§ 3, 4, 7, 8).

§ 3.

Grenzübergang bei beliebigen rechteckigen Gittern.

Wir betrachten zunächst wieder die Schwingungsgleichung

$$(1) \qquad \frac{\partial^2 u}{\partial t^2} - \frac{\partial^2 u}{\partial x^2} = 0,$$

legen aber nunmehr ein rechteckiges achsenparalleles Gitter zugrunde, dessen zeitliche Maschenweite h, dessen Raummasche $\varkappa h$ ist. Die zugehörige Differenzengleichung lautet:

$$(6) \qquad L(u) = \frac{1}{h^2}(u_1 - 2u_0 + u_3) - \frac{1}{\varkappa^2 h^2}(u_2 - 2u_0 + u_4) = 0,$$

wobei sich die Indizes auf den Mittelpunkt P_0 und die Ecken P_1, P_2, P_3, P_4 eines „Elementarrhombus" (vgl. Fig. 8) beziehen. Vermöge der Gleichung

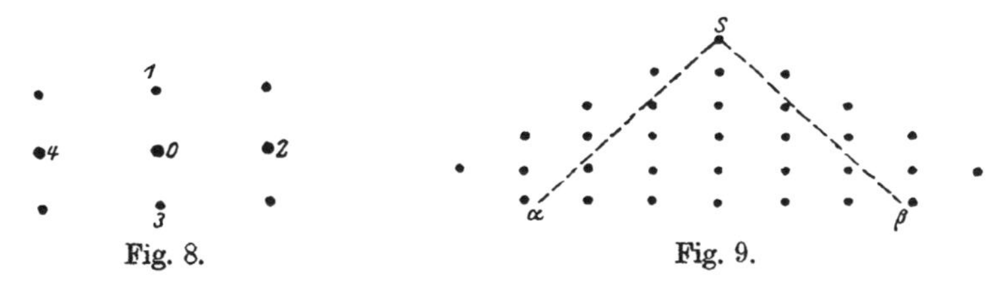

Fig. 8. Fig. 9.

$L(u) = 0$ können wir den Wert der Funktion u in einem Punkte S durch ihre Werte auf demjenigen Stück der beiden Anfangsreihen $t = 0$ und $t = h$ darstellen, das man erhält, wenn man vom Punkte S aus (vgl. Fig. 6, S. 59) die zu den Seiten eines Elementarrhombus parallelen „Bestimmtheitslinien“ zieht. Wir denken uns die Anfangswerte so vorgegeben, daß sie und die zwischen ihnen gebildeten ersten Differenzenquotienten bei abnehmender Maschenweite und bei festem $\varkappa$ gleichmäßig gegen stetige vorgegebene Funktionen auf der Geraden $t = 0$ konvergieren. Für die Lösungen der Differenzengleichungen läßt sich wohl eine explizite Darstellung durch ihre Anfangswerte aufstellen (entsprechend (3) in § 1); sie ist aber nicht so einfach, daß man unmittelbar den Grenzübergang zu verschwindender Maschenweite ausführen kann. Wir schlagen daher einen anderen Weg ein, der uns die Behandlung auch des allgemeinen Problem ermöglichen wird[24]).

Wir multiplizieren den Differenzenausdruck $L(u)$ mit $(u_1 - u_3)$ und formen das Produkt unter Beachtung der folgenden Indentitäten um:

$$(7) \qquad (u_1 - u_3)(u_1 - 2u_0 + u_3) = (u_1 - u_0)^2 - (u_0 - u_3)^2,$$

$$(8) \qquad (u_1 - u_3)(u_2 - 2u_0 + u_4) = (u_1 - u_0)^2 - (u_0 - u_3)^2 - \frac{1}{2}[(u_1 - u_2)^2 + (u_1 - u_4)^2 - (u_2 - u_3)^2 - (u_4 - u_3)^2].$$

Wir erhalten so

$$(9) \qquad 2(u_1 - u_3) L(u) = \frac{2}{h^2}\left(1 - \frac{1}{\varkappa^2}\right)[(u_1 - u_0)^2 - (u_0 - u_3)^2] + \frac{1}{h^2 \varkappa^2}[(u_1 - u_2)^2 + (u_1 - u_4)^2 - (u_2 - u_3)^2 - (u_4 - u_3)^2].$$

Wir summieren nun das Produkt (9) über alle Elementarrhomben eines Bestimmtheitsdreiecks $S\alpha\beta$. Die auf der rechten Seite von (9) auftretenden Differenzenquadrate kommen stets in zwei benachbarten Elementarrhomben vor, mit verschiedenen Vorzeichen versehen. Sie heben sich bei der Summation fort, sobald beide Elementarrhomben zum Dreieck $S\alpha\beta$ gehören; es bleibt also nur eine vom Rande des Dreiecks berührende Summe von Differenzenquadraten übrig. Wir erhalten so die Relation:

$$(10) \qquad h^2 \sum_{S\alpha\beta}\sum 2\frac{u_1 - u_3}{h} L(u)$$
$$= h\sum_{S\alpha}\left[2\left(1 - \frac{1}{\varkappa^2}\right)\left(\frac{\dot{u}}{h}\right)^2 + \frac{1}{\varkappa^2}\left(\frac{u'}{h}\right)^2\right]$$
$$+ h\sum_{S\beta}\left[2\left(1 - \frac{1}{\varkappa^2}\right)\left(\frac{\dot{u}}{h}\right)^2 + \frac{1}{\varkappa^2}\left(\frac{u'}{h}\right)^2\right]$$
$$- h\sum_{\mathrm{I\,II}}\left[2\left(1 - \frac{1}{\varkappa^2}\right)\left(\frac{\dot{u}}{h}\right)^2 + \frac{1}{\varkappa^2}\left(\frac{u'}{h}\right)^2 + \frac{1}{\varkappa^2}\left(\frac{u'}{h}\right)^2\right].$$

---

[24]) Zum folgenden vgl. K. Friedrichs und H. Lewy, Über die Eindeutigkeit usw., Math. Annalen (98 1928), S. 192 ff., wo analoge Umformungen für Integrale benutzt werden.

Hier bedeuten u' und $u^{\backprime}$ Differenzen in Bestimmtheitsrichtungen wie in § 1, während $\dot{u}$ die Differenz der Funktionswerte in zwei Nachbarpunkten bezeichnet, deren Verbindungslinie zur t-Achse parallel ist. Die Summen sind über alle aus zwei Parallelreihen bestehenden Randstreifen zu erstrecken, so daß sämtliche vorkommenden Differenzen u', $u^{\backprime}$, $\dot{u}$ einmal und nur einmal auftreten.

Für eine Lösung von $L(u) = 0$ verschwindet also die rechte Seite von (10). Die dort auftretende Summe über die Anfangsreihen I und II bleibt beschränkt, wenn wir die Maschenweite h (bei festgehaltenem $\varkappa$) zu Null abnehmen lassen; sie geht nämlich in ein Integral über vorgegebene Funktionen auf der Anfangslinie über. Infolgedessen bleiben auch die in (10) über $S\alpha$ und $S\beta$ erstreckten Summen beschränkt. Ist nun, wie wir fordern müssen (vgl. S. 61), $\varkappa \geqq 1$, also $1 - \frac{1}{\varkappa^2}$ nicht negativ, so folgt weiter die Beschränktheit der einzelnen Summen

$$h \sum_{S\alpha} \left(\frac{u'}{h}\right)^2, \qquad h \sum_{S\beta} \left(\frac{u^{\backprime}}{h}\right)^2,$$

die wir über irgendwelche Bestimmtheitslinien erstreckt denken können.

Hieraus können wir die „gleichartige Stetigkeit“ (vgl. 1. Teil § 4) der Folge der Gitterfunktionen in allen Richtungen der Ebene ableiten[25]); da die Werte von u auf der Anfangslinie beschränkt sind, folgt die Existenz einer gleichmäßig gegen eine Grenzfunktion $u(x, t)$ konvergierenden Teilfolge.

Zugleich mit der Funktion u genügen auch ihre ersten und zweiten Differenzenquotienten der Differenzengleichung $L(u) = 0$. Die Anfangswerte dieser Differenzenquotienten lassen sich vermittelst der Gleichung $L(u) = 0$ durch solche erste, zweite und dritte Differenzenquotienten von u ausdrücken, in denen nur Punkte der beiden Anfangsreihen I und II auftreten. Wir verlangen von ihnen, daß sie gegen stetige Grenzfunktionen streben, d. h. daß etwa die vorgegebenen Anfangswerte $u(x, 0)$, $u_t(x, 0)$ drei- bzw. zweimal stetig nach x differenzierbar sind.

Danach können wir unsere oben angestellte Konvergenzbetrachtung anstatt auf u auch auf seine ersten und zweiten Differenzenquotienten anwenden, also eine Teilfolge auswählen, so daß diese Differenzenquotienten gleichmäßig gegen Funktionen streben, die dann die ersten bzw. zweiten Ableitungen der Grenzfunktion $u(x, t)$ sein müssen. Die Grenzfunktion u

---

[25]) Sind S_1 und S_2 zwei Punkte im Abstand δ, so verbinde man sie durch einen Streckenzug aus zwei Strecken $S_1 S$ und $S S_2$: von denen die erste der einen, die zweite der anderen Bestimmtheitsrichtung parallel ist. Es gilt dann die Abschätzung

$$|u_{S_1} - u_{S_2}| \leqq |u_{S_1} - u_S| + |u_S - u_{S_2}| \leqq \sqrt{\delta} \sqrt{h \sum_{S_1 S} \left(\frac{u'}{h}\right)^2} + \sqrt{\delta} \sqrt{h \sum_{S S_2} \left(\frac{u^{\backprime}}{h}\right)^2}.$$

genügt infolgedessen der Differentialgleichung $\frac{\partial^2 u}{\partial t^2} - \frac{\partial^2 u}{\partial x^2} = 0$, die in der Grenze aus der Differenzengleichung $L(u) = 0$ entsteht; sie stellt also die Lösung des Anfangswertproblems dar. Da diese Lösung eindeutig bestimmt ist, konvergiert jede Teilfolge der Gitterfunktionen und damit die Folge selbst gegen die Grenzfunktion.

§ 4.

Die Schwingungsgleichung in drei Variablen.

Wir behandeln nun die Schwingungsgleichung

$$2\frac{\partial^2 u}{\partial t^2} - \frac{\partial^2 u}{\partial x^2} - \frac{\partial^2 u}{\partial y^2} = 0 \tag{11}$$

und knüpfen an die im § 2 gemachten Bemerkungen über die Beziehung der Abhängigkeitsbereiche an. Das Abhängigkeitsgebiet der Differentialgleichung (11) ist der Kreiskegel mit einer zur t-Richtung parallelen Achse und dem Öffnungswinkel α, mit $\operatorname{tg}\alpha = \frac{1}{\sqrt{2}}$. In irgendeinem rechtwinkligen achsenparallelen Gitter setzen wir entsprechend die Differenzengleichung

$$2\,u_{t\bar{t}} - u_{x\bar{x}} - u_{y\bar{y}} = 0 \tag{12}$$

an. Durch sie werden die Funktionswerte u in den Punkten eines „Elementaroktaeders" miteinander verknüpft. Sie gestattet, den Funktionswert in einem Punkte S eindeutig durch die Funktionswerte in gewissen Punkten der beiden Anfangsebenen $t = 0$ und $t = h$ auszudrücken. Wir erhalten zu jedem Punkte S eine Pyramide der Bestimmtheit, die aus den beiden Grundlinien als Abhängigkeitsgebiet zwei Rhomben ausschneidet.

Lassen wir die Maschenweiten etwa unter Festhaltung ihrer Verhältnisse gegen Null streben, so können wir eine Konvergenz der Folge der Gitterfunktionen gegen die Lösung der Differentialgleichung höchstens dann erwarten, wenn die Bestimmtheitspyramide den Bestimmtheitskegel der Differentialgleichung im Inneren enthält. Das einfachste Gitter dieser Eigenschaft wird dasjenige sein, das so liegt, daß die Bestimmungspyramide den Bestimmungskegel von außen berührt. Unsere Differentialgleichung ist gerade so gewählt, daß dies für ein *kubisches* achsenparalleles Gitter eintritt.

Die Differenzengleichung (12) nimmt in diesem Gitter in den Bezeichnungen der Figur 10 die Gestalt an:

$$L(u) = \frac{2}{h^2}(u_{a'} - 2u_0 + u_a) - \frac{1}{h^2}(u_1 - 2u_0 - u_3) - \frac{1}{h^2}(u_2 - 2u_0 - u_4), \tag{13}$$

in die der Funktionswert u_0 im Mittelpunkt P übrigens nicht mehr eingeht. Die Werte der Lösung auf den beiden Anfangsebenen seien die Werte einer viermal stetig nach x_1, y_2, t differenzierbaren Funktion.

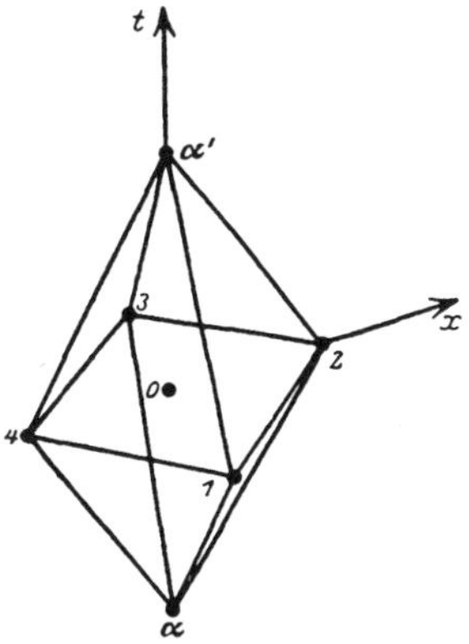

Fig. 10.

Wir benutzen zum Konvergenzbeweise wieder die im § 3 entwickelte Methode, indem wir für die Lösung unserer Differenzengleichung die dreifache Summe

$$h^3 \sum \sum \sum 2 \frac{u_{\alpha'} - u_\alpha}{h} L(u) = 0$$

bilden, die über alle Elementaroktaeder der vom Punkte S ausstrahlenden Bestimmtheitspyramide zu erstrecken ist. Auf Grund der fast wörtlich zu übernehmenden Schlußweise erkennen wir, daß die Werte der Funktion u in inneren Punkten der Bestimmtheitspyramide herausfallen und daß nur noch Flächensummen über die vier Seitendoppelflächen F und die beiden Grundflächen I II der Pyramide übrigbleiben.

Bezeichnen wir mit u' die Differenz der Funktionswerte in zwei Punkten, die durch eine Seitenlinie eines Elementaroktaeders verbunden wird, so lautet die entstehende Formel

$$(14) \qquad \sum_F \sum (u')^2 - \sum_{\mathrm{I}} \sum_{\mathrm{II}} (u')^2 = 0,$$

in der über sämtliche auf diesen Flächen enthaltenen Differenzen u' zu summieren ist, so daß jede solche Differenz nur einmal auftritt[26]). Da die Doppelsumme über die beiden Anfangsflächen beschränkt bleibt, weil sie ja in ein Integral über Anfangswerte übergeht, so bleibt auch die Summe über die „Bestimmtheitsflächen" F beschränkt.

Wir wenden unsere Betrachtung anstatt auf u selbst wieder auf ihre ersten, zweiten und dritten Differenzenquotienten, die ja selbst der Differenzengleichung (13) genügen und deren Anfangswerte vermittels (13) sich durch erste bis vierte Differenzenquotienten allein aus Werten auf den ersten beiden Anfangsebenen ausdrücken lassen. Ist $w = w_h$ einer der Differenzenquotienten bis zur dritten Ordnung, so wissen wir, daß die über eine Bestimmtheitsfläche F erstreckte Summe $h^2 \sum \sum \left(\frac{w'}{h}\right)^2$ beschränkt bleibt. Hieraus ergibt sich aber durch genau denselben Schluß, den wir im ersten Teil, § 4 angewandt haben, daß die Funktion u mit ihren ersten und

---

[26]) Es ist das Gitter gerade so gewählt, daß die Differenzen von u zwischen den beiden Flächen F nicht mehr auftreten.

zweiten Differenzenquotienten gleichartig stetig ist. Es gibt also eine gegen Null abnehmende Folge von Maschenweiten, so daß diese Größen, die ja am Anfang beschränkt sind, gegen stetige Grenzfunktionen konvergieren, und zwar offenbar gegen die Lösung der Differentialgleichung einschließlich ihrer ersten und zweiten Ableitungen, was genau wie früher (§ 3) folgt.

Anhang.

Ergänzungen und Verallgemeinerungen.

§ 5.

Beispiel einer Differentialgleichung erster Ordnung.

Wir haben im § 2 gesehen, daß unter Umständen das Abhängigkeitsgebiet der Differentialgleichung nur einen Teil des Abhängigkeitsgebietes der Differenzengleichung ausmacht, und daß also der Einfluß des Restgebietes in der Grenze herausfällt. Dies Phänomen können wir an dem Beispiel der Differentialgleichung erster Ordnung $\frac{\partial u}{\partial t} = 0$ explizite verfolgen, wenn wir sie durch die Differenzengleichung

$$2\, u_t - u_x + u_{\bar{x}} = 0 \tag{15}$$

ersetzen. Sie lautet ausgeschrieben in den Bezeichnungen der Fig. 5 (S. 59)

$$u_1 = \frac{u_2 + u_4}{2}. \tag{16}$$

Die Differenzengleichung verbindet wieder nur Punkte eines Teilgitters untereinander. Das Anfangswertproblem besteht darin, daß man auf der Reihe $t = 0$ der Funktion u in den Punkten $x = 2ih$ diejenigen Werte f_{2i} vorschreibt, die dort eine stetige Funktion $f(x)$ annimmt.

Wir betrachten etwa den Punkt S auf der t-Achse im Abstande $2nh$. Man verifiziert leicht die Darstellung der Lösung u in S:

$$u_S = \sum_{i=-n}^{n} \frac{1}{2^{2n}} \binom{2n}{n+i} f_{2i}. \tag{17}$$

Die Summe auf der rechten Seite strebt bei Verfeinerung der Maschenweite, d. h. bei $n \to \infty$ einfach gegen den Wert f_0. Man entnimmt das aus der Stetigkeit von $f(x)$ und dem Verhalten der Binomialkoeffizienten bei wachsendem n. (Vgl. den nächsten Paragraphen.)

§ 6.

Die Wärmeleitungsgleichung.

Die Differenzengleichung (16) des § 5 läßt sich auch als Analogon einer ganz anderen Differentialgleichung auffassen, nämlich der Wärmeleitungsgleichung

$$2\frac{\partial u}{\partial t} - \frac{\partial^2 u}{\partial x^2} = 0. \tag{18}$$

In irgendeinem rechteckigen achsenparallelen Gitter lautet die entsprechende Differenzengleichung

$$2\left(\frac{u_1 - u_0}{l}\right) = \left(\frac{u_2 + u_4 - 2u_0}{h^2}\right). \tag{19}$$

wo l die Zeit, h die Raummasche ist. Beim Grenzübergang zu verschwindender Maschenweite behält die Differenzengleichung nur dann ihre Form, wenn l proportional mit h^2 abnimmt. Setzen wir insbesondere $l = h^2$, so fällt der Wert u_0 aus der Gleichung heraus und es entsteht die Differenzengleichung

$$u_1 = \frac{u_2 + u_4}{2}, \tag{16}$$

deren Auflösung durch die Formel (17) gegeben wird:

$$u(0, t) = \sum_{i=-n}^{n} \frac{1}{2^{2n}} \binom{2n}{n+i} f_{2i}. \tag{17}$$

Ein Punkt ξ der x-Achse ist bei abnehmender Maschenweite immer durch den Index

$$2i = \frac{\xi}{h} \tag{20}$$

gekennzeichnet. Die Maschenweite h ist mit der Ordinate t des Aufpunktes S durch die Gleichung

$$2nh^2 = t \tag{21}$$

festgelegt.

Wir wollen untersuchen, was aus der Formel (17) entsteht, wenn h gegen Null, d. h. n gegen unendlich strebt. Unter Benutzung der Formel (21) schreiben wir die Gleichung (17) in die Gestalt

$$u(0, t) = \sum_{i=-n}^{n} \frac{\sqrt{2n}}{2 \cdot 2^{2n}\sqrt{t}} \binom{2n}{n+i} f_{2i} \cdot 2h. \tag{22}$$

Für den Koeffizienten von $2hf_{2i} = 2hf(\xi)$ verwenden wir die Abkürzung

$$\frac{1}{2\sqrt{t}} g_{2n}(\xi) = \frac{\sqrt{2n}}{2 \cdot 2^{2n}\sqrt{t}} \binom{2n}{n + \frac{\xi}{\sqrt{2t}}\sqrt{n}}.$$

Den Grenzwert dieses Koeffizienten, den man gewöhnlich mit Hilfe der Stirlingschen Formel bestimmt, wollen wir hier berechnen, indem wir die

Funktion $g_{2n}(\xi)$ als Lösung einer gewöhnlichen Differenzengleichung auffassen und den Grenzübergang zu verschwindender Maschenweite h und damit zur Differentialgleichung ausführen. Als diese Differenzengleichung findet man

$$\frac{1}{2h}(g_h(\xi + 2h) - g_h(\xi)) = -\frac{1}{2h} g_h(\xi) \frac{2i+1}{n+i+1}$$

(indem wir $g_h(\xi)$ anstatt $g_{2n}(\xi)$ schreiben). Oder

$$\frac{1}{2h}(g_h(\xi + 2h) - g_h(\xi)) = -g_h(\xi) \frac{\xi + h}{t + h\xi + 2h^2}.$$

$g_h(\xi)$ genügt außerdem der Normierungsbedingung

$$\sum_{i=-n}^{n} g_h(\xi) \cdot 2h = 2\sqrt{t}.$$

Diese Summe ist über das Abhängigkeitsgebiet der Differenzengleichung zu erstrecken, das, wenn $h \to 0$ strebt, in der Grenze die ganze x-Achse erfüllt.

Durch einfache Überlegungen erkennt man, daß $g_h(\xi)$ gleichmäßig gegen die Lösung $g(x)$ der Differentialgleichung

$$g'(x) = -g(x)\frac{x}{t}$$

mit der Nebenbedingung

$$\int_{-\infty}^{+\infty} g(x)\,dx = 2\sqrt{t}$$

konvergiert. Aus der Formel (22) entsteht dann durch Grenzübergang

$$u(0, t) = \int_{-\infty}^{+\infty} \frac{1}{\sqrt{2\pi t}} e^{-\frac{\xi^2}{2t}} f(\xi)\,d\xi,$$

die bekannte Lösung der Wärmeleitungsgleichung.

Die Betrachtungen dieses Paragraphen übertragen sich ohne weiteres auf den Fall von Differentialgleichungen

$$4\frac{\partial u}{\partial t} - \frac{\partial^2 u}{\partial x^2} - \frac{\partial^2 u}{\partial y^2} = 0$$

usw. bei mehr unabhängigen Veränderlichen.

§ 7.

Die allgemeine lineare homogene Differentialgleichung zweiter Ordnung in der Ebene.

Wir behandeln die Differentialgleichung

$$(23) \qquad \frac{\partial^2 u}{\partial t^2} - k^2 \frac{\partial^2 u}{\partial x^2} + \alpha \frac{\partial u}{\partial t} + \beta \frac{\partial u}{\partial x} + \gamma u = 0.$$

Die Koeffizienten sind zweimal stetig nach x, t differenzierbar, während die Anfangswerte auf der Geraden $t = 0$ dreimalig stetig nach x differenzierbar sind. Wir ersetzen die Differentialgleichung in einem Gitter mit der Zeitmaschenweite h und der Raummaschenweite $\varkappa h$, so daß in einer Umgebung des zu betrachtenden Stückes der Anfangsgeraden $1 - \frac{k^2}{\varkappa^2} > \varepsilon > 0$ für unser konstantes $\varkappa$ gilt, durch die Differenzengleichung

$$(24) \qquad L(u) = u_{t\bar{t}}(x, t) - k^2 u_{x\bar{x}}(x, t) + \alpha u_t + \beta u_x + \gamma u = 0$$

und wählen die Anfangswerte wie in § 3 (vgl. S. 63).

Zum Konvergenzbeweis formen wir wieder die Summe

$$h^2 \sum_{S\alpha\beta}\sum 2 \frac{u_1 - u_3}{h} L(u)$$

unter Verwendung der Identitäten (7), (8) um. Außer einer Summe (vgl. (10)) über den Rand des Dreiecks $S\alpha\beta$ (vgl. Fig. 6) tritt dann noch eine über das ganze Dreieck $S\alpha\beta$ erstreckte Summe auf, deren absoluter Betrag sich nach oben mit Hilfe der Schwarzschen Ungleichung durch

$$C h^2 \sum_{S\alpha\beta}\sum \left[\left(\frac{u^{\backprime}}{h}\right)^2 + \left(\frac{u'}{h}\right)^2 + \left(\frac{\dot{u}}{h}\right)^2 + u^2\right]$$

abschätzen läßt, wo die Konstante C nicht von der Funktion u, der Maschenweite h und in einer gewissen Umgebung der Anfangslinie auch nicht vom Punkte S abhängt.

Hier können wir weiter $h^2 \sum_{S\alpha\beta}\sum u^2$ nach oben durch

$$C_1 h^2 \sum_{S\alpha\beta}\sum \left(\frac{\dot{u}}{h}\right)^2 + C_2 h \sum_{\mathrm{I\,II}} u^2$$

abschätzen[27]), wo für die Konstanten C_1, C_2 dasselbe gilt, was für C gesagt wurde.

Wir erhalten so eine Ungleichung von der Form

$$(25) \qquad h \sum_{S\alpha} \left[2\left(1 - \frac{k^2}{\varkappa^2}\right)\left(\frac{\dot{u}}{h}\right)^2 + \frac{k^2}{\varkappa^2}\left(\frac{u'}{h}\right)^2\right]$$
$$+ h \sum_{S\beta} \left[2\left(1 - \frac{k^2}{\varkappa^2}\right)\left(\frac{\dot{u}}{h}\right)^2 + \frac{k^2}{\varkappa^2}\left(\frac{u^{\backprime}}{h}\right)^2\right]$$
$$\leqq C_3 h^2 \sum_{S\alpha\beta}\sum \left[\left(\frac{\dot{u}}{h}\right)^2 + \left(\frac{u'}{h}\right)^2 + \left(\frac{u^{\backprime}}{h}\right)^2\right] + D,$$

wo D eine für alle Punkte S und Maschenweiten h feste Schranke für die auftretenden Summen über die Anfangsgerade ist.

---

[27]) Vgl. zum Beweise die verwandte Ungleichung S. 64 unten.

Als Spitzen S unserer Dreiecke wählen wir nun von der Anfangsgeraden ausgehend der Reihe nach die Punkte $S_0, S_1, \ldots, S_n = S$ auf einer Parallelen zur t-Achse. Durch Summation der zugehörigen Ungleichungen (25) können wir die Ungleichung

$$
(26) \qquad \begin{aligned}
&h^2 \sum_{S S_0 \alpha}\sum \left[2\left(1-\frac{k^2}{\varkappa^2}\right)\left(\frac{\dot u}{h}\right)^2 + \frac{k^2}{\varkappa^2}\left(\frac{u'}{h}\right)^2\right] \\
&+ h^2 \sum_{S S_0 \beta}\sum \left[2\left(1-\frac{k^2}{\varkappa^2}\right)\left(\frac{\dot u}{h}\right)^2 + \frac{k^2}{\varkappa^2}\left(\frac{u^{\backprime}}{h}\right)^2\right] \\
&\leqq n h C_3 \sum_{S\alpha\beta}\sum \left[\left(\frac{\dot u}{h}\right)^2 + \left(\frac{u'}{h}\right)^2 + \left(\frac{u^{\backprime}}{h}\right)^2\right] + n h D
\end{aligned}
$$

erhalten. Beachten wir nun, daß man eine Differenz u' bzw. $u^{\backprime}$ durch zwei Differenzen $\dot u$ und eine Differenz $u^{\backprime}$ bzw. u' ausdrücken kann, so ergibt sich, daß wir die linke Seite von (26) höchstens verkleinern, wenn wir sie durch

$$
C_4 h^2 \sum_{S\alpha\beta}\sum \left[\left(\frac{\dot u}{h}\right)^2 + \left(\frac{u'}{h}\right)^2 + \left(\frac{u^{\backprime}}{h}\right)^2\right]
$$

ersetzen. Beschränken wir uns nun auf eine solche Umgebung $t \leqq n h$ der Anfangsgeraden, in der

$$
C_4 - n h C_3 = C_5 > 0
$$

ist, so erhalten wir aus (26)

$$
(27) \qquad C_5 h^2 \sum_{S\alpha\beta}\sum \left[\left(\frac{\dot u}{h}\right)^2 + \left(\frac{u'}{h}\right)^2 + \left(\frac{u^{\backprime}}{h}\right)^2\right] \leqq \frac{C_4}{C_3} D .
$$

Aus der in (27) ausgedrückten Beschränktheit der linken Seite ergibt sich nach (25) die Beschränktheit von

$$
h \sum_{S\alpha} \left(\frac{u'}{h}\right)^2 + h \sum_{S\beta} \left(\frac{u^{\backprime}}{h}\right)^2,
$$

woraus sich wie in § 3 die gleichartige Stetigkeit von u ergibt.

Wir wenden die Ungleichung (25) anstatt auf die Funktion u selbst auf deren erste und zweite Differenzenquotienten w an, die auch Differenzengleichungen genügen, deren Glieder zweiter Ordnung wie in (24) lauten. In den Zusatzgliedern können zwar noch Ableitungen von u auftreten, die sich nicht durch w ausdrücken lassen, aber deren mit h^2 multiplizierte Flächenquadratsumme schon als beschränkt angenommen werden kann. Das aber genügt, um auf diese Differenzengleichung für w denselben Schluß anzuwenden, den wir oben auf u angewandt haben. Wir können somit die gleichartige Stetigkeit und Beschränktheit der Funktionen u und ihrer ersten und zweiten Ableitungen folgern, die infolgedessen eine Teilfolge besitzen, die gleichmäßig gegen die Lösung des

Anfangswertproblems der Differentialgleichung konvergiert. Aus deren Eindeutigkeit folgt wieder, daß die Funktionenfolge selber konvergiert.

Wir müssen dabei allerdings voraussetzen, daß die Differenzenquotienten bis zur dritten Ordnung auf und zwischen den beiden Anfangsreihen gleichmäßig gegen stetige Grenzfunktionen konvergieren[28]).

§ 8.

Das Anfangswertproblem einer beliebigen hyperbolischen linearen Differentialgleichung zweiter Ordnung.

Wir wollen nun zeigen, daß die bisher entwickelten Methoden dazu ausreichen, das Anfangswertproblem einer beliebigen linearen homogenen hyperbolischen Differentialgleichung zweiter Ordnung zu lösen. Es genügt dabei, sich auf den Fall von drei Variablen zu beschränken. Der Gedankengang läßt sich unmittelbar auf mehr Variable übertragen. Man kann leicht einsehen, daß das allgemeinste derartige Problem durch eine Variablentransformation auf folgendes zurückgeführt werden kann: Eine Funktion $u(x, y, t)$ zu finden, die der Differentialgleichung

$$u_{tt} - (a u_{xx} + 2 b u_{xy} + c u_{yy}) + \alpha u_t + \beta u_x + \gamma u_y + \delta u = 0 \tag{28}$$

genügt und die mit ihren ersten Ableitungen auf der Fläche $t = 0$ vorgegebene Werte annimmt. Dabei sollen die Koeffizienten Funktionen der Variablen x, y, t sein und den Bedingungen

$$a > 0, \quad c > 0, \quad ac - b^2 > 0$$

genügen.

Die Koeffizienten setzen wir dabei als dreimal nach x, y, t, und die Anfangswerte u als viermal bzw. u_t als dreimal nach x, y stetig differenzierbar voraus.

Wir denken uns ferner die Koordinaten x und y um einen Punkt der Anfangsebene so gedreht, daß in ihm $b = 0$ ist. Dann ist in einer gewissen Umgebung G dieses Punktes sicher die Bedingung

$$a - |b| > 0, \quad c - |b| > 0$$

erfüllt. Auf diese Umgebung beschränken wir unsere Betrachtungen. Wir können dann eine dreimal stetig differenzierbare Funktion $d > 0$ so wählen, daß

$$\left.\begin{matrix} a - d \\ c - d \\ d - |b| \end{matrix}\right\} > \varepsilon > 0 \tag{29}$$

---

[28]) Diese Voraussetzung und die über die Differenzierbarkeit der Koeffizienten der Differentialgleichung und ferner die Beschränkung auf eine genügend kleine Umgebung der Anfangsgeraden lassen sich in Sonderfällen mildern.

mit konstantem ε gilt. Dann setzen wir die Differentialgleichung in die Form

$$(30)\quad u_{tt}-(a-d)u_{xx}-(c-d)u_{yy}-\frac{1}{2}(d+b)(u_{xx}+2u_{xy}+u_{yy})$$
$$-\frac{1}{2}(d-b)(u_{xx}-2u_{xy}+u_{yy})+\alpha u_t+\beta u_x+\gamma u_y+\delta u=0.$$

Wir legen nun in den Raum das Gitter der Punkte

$$t=lh,\quad x+y=m\varkappa h,\quad x-y=n\varkappa h\qquad (l,m,n=\ldots-1,0,1,2,\ldots)$$

und ersetzen die Gleichung (30) durch eine Differenzengleichung $L(u)=0$ in diesem Gitter. Wir ordnen zu dem Zweck jedem Gitterpunkt P_0 folgende Nachbarpunkte zu: Die Punkte $P_{\alpha'}$ bzw. P_α, die aus P_0 durch Verschiebung um h bzw. $-h$ in Richtung der t-Achse entstehen; ferner die Punkte $P_1,\ldots,P_8$, die mit P_0 in derselben Parallelebene zur (x,y)-Ebene liegen; vgl. Fig. 11. Diese Punkte bilden ein „Elementaroktaeder" mit den Eckpunkten $P_{\alpha'}, P_\alpha, P_1, P_2, P_3, P_4$.

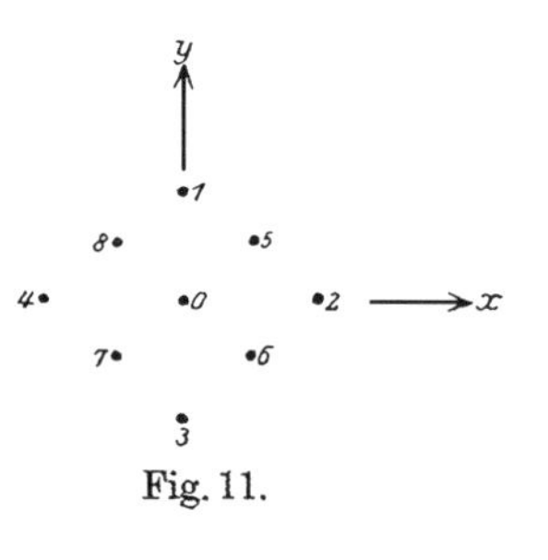

Fig. 11.

Für jeden Gitterpunkt P_0, der innerhalb von G liegt, ersetzen wir die in (30) auftretenden zweiten Differentialquotienten folgendermaßen durch Differenzenquotienten aus dem zu P_0 gehörigen Elementaroktaeder. Wir ersetzen

$$u_{tt}\quad\text{durch}\quad \frac{1}{h^2}(u_{\alpha'}-2u_0+u_\alpha)$$

$$u_{xx}\quad\text{durch}\quad \frac{1}{\varkappa^2h^2}(u_2-2u_0+u_4)$$

$$u_{yy}\quad\text{durch}\quad \frac{1}{\varkappa^2h^2}(u_1-2u_0+u_3)$$

$$u_{xx}+2u_{xy}+u_{yy}\quad\text{durch}\quad \frac{4}{\varkappa^2h^2}(u_6-2u_0+u_8)$$

$$u_{xx}-2u_{xy}+u_{yy}\quad\text{durch}\quad \frac{4}{\varkappa^2h^2}(u_5-2u_0+u_7).$$

Die in (30) auftretenden ersten Differentialquotienten ersetzen wir durch irgendwelche entsprechende Differenzenquotienten in dem Elementaroktaeder. Den Koeffizienten in der Differenzengleichung geben wir die Werte, die die Koeffizienten der Differentialgleichung im Punkte P_0 annehmen.

Auf den ersten beiden Anfangsebenen $t=0$ und $t=h$ denken wir uns die Funktionswerte so vorgegeben, daß sie bei Verfeinerung der Maschenweiten unter Festhaltung des Verhältnisses $\varkappa$ der Raum- zur Zeitmaschenweite gegen die vorgegebenen Anfangswerte auf $t = 0$ streben, wobei die zwischen den beiden Ebenen $t = 0$ und $t = h$ gebildeten Differenzenquotienten bis zur vierten Ordnung gegen die entsprechenden vorgegebenen Differentialquotienten gleichmäßig konvergieren sollen.

Die Lösung der Differenzengleichung $L(u) = 0$ in einem Punkte ist eindeutig durch die Werte auf den beiden Grundflächen der durch ihn gehenden Bestimmtheitspyramide bestimmt.

Für den Konvergenzbeweis bilden wir die über alle Elementaroktaeder einer Bestimmtheitspyramide erstreckte Summe

$$h^3 \sum \sum \sum 2 \frac{u_{a'} - u_a}{h} L(u)$$

und formen sie vermittels der Identitäten (7), (8) um. Dadurch entsteht einmal eine mit h^3 multiplizierte Raumsumme, die in den ersten Differenzenquotienten quadratisch ist, und ferner eine mit h^2 multiplizierte Summe über die Seitendoppelflächen, in denen die Quadrate *aller* auf und zwischen den Doppelflächen vorkommenden Differenzenquotienten vom Typus $u_a - u_0, u_a - u_1, \ldots, u_a - u_8$ auftreten, wobei ihre Koeffizienten wegen (29) größer als eine feste positive Konstante sind, wenn wir überdies noch das Verhältnis $\frac{1}{\varkappa}$ von Zeit- zu Raummaschenweite genügend klein wählen.

Von hier aus können wir ganz ebenso vorgehen wie in den §§ 7, 4 und nachweisen, daß die Lösungen unserer Differenzengleichung gegen die Lösung der Differentialgleichung konvergieren.

(Eingegangen am 1. 9. 1927.)

Appendix C

Reprint of CFL paper English translation[1]

[1] Special permission granted by *IBM Journal of Research and Development*.

C.A. de Moura, C.S. Kubrusly (eds.), *The Courant–Friedrichs–Lewy (CFL) Condition*, DOI 10.1007/978-0-8176-8394-8, © Springer Science+Business Media New York 2013

R. Courant*
K. Friedrichs*
H. Lewy†

On the Partial Difference Equations of Mathematical Physics

Editor's note: This paper, which originally appeared in *Mathematische Annalen* 100, 32–74 (1928), is republished by permission of the authors. We are also grateful to the Atomic Energy Commission for permission to republish this translation, which had appeared as AEC Report NYO-7689, and to Phyllis Fox, the translator, who did the work at the AEC Computing Facility at New York University under AEC Contract No. AT(30-1)-1480. Professor Eugene Isaacson had made suggestions on this translation.

Introduction

Problems involving the classical linear partial differential equations of mathematical physics can be reduced to algebraic ones of a very much simpler structure by replacing the differentials by difference quotients on some (say rectilinear) mesh. This paper will undertake an elementary discussion of these algebraic problems, in particular of the behavior of the solution as the mesh width tends to zero. For present purposes we limit ourselves mainly to simple but typical cases, and treat them in such a way that the applicability of the method to more general difference equations and to those with arbitrarily many independent variables is made clear.

Corresponding to the correctly posed problems for partial differential equations we will treat boundary value and eigenvalue problems for elliptic difference equations, and initial value problems for the hyperbolic or parabolic cases. We will show by typical examples that the passage to the limit is indeed possible, i.e., that the solution of the difference equation converges to the solution of the corresponding differential equation; in fact we will find that for elliptic equations in general a difference quotient of arbitrarily high order tends to the corresponding derivative. Nowhere do we assume the existence of the solution to the differential equation problem—on the contrary, we obtain a simple existence proof by using the limiting process.[1] For the case of elliptic equations convergence is obtained independently of the choice of mesh, but we will find that for the case of the initial value problem for hyperbolic equations, convergence is obtained only if the ratio of the mesh widths in different directions satisfies certain inequalities which in turn depend on the position of the characteristics relative to the mesh.

We take as a typical case the boundary value problem of potential theory. Its solution and its relation to the solution of the corresponding difference equation has been extensively treated during the past few years.[2] However in contrast to the present paper, the previous work has involved the use of quite special characteristics of the potential equation so that the applicability of the method used there to other problems has not been immediately evident.

In addition to the main part of the paper, we append an elementary algebraic discussion of the connection of the boundary value problem of elliptic equations with the random walk problem arising in statistics.

[1] Our method of proof may be extended without difficulty to cover boundary value and eigenvalue problems for arbitrary linear elliptic differential equations and initial value problems for arbitrary linear hyperbolic differential equations.

* Now at Courant Institute of Mathematical Sciences, New York University.

† Now at University of California, Berkeley.

[2] J. le Roux, "Sur le problem de Dirichlet", *Journ. de mathém. pur. et appl.* (6) **10**, 189 (1914). R. G. D. Richardson, "A new method in boundary problems for differential equations", *Trans. of the Am. Math. Soc.* **18**, p. 489 ff, (1917). H. B. Philips and N. Wiener, *Nets and the Dirichlet Problem*, Publ. of M.I.T. (1925). Unfortunately these papers were not known by the first of the three authors when he prepared his note "On the theory of partial difference equations," *Gött. Nachr.* 23, X, 1925, on which the present work depends. See also L. Lusternik, "On an application of the direct method in variation calculus," Recueil de la Société Mathém. de Moscou, 1926. G. Bouligand, "Sur le probléme de Dirichlet," *Ann. de la soc. polon. de mathém.* **4**, Cracow (1926). On the meaning of the difference expressions and on further applications of them, see R. Courant, "Über direkte Methoden in der Variationsrechnung," *Math. Ann.* **97**, p. 711, and the references given therein.

I. The elliptic case

Section 1. Preliminary remarks

• 1. *Definitions*

Consider a rectangular array of points in the (x, y)-plane, such that for mesh width $h > 0$ the points of the lattice are given by

$$\left.\begin{aligned} x &= nh \\ y &= mh \end{aligned}\right\} \quad m, n = 0, \pm 1, \pm 2, \cdots .$$

Let G be a region of the plane bounded by a continuous closed curve which has no double points. Then the corresponding mesh region, G_h—which is uniquely determined for sufficiently small mesh width—consists of all those mesh points lying in G which can be connected to any other given point in G by a connected chain of mesh points. By a connected chain of mesh points we mean a sequence of points such that each point follows in the sequence one of its four neighboring points. We denote as a *boundary point* of G_h a point whose four neighboring points do not all belong to G_h. All other points of G_h we call *interior points.*

We shall consider functions $u, v, \cdots$ of position on the grid, i.e., functions which are defined only for grid points, but we shall denote them as $u(x, y), v(x, y), \cdots$. For their forward and backward difference quotients we employ the following notation,

$$u_x = \frac{1}{h}[u(x+h, y) - u(x, y)],$$

$$u_y = \frac{1}{h}[u(x, y+h) - u(x, y)].$$

$$u_{\bar{x}} = \frac{1}{h}[u(x, y) - u(x-h, y)],$$

$$u_{\bar{y}} = \frac{1}{h}[u(x, y) - u(x, y-h)].$$

In the same way the difference quotients of higher order are formed, e.g.,

$$(u_x)_{\bar{x}} = u_{x\bar{x}} = u_{\bar{x}x}$$
$$= \frac{1}{h^2}[u(x+h, y) - 2u(x, y) + u(x-h, y)], \text{ etc.}$$

• 2. *Difference expressions and Green's function*

In order to study linear difference expressions of second order, we form (using as a model the theory of partial differential equations), a bilinear expression from the forward difference quotients of two functions, u and v,

$$B(u, v) = au_xv_x + bu_xv_y + cu_yv_x + du_yv_y + \alpha u_xv$$
$$+ \beta u_yv + \gamma uv_x + \delta uv_y + guv,$$

where $a = a(x, y), \cdots, \alpha = \alpha(x, y), \cdots, g = g(x, y)$ are functions on the mesh.

From the bilinear expression of first order we derive a difference expression of second order in the following way: we form the sum

$$h^2 \sum_{G_h}\sum B(u, v)$$

over all points of a region G_h in the mesh, where in $B(u, v)$ the difference quotients between boundary points and points not belonging to the mesh are to be set equal to zero. We now transform the sum through partial summation, i.e., we arrange the sum according to v, and split it up into a sum over the set of interior points, G_h' and a sum over the set of boundary points, Γ_h. Thus we obtain:

$$h^2 \sum_{G_h}\sum B(u, v)$$
$$= -h^2 \sum_{G_h'}\sum vL(u) - h\sum_{\Gamma_h} v\Re(u). \quad (1)$$

$L(u)$ is a linear difference expression of second order defined for all interior points of G_h:

$$L(u) = (au_x)_{\bar{x}} + (bu_x)_{\bar{y}} + (cu_y)_{\bar{x}} + (du_y)_{\bar{y}}$$
$$- \alpha u_x - \beta u_y + (\gamma u)_{\bar{x}} + (\delta u)_{\bar{y}} - gu.$$

$\Re(u)$ is, for every boundary point, a linear difference expression whose exact form will not be given here.

If we arrange $\sum_{G_h}\sum B(u, v)$ according to u, we find

$$h^2 \sum_{G_h}\sum B(u, v)$$
$$= -h^2 \sum_{G_h'}\sum uM(v) - h\sum_{\Gamma_h'} u\mathcal{S}(v). \quad (2)$$

$M(v)$ is called the adjoint difference expression of $L(u)$ and has the form

$$M(v) = (av_x)_{\bar{x}} + (bv_y)_{\bar{x}} + (cv_x)_{\bar{y}} + (dv_y)_{\bar{y}}$$
$$+ (\alpha v)_{\bar{x}} + (\beta v)_{\bar{y}} - \gamma v_x - \delta v_y - gv,$$

while $\mathcal{S}(v)$ is a difference expression corresponding to $\Re(u)$ for the boundary.

Formulas (1) and (2) give

$$h \sum_{G_h'}\sum [vL(u) - uM(v)]$$
$$+ h\sum_{\Gamma_h'} [v\Re(u) - u\mathcal{S}(v)] = 0. \quad (3)$$

Formulas (1), (2), and (3) are called *Green's formula.*

The simplest and most important case results if the bilinear form is symmetric, i.e., if the relations $b = c$, $\alpha = \gamma$, $\beta = \delta$ hold. In this case $L(u)$ is identical with $M(u)$—the self-adjoint case—and it can be derived from the quadratic expression

$$B(u, u) = au_x^2 + 2bu_xu_y + du_y^2$$
$$+ 2\alpha u_xu + 2\beta u_yu + gu^2.$$

In the following we shall limit ourselves mainly to expressions $L(u)$ which are self-adjoint. The character of the difference expression $L(u)$ depends principally on the nature of those terms in the quadratic form $B(u, u)$ which are quadratic in the first difference quotients. We call this part of $B(u, u)$ the *characteristic form*:

$$P(u, u) = au_x^2 + 2bu_xu_y + du_y^2.$$

We call the corresponding difference expression $L(u)$ elliptic or hyperbolic, depending on whether the function $P(u, u)$ of the difference quotients is (positive) definite or indefinite.

The difference expression $\Delta u = u_{x\bar{x}} + u_{y\bar{y}}$ with which we shall concern ourselves in the following paragraph is elliptic, i.e., it comes from the quadratic expression

$$B(u, u) = u_x^2 + u_y^2 \quad \text{or} \quad u_{\bar{x}}^2 + u_{\bar{y}}^2.$$

The corresponding Green's formulas are

$$h^2 \sum_{G_h}\sum (u_x^2 + u_y^2) = -h^2 \sum_{G_h'}\sum u\Delta u - h \sum_{\Gamma_h} u\mathfrak{R}(u) \qquad \text{[Note 3]} \qquad (4)$$

$$h^2 \sum_{G_h'}\sum (v\Delta u - u\Delta v) + h \sum_{\Gamma_h} [v\mathfrak{R}(u) - u\mathfrak{R}(v)] = 0. \qquad (5)$$

The difference expression, $\Delta u = u_{x\bar{x}} + u_{y\bar{y}}$, is obviously the analogue of the differential expression $(\partial^2 u/\partial x^2) + (\partial^2 u/\partial y^2)$ for a function $u(x, y)$ of the continuous variables x and y. Written out explicitly the difference expression is

$$\Delta u = \frac{1}{h^2}[u(x + h, y) + u(x, y + h) + u(x - h, y) + u(x, y - h) - 4u(x, y)].$$

Therefore $(h^2/4)\Delta u$ is the excess of the arithmetic mean of the functional values at the four neighborhood points over the functional value at the point in question.

Completely similar considerations lead to linear difference expressions of the fourth order and corresponding Green's formula, provided one starts from bilinear difference expressions which are formed from the difference quotients of second order. Consider for example the difference expression

$$\Delta\Delta u = u_{xx\bar{x}\bar{x}} + 2u_{x\bar{x}y\bar{y}} + u_{yy\bar{y}\bar{y}}.$$

This corresponds to the quadratic expression

$$B(u, u) = (u_{x\bar{x}} + u_{y\bar{y}})^2 = (\Delta u)^2,$$

provided one orders the sum

$$h^2 \sum_{G_h'}\sum \Delta u \Delta v$$

according to v, or equivalently replaces u by Δu in Eq. (5). One must notice however that in the expression $\Delta\Delta u$, the functional value at a point is connected with the values at its neighboring points and at their neighboring points, and accordingly is defined only for such points of the region G_h as are also interior points of the region G_h' (See Section 5). The entirety of such points we designate as G_h''. We obtain then Green's formula

$$h^2 \sum_{G_h'}\sum \Delta u \cdot \Delta v = h^2 \sum_{G_h''}\sum v \cdot \Delta\Delta u + h \sum_{\Gamma_h + \Gamma_h'} v \cdot \mathfrak{R}(u), \qquad (6)$$

where $\mathfrak{R}(u)$ is a definable linear difference expression for each point of the boundary strip $\Gamma_h + \Gamma_h'$. Γ_h' indicates the set of boundary points of G_h'.

Section 2. Boundary value and eigenvalue problems

• 1. *The theory of boundary value problems*

The boundary value problem for linear elliptic homogeneous difference equations of second order, which corresponds to the classical boundary value problem for partial differential equations, can be formulated in the following way.

Let there be given a self-adjoint elliptic linear difference expression of second order, $L(u)$, in a mesh region, G_h. $L(u)$ results from a quadratic expression $B(u, u)$ which is positive definite in the sense that it cannot vanish if u_x and u_y do not themselves vanish.

A function, u, is to be determined satisfying in G_h the difference equation $L(u) = 0$, and assuming prescribed values at the boundary points.

Under these requirements there will be exactly as many linear equations as there are interior points of the mesh at which the function u is to be determined.[4] Some of these equations which involve only mesh points whose neighbors also lie in the interior of the region are homogeneous; others which involve boundary points of the mesh region are inhomogeneous. If the right-hand side of the inhomogeneous equations is set equal to zero, that is if $u = 0$ on the boundary, then it follows from Green's formula (1), by setting $u = v$ that $B(u, u)$ vanishes, and further, from the definiteness of $B(u, u)$, that u_x and u_y vanish, and hence that u itself vanishes. Thus the difference equation for zero boundary values has the solution $u = 0$, or in other words the solution is uniquely determined since the difference of two solutions with the

[3] The boundary expression $\mathfrak{R}(u)$ may be written as follows: Let $u_0, u_1, \ldots, u_\nu$ be values of the function at a boundary point and at its ν neighboring points ($\nu \leq 3$), then

$$\mathfrak{R}(u) = \frac{1}{h}(u_1 + \cdots + u_\nu - \nu u_0).$$

[4] If the matrix of the linear system of equations corresponding to an arbitrary difference equation of second order, $L(u) = 0$, is transposed, then the transposed set of equations corresponds to the adjoint difference equation $M(v) = 0$. Thus the above self-adjoint system gives rise to a set of linear equations with symmetric coefficients.

same boundary value must vanish. Further, if a linear system of equations with as many unknowns as equations is such that for vanishing right-hand side the unknowns must vanish, then the fundamental theorem of the theory of equations asserts that for an arbitrary right-hand side exactly one solution must exist. In our case there follows at once the existence of a solution of the boundary value problem.

Therefore we see that for elliptic difference equations the uniqueness and existence of the solution of the boundary value problem are related to each other through the fundamental theorem of the theory of linear equations, whereas for partial differential equations both facts must be proved by quite different methods. The basis for this difficulty in the latter case is to be found in the fact that the differential equations are no longer equivalent to a finite number of equations, and so one can no longer depend on the equality of the number of unknowns and equations.

Since the difference expression $\Delta u = 0$ can be derived from the positive definite quadratic expression

$$h^2 \sum_{G_h} \sum (u_x^2 + u_y^2),$$

the boundary value problem for the difference expression is uniquely solvable.

The theory for difference equations of higher order is developed in exactly the same way as that for difference equations of second order; for example one can treat the fourth order difference equation $\Delta\Delta u = 0$. In this case the values of the function must be prescribed on the boundary strip $\Gamma_h + \Gamma_h'$. Evidently here also the difference equation yields just as many linear equations as there are unknown functional values at the points of G_h''. In order to demonstrate the existence and uniqueness of the solution one needs only to show that a solution which has the value zero in the boundary strip $\Gamma_h + \Gamma_h'$ necessarily vanishes identically. To this end we note that the sum over the corresponding quadratic expression

$$h^2 \sum_{G_h''} \sum (\Delta u)^2 \tag{7}$$

for such a function vanishes, as can be seen by transforming the sum according to Green's formula (6). The vanishing of the sum (7) however implies that Δu vanishes at all points of G_h', and according to the above proof this can only happen for vanishing boundary values if the function u assumes the value zero throughout the region. Thus our assertion is proved, and both the existence and uniqueness of the solution to the boundary value problem for the difference equations are guaranteed.[5]

[5] For the actual process of carrying through the solution of the boundary value problem by an iterative method, see among others the treatment: "Über Randwertaufgaben bei partiellen Differenzengleichungen" by R. Courant, *Zeitschr. f. angew. Mathematik u. Mechanik* **6**, 322–325 (1925). Also there is a report by H. Henky, in *Zeitschr. f. angew. Math. u. Mech.* **2**, 58 ff (1922).

• 2. *Relation to the minimum problem*

The above boundary value problem is related to the following minimum problem: among all functions $\varphi(x, y)$ defined in the mesh region G_h and assuming given values at the boundary points, that function $\varphi = u(x, y)$ is to be found for which the sum

$$h^2 \sum_{G_h} \sum B(\varphi, \varphi)$$

over the mesh region assumes the least possible value. We assume that the quadratic difference expression of first order, $B(u, u)$ is positive definite in the sense described in Section 1, Part 2. One can show that the difference equation $L(\varphi) = 0$ results from this minimum requirement on the solution $\varphi = u(x, y)$, where $L(\varphi)$ is the difference expression of second order derived previously from $B(\varphi, \varphi)$. In fact this can be seen either by applying the rules of differential calculus to the sums $h^2 \sum_{G_h} \sum B(\varphi, \varphi)$ as a function of a finite number of values of φ at the grid points, or by employing the usual methods from the calculus of variations.

By way of example, solving the boundary value problem of finding the solution to the equation $\Delta\varphi = 0$ which assumes given boundary values, is equivalent to minimizing the sum $h^2 \sum_{G_h} \sum (\varphi_x^2 + \varphi_y^2)$ over all functions which take on the boundary values.

There is a similar correspondence for the fourth-order difference equations, where we limit ourselves to the example $\Delta\Delta\varphi = 0$. The boundary value problem corresponding to this difference equation is equivalent to the problem of minimizing the sum

$$h^2 \sum_{G_h'} \sum (\Delta\varphi)^2$$

for functions that take on given values on the boundary strip Γ_h'. Besides this sum there are yet other quadratic expressions in the second derivatives which give rise to the equation $\Delta\Delta u = 0$ under the process of being minimized. For example this is true in G_h for the sum

$$h^2 \sum_{G_h'} \sum (u_{xx}^2 + 2u_{xy}^2 + u_{yy}^2).$$

That the minimum problem posed above always has a solution follows from the theorem that a continuous function of a finite number of variables (the functional values of φ at the grid points) always has a minimum if it is bounded from below and if it tends to infinity as soon as any of the independent variables goes to infinity.[6]

• 3. *Green's function*

It is possible to treat the boundary value problem for the inhomogeneous equation, $L(u) = -f$, in much the same way as the homogeneous case, $L(u) = 0$. For the inhomogeneous case it is sufficient to consider only the case of

[6] It can easily be verified that the hypotheses for the application of this theorem are satisfied.

zero boundary conditions, since different boundary conditions can be taken care of by adding a suitable solution of the homogeneous equation. To solve the linear system of equations representing the boundary value problem, $L(u) = -f$, we first choose as the function $f(x, y)$ a function which assumes the value $-1/h^2$ at the point $x = \xi$, $y = \eta$ of the mesh. If $K(x, y; \xi, \eta)$ is the solution (vanishing on the boundary) of this difference equation which depends on the parametric point (ξ, η), then the solution for an arbitrary boundary condition can be represented by the sum

$$u(x, y) = h^2 \sum_{(\xi,\eta) \text{ in } G_h} K(x, y; \xi, \eta) f(\xi, \eta).$$

The function $K(x, y; \xi, \eta)$ which depends on the points (x, y) and (ξ, η) is called the Green's function of the differential expressions $L(u)$. If we call the Green's function for the adjoint expression $M(v)$, $\bar{K}(x, y; \xi, \eta)$, then the equivalence

$$K(\bar{\xi}, \bar{\eta}; \xi, \eta) = \bar{K}(\xi, \eta; \bar{\xi}, \bar{\eta}) \quad \text{holds,}$$

as can be seen to follow immediately from Green's formula (3) when $u = K(x, y; \xi, \eta)$, and $v = \bar{K}(x, y; \bar{\xi}, \bar{\eta})$. For a self-adjoint difference expression the above relation gives the symmetric expression

$$K(\bar{\xi}, \bar{\eta}; \xi, \eta) = K(\xi, \eta; \bar{\xi}, \bar{\eta}).$$

• 4. *Eigenvalue problems*

Self-adjoint difference expressions, $L(u)$, give rise to eigenvalue problems of the following type: find the value of a parameter λ, the eigenvalue, such that in G_h, a solution, the eigenfunction can be found for the difference equation $L(u) + \lambda u = 0$, where u is to be zero on the boundary, Γ_h.

The eigenvalue problem is equivalent to finding the principle axes of the quadratic form $B(u, u)$. Exactly as many eigenvalues and corresponding eigenfunctions exist as there are interior mesh points of the region G_h. The system of eigenfunctions and eigenvalues (and a proof of their existence) is given by the minimum problem:

Among all functions, $\varphi(x, y)$, vanishing on the boundary, and satisfying the orthogonality relation

$$h^2 \sum_{G_h} \sum \varphi u^{(\nu)} = 0, \qquad (\nu = 1, \cdots, m - 1)$$

and normalized such that

$$h^2 \sum_{G_h} \sum \varphi^2 = 1,$$

the function, $\varphi = u$, is to be found for which the sum

$$h^2 \sum_{G_h} \sum B(\varphi, \varphi)$$

assumes its minimum value. The value of this minimum is the mth eigenvalue, and the function for which it is assumed is the mth eigenfunction.[7]

Section 3.[8] *Connections with the problem of the random walk*
The theme of the following is related to a question from the theory of probability, namely the problem of the random walk in a bounded region.[9] We consider the lines of a mesh region G_h as paths along which a particle can move from one grid point to a neighboring one. In this net of streets the particle can wander aimlessly, and it can choose at random one of the four directions leading from each intersection of paths of the net—all four directions being equally probable. The walk ends as soon as a boundary point of G_h is reached because here the particle must be absorbed.

We ask:

1) What is the probability $w(P; R)$ that a random walk starting from a point P reaches a particle point R of the boundary?
2) What is the mathematical expectation $v(P; Q)$ that a random walk starting from P reaches a point Q of G_h without touching the boundary?

This probability or mathematical expectation, respectively, will be defined more precisely by the following process. Assume that at the point P there is a unit concentration of matter. Let this matter diffuse into the mesh with constant velocity, traveling say a mesh width in unit time. At each meshpoint let exactly one-fourth of the matter at the point diffuse outwards in each of the four possible directions. The matter which reaches a boundary point is to remain at that point. If the point of origin P is itself a boundary point, then the matter never leaves that point.

We define the probability $w(P; R)$ that a random walk starting from P reaches the boundary point R (without

[7] From the orthogonality condition on the eigenfunctions,

$$h^2 \sum_{G_h} \sum u^{(\nu)} u^{(\mu)} = 0, \qquad (\nu \neq \mu)$$

it follows that each function, $g(x, y)$, which vanishes on the boundary of the mesh can be expanded in terms of the eigenfunctions in the form

$$y = \sum_{\nu=1}^{N} c^{(\nu)} u^{(\nu)},$$

where the coefficients are determined from the equations

$$c^{(\nu)} = h^2 \sum_{G_h} \sum g u^{(\nu)}.$$

In this way in particular the following representation for the Green's functions may be derived,

$$K(x, y; \xi, \eta) = -\sum_{\nu=1}^{N} \frac{u^{(\nu)}(x, y) \cdot u^{(\nu)}(\xi, \eta)}{\lambda^{(\nu)}}.$$

[8] Section 3 is not prerequisite to Section 4.
[9] The present treatment is essentially different from the familiar treatments which can be carried through, say for example in the case of Brownian motion for molecules. The difference lies precisely in the way in which the boundary of the region enters.

previously attaining the boundary), as the amount of matter which accumulates at this boundary point over an infinite amount of time.

We define the probability $E_n(P; Q)$ that the walk starting from the point P reaches the point Q in exactly n steps without touching the boundary by the amount of matter which accumulate in n units of time provided P and Q are both interior points. If either P or Q are boundary points then E_n is set equal to zero.

The value $E_n(P; Q)$ is exactly equal to $1/4^n$ times the number of paths of n steps leading from P to Q without touching the boundary. Thus $E_n(P; Q) = E_n(Q; P)$.

We define the mathematical expectation $v(P; Q)$ that one of the paths considered above leads from P to the point Q by the infinite sum of all of these possibilities,

$$v(P; Q) = \sum_{\nu=0}^{\infty} E_\nu(P; Q) \qquad \text{(Note 10)},$$

i.e., for the interior points P and Q, the sum of all the concentrations which have passed through the point Q at different times. This will be assigned the expected value 1 for a concentration originating at Q.

If the amount arriving at the boundary point R in exactly n steps is designated as $F_n(P; R)$, then the probability $w(P; R)$ is given by the series

$$w(P; R) = \sum_{\nu=0}^{\infty} F_\nu(P; R).$$

All the terms of this series are positive and the partial sum is bounded by one (since the concentration reaching the boundary can be made up of only part of the initial concentration), and therefore the convergence of the series is assured.

Now it is easy to see that the probability $E_n(P; Q)$, that is, the concentration reaching the point Q in exactly n steps tends to zero as n increases. For if at any point Q, from which the boundary point R can be reached in m steps, we have $E_n(P; Q) > \alpha > 0$, then at least $\alpha/4^m$ of the concentration at Q will reach the point R after m steps. However, since the sum $\sum_{\nu=0}^{\infty} F_\nu(P; R)$ converges, the concentration reaching R goes to zero with increasing time, where the value of $E_n(P; Q)$ must itself vanish as time increases; that is, the probability of an infinitely long walk remaining in the interior of the region is zero.

From this it follows that the entire concentration eventually reaches the boundary; or in other words that the sum $\sum_R w(P; R)$ over all the boundary points R is equal to one.

The convergence of the infinite series for the mathematical expectation

$$v(P; Q) = \sum_{\nu=0}^{\infty} E_\nu(P; Q)$$

remains to be shown.

[10] The convergence will be shown below.

To this end we remark that the quantity $E_n(P; Q)$ satisfies the following relations

$$E_{n+1}(P; Q) = \tfrac{1}{4}\{E_n(P; Q_1) + E_n(P; Q_2) + E_n(P; Q_3) + E_n(P; Q_4)\}, \quad [n \geq 1],$$

where Q_1 through Q_4 are the four neighboring points of the interior point Q. That is, the concentration at the point Q at the $n + 1^{\text{st}}$ step consists of 1/4 times the sum of the concentrations at the four neighboring points at the nth step. If one of the neighbors of Q, for example $Q_1 = R$, is a boundary point then it follows that no concentration flows from this boundary point to Q since the expression $E_n(P; R)$ is zero in this case. Furthermore, for an interior point, $E_0(P; P) = 1$ and of course $E_1(P; Q) = 0$.

From these relationships we find for the partial sum

$$v_n(P; Q) = \sum_{\nu=0}^{n} E_\nu(P; Q)$$

the equation

$$v_{n+1}(P; Q) = \tfrac{1}{4}\{v_n(P; Q_1) + v_n(P; Q_2) + v_n(P; Q_3) + v_n(P; Q_4)\},$$

for $P \neq Q$. For the case of $P = Q$,

$$v_{n+1}(P; P) = 1 + \tfrac{1}{4}\{v_n(P; P_1) + v_n(P; P_2) + v_n(P; P_3) + v_n(P; P_4)\},$$

that is, the expectation that a point goes back into itself consists of the expectation that a nonvanishing path leads from P back again to itself—namely,

$$\tfrac{1}{4}\{v_n(P; P_1) + v_n(P; P_2) + v_n(P; P_3) + v_n(P; P_4)\},$$

together with the expectation unity that expresses the initial position of the concentration itself at this point.

The quantity $v_n(P; Q)$ thus satisfies the following difference equation[11]

$$\Delta v_n(P; Q) = \frac{4}{h^2} E_n(P; Q), \qquad \text{for} \quad P \neq Q,$$

$$\Delta v_n(P; Q) = \frac{4}{h^2} \{E_n(P; Q) - 1\}, \quad \text{for} \quad P = Q.$$

$v_n(P; Q)$ is equal to zero when Q is a boundary point.

[11] This defines the Δ-operation for the variable point Q. This equation can be interpreted as an equation of the heat conduction type. That is, if the function $v_n(P; Q)$ is considered, not as a function of the index n as in our presentation above, but rather as a function of time, t, which is proportional to n, so that $t = n\tau$ and $v_n(P; Q) = v(P; Q; t) = v(t)$, then the above equations can be written in the following form:

$$\Delta v(t) = \frac{4\tau}{h^2} \cdot \frac{v(t + \tau) - v(t)}{\tau} \qquad \text{for} \quad P \neq Q,$$

$$\Delta v(t) = \frac{4\tau}{h^2} \cdot \left(\frac{v(t + \tau) - v(t)}{\tau} - \frac{1}{\tau}\right) \quad \text{for} \quad P = Q.$$

For a similar difference equation which has a parabolic differential equation as its limiting form, see Section 6 of the second half of the paper.

The solution of this boundary value problem for arbitrary right-hand side is uniquely determined as we have explained earlier (Section 2, Part 1), and depends continuously on the right-hand side. Since the variables $E_n(P; Q)$ tend to zero, the solution $v_n(P; Q)$ converges to the solution $v(P; Q)$ of the difference equation

$$\Delta v(P; Q) = 0 \qquad \text{for} \quad P \neq Q$$

$$\Delta v(P; Q) = -\frac{4}{h^2} \quad \text{for} \quad P = Q,$$

with boundary values $v(P; R) = 0$.

Thus we see that the mathematical expectation $v(P; Q)$ exists and is none other than the Green's function for the difference equation $\Delta u = 0$, except for a factor of 4. The symmetry of the Green's function is an immediate consequence of the symmetry of the quantity $E_n(P; Q)$ which was used to define it.

The probability $w(P; R)$ satisfies, with respect to P, the relation

$$w(P; R) = \tfrac{1}{4}\{w(P_1; R) + w(P_2; R) + w(P_3; R) + w(P_4; R)\},$$

and thus the difference equation $\Delta w = 0$. That is, if P_1, P_2, P_3, P_4, are the four neighboring points of the interior point P, then each path from P to R must pass through one of these four directions, and each of the four is equally likely. Furthermore, the probability of going from one boundary point R to another R' is $w(R; R') = 0$ unless the two points R and R' coincide, in which case $w(R; R) = 1$. Thus $w(P; R)$ is that solution of the boundary value problem $\Delta w = 0$ which assumes the value 1 at the boundary point R and the value 0 at all other points of the boundary. Therefore the solution of the boundary value problem for an arbitrary boundary value $u(R)$ has the simple form $u(P) = \sum_R w(P; R)u(R)$, where the sum is to be extended over all the boundary points.[12] If the function $u \equiv 1$ is substituted for u in this expression, then we check the relation $1 = \sum_R w(P; R)$.

The interpretation given above for Green's function as an expectation yields immediately further properties. We mention only the fact that the Green's function decreases if one goes from the region G to a subregion $\tilde{G}$ lying within G; that is, the number of possible paths for steps on the mesh leading from one point P to another Q (without touching the boundary), decreases as the region decreases.

Of course, corresponding relationships hold for more than two independent variables. We note only that other elliptic difference equations admit a similar probability interpretation.

If the limit for vanishing mesh width is considered by methods given in the following section, then the Green's function on the mesh goes over to the Green's function of the potential equation except for a numerical factor; a similar relationship holds between the expression $w(P; R)/h$ and the normal derivative of the Green's function at the boundary of the region. In this way, for example, the Green's function for the potential equation could be interpreted as the specific mathematical expectation of going from one point to another,[13] without reaching the boundary.

In going over to the limit of a continuum from the mesh, the influence of the direction in the mesh prescribed for the random walk vanishes. This fact suggests that it would be of some interest to consider limiting cases of more general random walks for which the limitations on the direction of the walk are relaxed. This lies outside of the scope of this presentation, however, and we can only hope to renew the question at some other opportunity.

Section 4. The solution of the differential equation as a limiting form of the solution of the difference equation

- 1. *The boundary value problem of potential theory*

In letting the solution of the difference equation tend to the solution of the corresponding differential equation we shall relinquish the greatest possible degree of generality with regard to the boundary and boundary values in order to demonstrate more clearly the character of our method.[14] Therefore we assume that we are given a simply connected region G with a boundary formed of a finite number of arcs with continuously turning tangents. Let $f(x, y)$ be a given function which is continuous and has continuous partial derivatives of first and second order in a region containing G. If G_h is the mesh region with mesh width h belonging to the region G, then let the boundary value problem for the difference equation $\Delta u = 0$ be solved for the same boundary values which the function $f(x, y)$ assumes on the boundary; let $u_h(x, y)$ be the solution. We shall prove that as $h \to 0$ the function $u_h(x, y)$ converges to a function $u(x, y)$ which satisfies in G the partial differential equation $(\partial^2 u/\partial x^2) + (\partial^2 u/\partial y^2) = 0$ and takes the value of $f(x, y)$ at each of the points of the boundary. We shall show further that for any region lying entirely within G the difference quotients of u_h of arbitrary order tend uniformly towards the corresponding partial derivatives of $u(x, y)$.

In the convergence proof it is convenient to replace the boundary condition $u = f$ by the weaker requirement that

$$\frac{1}{r}\iint_{S_r} (u - f)^2 \, dxdy \to 0 \quad \text{as} \quad r \to 0,$$

[12] Moreover it is easy to show that the probability $w(P; R)$ of reaching the boundary is the boundary expression $\mathfrak{R}(K(P, Q))$, constructed from the Green's function $K(P; Q)$ in terms of Q, where $u(x, y)$ is to be identified with $w(P, Q)$, and $v(x, y)$ with $v(P, Q)$ in Green's formula (5).

[13] Here the *a priori* expectation of reaching a certain area element is understood to be equal to the area of the element.

[14] We mention however that carrying through our method for more general boundaries and boundary values in no way causes any fundamental difficulty.

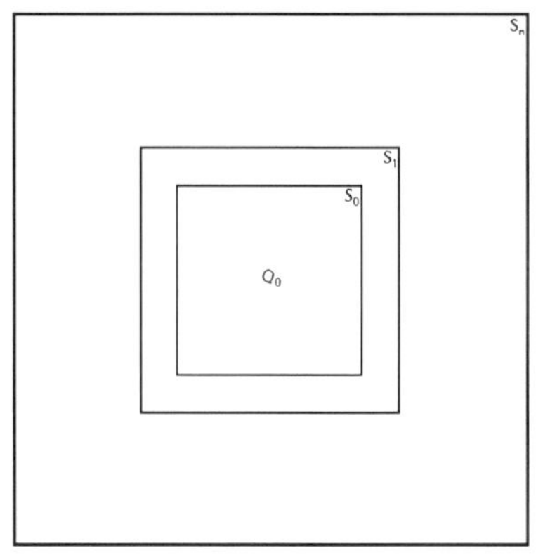

Figure 1

where S_r is that strip of G whose points are at a distance from the boundary smaller than r.[15] The convergence proof depends on the fact that for any subregion G^* lying entirely within G, the function $u_h(x, y)$ and each of its difference quotients is bounded and uniformly continuous as $h \to 0$ in the following sense: For each of these functions, say $w_h(x, y)$, there exists a $\delta(\varepsilon)$ depending only on the subregion and not on h such that

$$|w_h(P) - w_h(P_1)| < \varepsilon$$

provided the mesh points P and P_1 lie in the same subregion of G_h and are separated from each other by a distance less than $\delta(\varepsilon)$.

Once uniform continuity in this sense (equicontinuity) has been established we can in the usual way select from the functions u_h a subsequence of functions which tend uniformly in any subregion G^* to a limit function $u(x, y)$, while the difference quotients of u_h tend uniformly towards the partial derivatives of u. The limit function then possesses derivatives of arbitrarily high order in any proper subregion G^* of G and satisfies $\nabla^2 u = 0$ in this region. If we can show also that u satisfies the boundary condition we can regard it as the solution of our boundary value problem for the region G. Since this solution is uniquely determined, it is clear that not only a partial sequence of the functions u_h, but this sequence of functions itself possesses the required convergence properties.

The uniform continuity (equicontinuity) of our quantities may be established by proving the following lemmas.

1) As $h \to 0$ the sums over the mesh region $h^2 \sum_{G_h} \sum u^2$ and $h^2 \sum_{G_h} \sum (u_x^2 + u_y^2)$ remain bounded.[16]
2) If $w = w_h$ satisfies the difference equation $\Delta w = 0$ at a mesh point of G_h, and if, as $h \to 0$ the sum

   $$h^2 \sum_{G_h^*} \sum w^2,$$

   extended over a mesh region G_h^* associated with a subregion G^* of G remains bounded, then for any fixed subregion G^{**} lying entirely within G^* the sum

   $$h^2 \sum_{G_h^{**}} \sum (w_x^2 + w_y^2)$$

   over the mesh region G_h^{**} associated with G^{**} likewise remains bounded as $h \to 0$. Using 1) there follows from this, since all of the difference quotients w of the function u_h again satisfy the difference equation $\Delta w = 0$, that each of the sums $h^2 \sum_{G_h^*} \sum w^2$ is bounded.
3) From the boundedness of these sums there follows finally the boundedness and uniform continuity of all the difference quotients themselves.

• 2. *Proof of the lemmas*

The proof of 1) follows from the fact that the functional values u_h are themselves bounded. For the greatest (or least) value of the function is assumed on the boundary[17] and so is bounded by a prescribed finite value. The boundedness of the sum $h^2 \sum_{G_h} \sum (u_x^2 + u_y^2)$ is an immediate consequence of the minimum property of our mesh function formulated in Part 2 of Section 2 which gives in particular

$$h^2 \sum_{G_h} \sum (u_x^2 + u_y^2) \le h^2 \sum_{G_h} \sum (f_x^2 + f_y^2),$$

but as $h \to 0$ the sum on the right tends to the integral

$$\iint_G \left[\left(\frac{\partial f}{\partial x}\right)^2 + \left(\frac{\partial f}{\partial y}\right)^2 dxdy\right]$$

which, by hypothesis, exists.

[15] The weaker boundary value requirement does in fact provide the unique characterization of the solution, as can be seen from the easily proved theorem: If the boundary condition above is satisfied for $f(x, y) = 0$ for a function satisfying the equation

$$\frac{\partial^2 u}{\partial x^2} + \frac{\partial^2 u}{\partial y^2} = 0$$

in the interior of G and if

$$\iint_G \left[\left(\frac{\partial u}{\partial x}\right)^2 + \left(\frac{\partial u}{\partial y}\right)^2\right] dxdy$$

exists, then $u(x, y)$ is identically zero. (See Courant, "Über die Lösungen der Differentialgleichungen der Physik," *Math Ann.* **85**, 296 ff.)

In the case of two independent variables the boundary values are actually attained, as follows from the weaker requirement; but in the case of more variables the corresponding result cannot in general be expected since there can exist exceptional points on the boundary at which the boundary value is not taken on even though a solution exists under the weaker requirement.

[16] Here and in the following we drop the index h from the grid functions.

[17] We note, however, with a view to carrying over the method to other differential equations, that we can relax these conditions. To this end we need only to bring into play the inequality (15) or to use the reasoning of the alternative (see Part 4, Section 4).

To prove 2) we consider the quadratic sum

$$h^2 \sum_{\text{int.}Q_1}\sum (w_x^2 + w_{\bar{x}}^2 + w_y^2 + w_{\bar{y}}^2),$$

where the summation extends over all the interior points of a square Q_1, (see Fig. 1). We denote the values of the function on the boundary S_1 of the square Q_1, by w_1, and those on the boundary S_0, of Q_0, by w_0. Then Green's formula gives

$$h^2 \sum_{\text{int.}Q_1}\sum (w_x^2 + w_{\bar{x}}^2 + w_y^2 + w_{\bar{y}}^2) \tag{8}$$
$$= \sum_{S_1} w^2 - \sum_{S_0} w^2 - \sum_{C_1} w^2$$

where S_1 and S_0 are respectively the boundary of Q_1 and the square boundary of the lattice points Q_0 lying within S_1, while C_1 consists of the four corner points of the boundary of Q_1.

We now consider a sequence of concentric squares $Q_0, Q_1, \cdots, Q_n$ with boundaries $S_0, S_1, \cdots, S_n$, where each boundary is separated from the next by a mesh width. Applying the formula to each of these squares and observing that we have always

$$2h^2 \sum_{Q_0}\sum (w_x^2 + w_y^2)$$
$$\leq h^2 \sum_{Q_k}\sum (w_x^2 + w_{\bar{x}}^2 + w_y^2 + w_{\bar{y}}^2), \qquad (k \geq 1)$$

we obtain

$$2h^2 \sum_{Q_0}\sum (w_x^2 + w_y^2)$$
$$\leq \sum_{S_k} w^2 - \sum_{S_{k-1}} w^2 - \sum_{C_k} w^2 \qquad (1 \leq k < n),$$

where C_k consists of the four corner points of the boundary Q_k.

We strengthen the inequality by neglecting the last non-positive term on the right and we then add the n inequalities to obtain

$$2nh^2 \sum_{Q_0}\sum (w_x^2 + w_y^2) \leq \sum_{S_n} w^2 - \sum_{S_0} w^2 \leq \sum_{S_n} w^2.$$

Summing this inequality from $n = 1$ to $n = N$ we get

$$N^2h^2 \sum_{Q_0}\sum (w_x^2 + w_y^2) \leq \sum_{Q_N}\sum w^2.$$

Diminishing the mesh width h we can make the squares Q_0 and Q_n converge towards two fixed squares lying within G and having corresponding sides separated by a distance a. In this process $Nh \to a$ and we have, independent of the mesh width

$$h^2 \sum_{Q_0}\sum (w_x^2 + w_y^2) \leq \frac{h^2}{a^2} \sum_{Q_N}\sum w^2. \tag{9}$$

For sufficiently small h this inequality holds of course not only for two squares Q_0 and Q_N but with a change in the constant, a, for any two subregions of G such that one

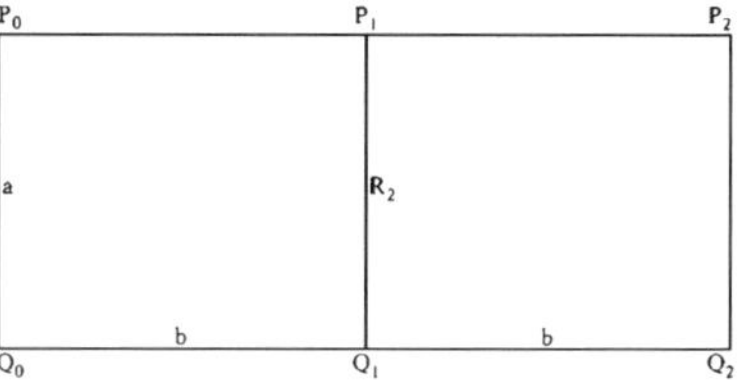

Figure 2

is contained entirely within the other. Thus lemma (2) is proved.[18]

In order to prove the third result, that u_h and all its partial difference quotients w_h remain bounded and uniformly continuous as $h \to 0$, we consider a rectangle R (Fig. 2) with corners P_0, Q_0, P, Q and with sides P_0Q_0 and PQ of length a parallel to the x-axis.

We start with the relation

$$w(Q_0) - w(P_0) = h \sum_{PQ} w_x - h^2 \sum_{R}\sum w_{xy},$$

and the inequality

$$|w(Q_0) - w(P_0)| \leq h \sum_{PQ} |w_x| + h^2 \sum_{R}\sum |w_{xy}|, \tag{11}$$

which is a consequence of it.

We then let the side PQ of the rectangle vary between an initial line P_1Q_1, a distance b from P_0Q_0 and a final line P_2Q_2 a distance $2b$ from P_0Q_0, and we sum the corresponding $(b/h) + 1$ inequalities (11). We obtain the estimate

$$|w(P_0) - w(Q_0)| \leq \frac{1}{b+h} h^2 \sum_{R_2}\sum |w_x|$$
$$+ h^2 \sum_{R_2}\sum |w_{xy}|$$

where the summations extend over the entire rectangle, $R_2 = P_0Q_0P_2Q_2$. From Schwarz's inequality it then follows that,

$$|w(P_0) - w(Q_0)| \leq \frac{1}{b} \sqrt{2ab} \sqrt{h^2 \sum_{R_2}\sum w_x^2} \tag{12}$$
$$+ \sqrt{2ab} \sqrt{h^2 \sum_{R_2}\sum w_{xy}^2}.$$

[18] If we do not assume that $\Delta w = 0$, then in place of the inequality (9) we find:

$$h^2 \sum_{G^{**}}\sum (w_x^2 + w_y^2)$$
$$\leq c_1h^2 \sum_{G^*}\sum w^2 + c_2h^2 \sum_{G^*}\sum (\Delta w)^2 \tag{10}$$

for suitable constants c_1 and c_2 independent of h, where G^{**} lies entirely within G^*, and G^* in turn is contained in the interior of G.

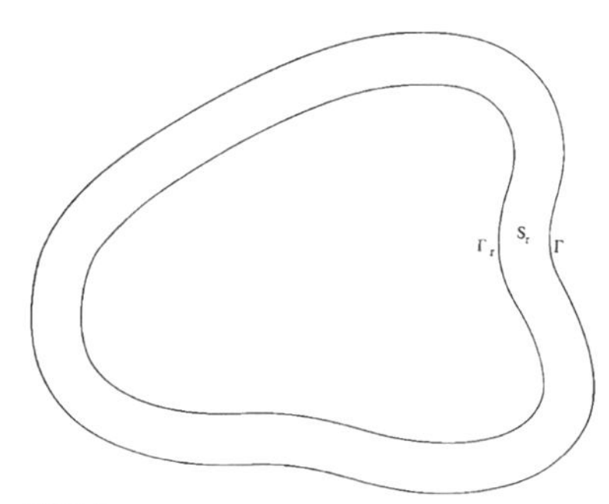

Figure 3

Since, by hypothesis, the sums which occur here multiplied by h^2 remain bounded, it follows that as $a \to 0$ the difference $|w(P_0) - w(Q_0)| \to 0$ independently of the mesh-width, since for each subregion G^* of G the quantity b can be held fixed. Consequently the uniform continuity (equicontinuity) of $w = w_h$ is proved for the x-direction. Similarly it holds for the y-direction and so also for any subregion G^* of G. The boundedness of the function w_h in G^* finally follows from its uniform continuity (equicontinuity) and the boundedness of $h^2 \sum_{G^*} \sum w_h^2$.

By this proof we establish the existence of a subsequence of functions u_h which converge towards a limit function $u(x, y)$ and which do so uniformly together with all their difference quotients, in the sense discussed above for every interior subregion of G. This limit function $u(x, y)$ has throughout G continuous partial derivatives of arbitrary order, and satisfies there the potential equation:

$$\frac{\partial^2 u}{\partial x^2} + \frac{\partial^2 u}{\partial y^2} = 0.$$

• 3. *The boundary condition*

In order to prove that the solution satisfies the boundary condition formulated above we shall first of all establish the inequality

$$h^2 \sum_{S_{r,h}} \sum v^2 \leq Ar^2h^2 \sum_{S_{r,h}} \sum (v_x^2 + v_y^2) + Brh \sum_{\Gamma_h} v^2 \qquad (13)$$

where $S_{r,h}$ is that part of the mesh region G_h which lies inside a boundary strip S_r. This boundary strip S_r consists of all points of G whose distance from the boundary is less than r; it is bounded by Γ and another curve Γ_r (Fig. 3). The constants A and B depend only on the region and not on the function v nor the mesh width h.

In order to prove the above inequality, we divide the boundary, Γ, of G into a finite number of pieces for which the angle of the tangent with one of the x- or y-axes is greater than some positive value (say 30°). Let γ, for instance, be a piece of Γ which is sufficiently steep (in the above sense) relative to the x-axis (see Fig. 4). Lines parallel to the x-axis intersecting γ will cut a section γ_r from the neighboring curve Γ_r, and will define together with γ and γ_r a piece s_r of the boundary strip S_r. We use the symbol $s_{r,h}$ to denote the portion of G_h contained in s_r and denote the associated portion of the boundary Γ_h by γ_h.

We now imagine a parallel to the x-axis to be drawn through a mesh point P_h of $s_{r,h}$. Let it meet the boundary γ_h in a point $\bar{P}_h$. The portion of this line which lies in $s_{r,h}$ we call $p_{r,h}$. Its length is certainly smaller than cr, where the constant c depends only on the smallest angle of inclination of a tangent γ to the x-axis.

Between the values of v at P_h and $\bar{P}_h$ we have the relation

$$v(P_h) = r(\bar{P}_h) \pm h \sum_{P_h\bar{P}_h} v_x.$$

Squaring both sides and applying Schwarz's inequality, we obtain

$$v(P_h)^2 \leq 2v(\bar{P}_h)^2 + 2cr \cdot h \sum_{p_{r,h}} v_x^2.$$

Summing with respect to P_h in the x-direction, we get

$$h \sum_{p_r} v^2 \leq 2crv(\bar{P}_h)^2 + 2c^2r^2h \sum_{p_r} v_x^2.$$

Summing again in the y-direction we obtain the relation

$$h \sum_{S_{r,h}} \sum v^2 \leq 2cr \sum_{\Gamma_h} v(\bar{P}_h)^2 + 2c^2r^2h \sum_{S_{r,h}} \sum v_x^2. \qquad (14)$$

Writing down the inequalities associated with the other portions of Γ and adding all the inequalities together we obtain the desired inequality (13).[19]

We next set $v_h = u_h - f_h$ so that $v_h = 0$ on Γ_h.

Then since $h^2 \sum_{G_h} \sum (v_x^2 + v_y^2)$ remains bounded as $h \to 0$, we obtain from (13)

$$\frac{h^2}{r} \sum_{S_{r,h}} \sum v^2 \leq \kappa r, \qquad (16)$$

where κ is a constant which does not depend on the function v or the mesh size. Extending the sum on the left to the difference $S_{r,h} - S_{\rho,h}$ of two boundary strips, the inequality (16) still holds with the same constant κ and we can pass to the limit $h \to 0$.

[19] By similar reasoning we can also establish the inequality

$$h^2 \sum_{G_h} \sum v^2 \leq c_1 h \sum_{\Gamma_h} v^2 + c_2 h^2 \sum_{G_h} \sum (v_x^2 + v_y^2) \qquad (15)$$

in which the constants c_1 and c_2 depend only on the region G and not on the mesh division.

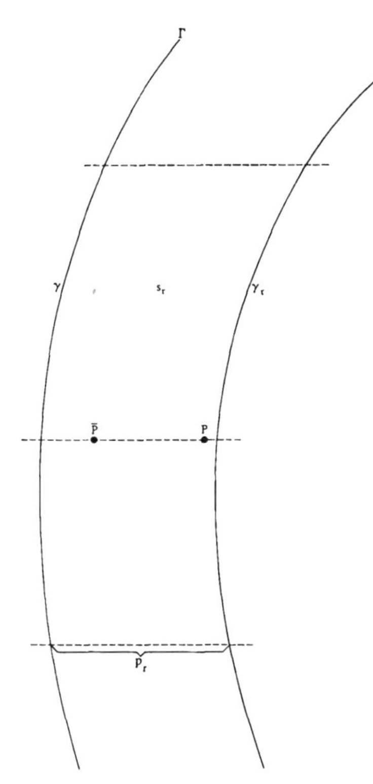

Figure 4

From the inequality (16) we then get

$$\frac{1}{r}\iint_{S_r - S_\rho} v^2\,dxdy \leq \kappa r, \qquad v = u - f.$$

Now letting the narrower boundary strip S_ρ approach the boundary we obtain the inequality

$$\frac{1}{r}\iint_{S_r} v^2\,dxdy = \frac{1}{r}\iint_{S_r} (u - f)^2\,dxdy \leq \kappa r$$

which states that the limit function u satisfies the prescribed boundary condition.

• 4. *Applicability of the method to other problems*

Our method is based essentially on the inequalities arising from the lemmas stated previously since the principal points of the proofs follow from the two last lemmas in Part 1 of Section 4.[20] No use is made of special fundamental solutions or special properties of the difference expression, and therefore the method can be carried over directly to the case of arbitrarily many independent variables as well as to the eigenvalue problem,

$$\frac{\partial^2 u}{\partial x^2} + \frac{\partial^2 u}{\partial y^2} + \lambda u = 0.$$

The same sort of convergence relations will obtain in this case as above.[21] Also the method applies to linear partial differential equations of other types, in particular its application to equations with variable coefficients requires only some minor modifications. The essential difference in this case lies only in proving the boundedness of $h^2 \sum \sum u_h^2$ which of course does not always hold for an arbitrary linear problem. However in case this sum is not bounded it can be shown that the general boundary value problem for the differential equation in question also possesses effectively no solutions, but that in this case there exist nonvanishing solutions of the corresponding homogeneous problem, i.e., eigenfunctions.[22]

• 5. *The boundary value problem* $\Delta\Delta u = 0$

In order to show that the method can be carried over to the case of differential equations of higher order, we will treat briefly the boundary value problem of the differential equation:

$$\frac{\partial^4 u}{\partial x^4} + 2\frac{\partial^4 u}{\partial x^2\,\partial y^2} + \frac{\partial^4 u}{\partial y^4} = 0.$$

We seek, in G, a solution of this equation for which the values of u and its first derivative are given on the boundary, being specified there by some function $f(x, y)$.

To this end we assume as previously that $f(x, y)$ together with its first and second derivatives is continuous in that region of the plane containing the region G.

We replace our differential equation problem by the new problem of solving the difference equation $\Delta u = 0$ in the mesh region G subject to the condition that at the points of the boundary strip $\Gamma_h + \Gamma_h'$ the function u takes on the values $f(x, y)$. From Section 2 we know that this boundary value problem has a unique solution. We will show that as the mesh size decreases, this solution, in each

[20] For an application of corresponding integral inequalities see K. Friedrichs, "Die Rand- und Eigenwertprobleme aus der Theorie der elastischen Platten", *Math. Ann.* **98**, 222 (1926).

[21] Similarly one proves at the same time that every solution of such a differential equation problem has derivatives of every order.

[22] See Courant-Hilbert, *Methoden der Mathematischen Physik* **1**, Ch. III, Section 3, where the theory of integral equations is handled with the help of the corresponding principle of the alternative. See also the Dissertation (Göttingen) of W. v. Koppenfels, which will appear soon.

interior subregion of G, converges to the solution of the differential equation, and that all of its difference quotients converge to the corresponding partial derivatives.

We note first that for the solution $u = u_h$, the sum

$$h^2 \sum_{G_h'}\sum (u_{xx}^2 + 2u_{xy}^2 + u_{yy}^2)$$

remains bounded as $h \to 0$. That is, by applying the minimum requirements on the solution (Part 2, Section 2) one finds that this sum is not larger than the corresponding sum

$$h^2 \sum_{G_h'}\sum (f_{xx}^2 + 2f_{xy}^2 + f_{yy}^2),$$

and this converges as $h \to 0$ to

$$\iint_G \left(\frac{\partial^2 f}{\partial x^2} + 2\frac{\partial^2 f}{\partial x \partial y} + \frac{\partial^2 f}{\partial y^2}\right) dxdy$$

which exists, by hypothesis.

From the boundedness of the sum

$$h^2 \sum_{G_h'}\sum (u_{xx}^2 + 2u_{xy}^2 + u_{yy}^2)$$

follows immediately the boundedness of $h^2 \sum_{G_h'} \sum (\Delta u)^2$ and also that of

$$h^2 \sum_{G_h'}\sum (u_x^2 + u_y^2) \quad \text{and} \quad h^2 \sum_{G_h}\sum u^2.$$

That is, for arbitrary w the following inequality holds (see Footnote 19),

$$h^2 \sum_{G_h}\sum w^2 \leq ch^2 \sum_{G_h}\sum (w_x^2 + w_y^2) + ch \sum_{\Gamma_h} w^2. \tag{15}$$

Then if one substitutes the first difference quotients of w for w itself in this inequality and applies the expression over the subregion of G_h for which they are defined, one finds the further inequality,

$$h^2 \sum_{G_h}\sum (w_x^2 + w_y^2) \leq ch^2 \sum_{G_h'}\sum (w_{xx}^2 + 2w_{xy}^2 + w_{yy}^2) + ch \sum_{\Gamma_h + \Gamma_h'} (w_x^2 + w_y^2),$$

where again the constant c is independent of the function and of the mesh size. We apply this inequality to $w = u_h$ and thus find the boundedness of the sum over $\Gamma_h + \Gamma_h'$ on the right-hand side; by definition these boundary sums converge to the corresponding integral containing $f(x, y)$. Thus from the boundedness of

$$h^2 \sum_{G_h'}\sum (u_{xx}^2 + 2u_{xy}^2 + u_{yy}^2)$$

follows the boundedness of $h^2 \sum_{G_h} \sum (u_x^2 + u_y^2)$ and thence that of $h^2 \sum_{G_h} \sum u^2$.

For the third step we substitute one after the other the expressions $\Delta u, \Delta u_x, \Delta u_y, \Delta u_{xx}, \cdots$, for w in the inequality

$$h^2 \sum_{G^{**}}\sum (w_x^2 + w_y^2) \leq ch^2 \sum_{G^*}\sum w^2 + ch^2 \sum_{G^*}\sum (\Delta w)^2$$

(see Part 2, Section 4) where G^* is a subregion of G containing G^{**} in its interior. The expressions introduced all satisfy the equation $\Delta w = 0$. It follows then that for each expression in turn and for all subregions G^* of G that the sums, $h^2 \sum_{G^*} \sum (w_x^2 + w_y^2)$, that is, $h^2 \sum_{G^*} \sum (\Delta u_x^2 + \Delta u_y^2)$, $h^2 \sum_{G^*} \sum (\Delta u_{xx}^2 + \Delta u_{xy}^2)$, $\cdots$ are bounded together with the sums:

$$h^2 \sum_{G_h}\sum u^2, \qquad h^2 \sum_{G_h}\sum (u_x^2 + u_y^2),$$

and $h^2 \sum_{G_h}\sum (\Delta u)^2$,

which are already known to be bounded.

Finally we substitute into the inequality (10), in place of w, the sequence of functions $u_{xx}, u_{xy}, u_{yy}, u_{xxx}, \cdots$, for which

$$h^2 \sum_{G_h^*}\sum (\Delta w)^2, \quad \text{i.e.} \quad h^2 \sum_{G_h^*}\sum (\Delta u_{xx})^2, \cdots$$

are bounded as shown above. We then find that for all subregions the sums

$$h^2 \sum_{G_h^*}\sum (u_{xxx}^2 + u_{xxy}^2), \quad h^2 \sum_{G_h^*}\sum (u_{xyx}^2 + u_{xyy}^2), \cdots$$

remain bounded.

From these facts we can now conclude as previously that from our sequence of mesh functions a subsequence can be chosen which in each interior subregion of G converges (together with all its difference quotients) uniformly to a limit function (or respectively its derivatives) which is continuous in the interior of G.

We have yet to show that this limit function which obviously satisfies the differential equation $\Delta\Delta u = 0$ also takes on the prescribed boundary conditions. For this purpose we say here only that, analogous to the foregoing, the expressions

$$\iint_{S_r} (u - f)^2 \, dxdy \leq cr^2,$$

$$\iint_{S_r} \left[\left(\frac{\partial u}{\partial x} - \frac{\partial f}{\partial x}\right)^2 + \left(\frac{\partial u}{\partial y} - \frac{\partial f}{\partial y}\right)^2\right] dxdy \leq cr^2$$

hold.[23] That the limit function fulfills these conditions may be seen by carrying over the treatment in Part 3, Section 4 to the function u and its first difference quotient.

[23] That the boundary values for the function and its derivatives actually are assumed is not difficult to prove. See for instance the corresponding treatment of K. Friedrichs, *loc. cit.*

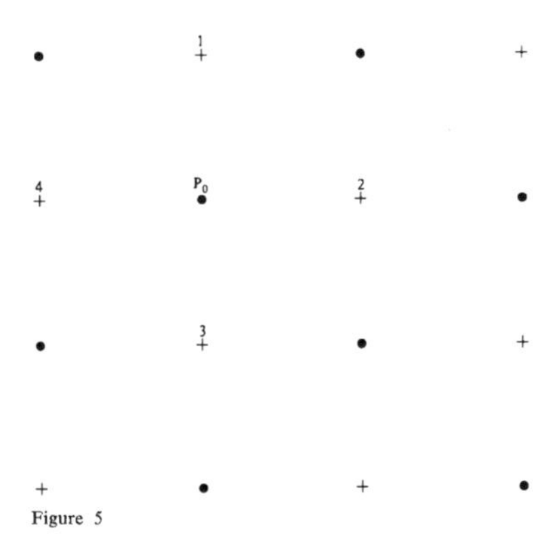

Figure 5

From the uniqueness of our boundary value problem we see furthermore that not only a selected subsequence, but also the original sequence of functions u possesses the asserted convergence properties.

II. The hyperbolic case

Section 1. The equation of the vibrating string

In the second part of this paper we shall consider the initial value problem for linear hyperbolic partial differential equations. We shall prove that under certain hypotheses the solutions of the difference equations converge to the solutions of the differential equations as the mesh size decreases.

We can discuss the situation most easily by considering the example of the approximation to the solution of the wave equation

$$\frac{\partial^2 u}{\partial t^2} - \frac{\partial^2 u}{\partial x^2} = 0 \tag{1}$$

We limit ourselves to the particular initial value problem where the value of the solution u, and its derivatives are given on the line $t = 0$.

In order to find the corresponding difference equation, we construct in the (x, t)-plane a square grid with lines parallel to the axes and with mesh width h. Following the notation of the first part of the paper we replace the differential equation (1) by the difference equation $u_{t\bar{t}} - u_{x\bar{x}} = 0$. If we select a grid point, P_0, then the corresponding difference equation relates the value of the function at this point to the values at four neighboring points. If we characterize the four neighboring values by the four indices 1, 2, 3, 4 (cf. Fig. 5), then the difference equation assumes the simple form

$$u_1 + u_3 - u_2 - u_4 = 0. \tag{2}$$

Note that the value of the function u at the point P_0 does not appear itself in the equation.

We consider the grid split up into two subgrids, indicated in Fig. 5 by dots and crosses respectively. The difference equation connects the values of the function over each of the subgrids separately, and so we shall consider only one of the two grids. As initial condition the values of the function are prescribed on the two rows of the grid, $t = 0$ and $t = h$. We can give the solution of this initial value problem explicitly; that is, we express the value of the solution at any point S in terms of the values given along the two initial rows. One can see at once that the value at a point of the row $t = 2h$ is uniquely determined by only the three values at the points close to it in the two first rows. The value at a point of the fourth row is uniquely determined by the values of the solution at three particular points in the second and third rows, and through them it is related to certain values in the first two rows. In general to a point S there will correspond a certain region of dependence in the first two rows; it may be found by drawing the lines $x + t =$ const. and $x - t =$ const., through the point S and extending them until they meet the second row at the points α and β respectively (cf. Fig. 6). The triangle $S\alpha\beta$ is called the *triangle of determination* because all the values of u in it remain unchanged provided the values on the first two rows of it are held fixed. The sides of the triangle are called *lines of determination.*

If one denotes the differences of u in the direction of the lines of determination by u' and $u^{\backprime}$, that is,

$$u_1' = u_1 - u_4, \qquad u_1^{\backprime} = u_1 - u_2$$
$$u_2' = u_2 - u_3, \qquad u_4^{\backprime} = u_4 - u_3,$$

Figure 6

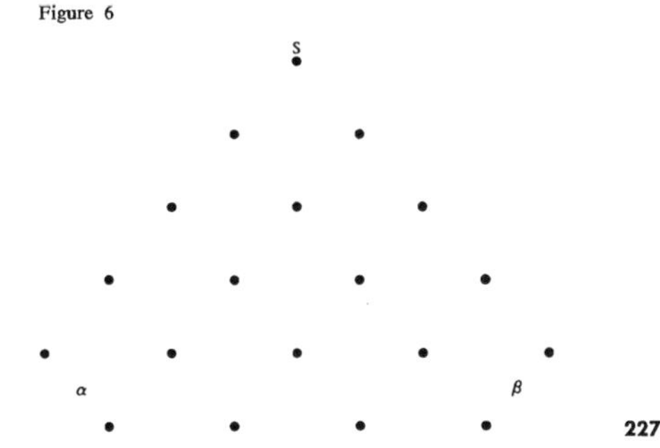

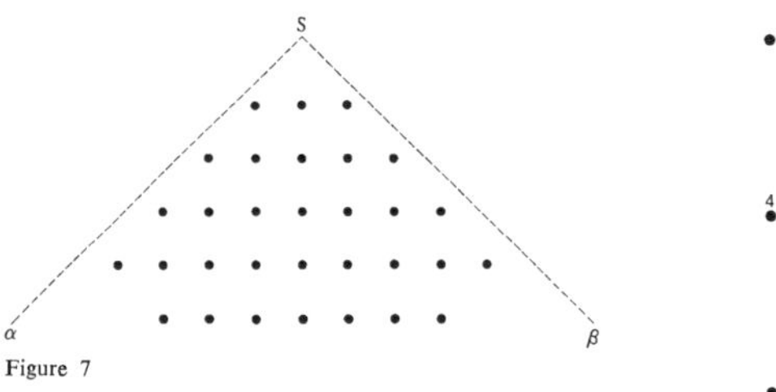

Figure 7

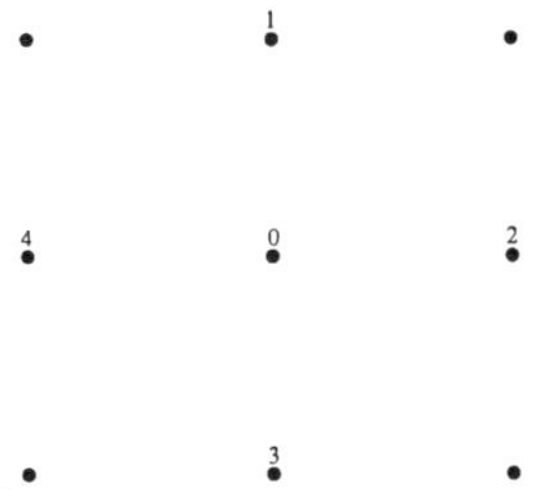

Figure 8

then the difference equation assumes the form

$$u_1' = u_2',$$

i.e., along a line of determination the differences with respect to the other direction of determination are constant, and thus are equal to one of the *given* differences between the value at two points on the first two rows. Moreover the difference $u_S - u_\alpha$ is a sum of differences u' along the determining line $\overline{\alpha S}$, so that using the remark above, we can obtain the final result (in obvious notation):

$$u_S = u_\alpha + \sum_{\alpha_1}^{\beta_1} u'. \tag{3}$$

We now let h go to zero, and let the prescribed values on the second and first rows converge uniformly to a twice continuously differentiable function, $f(x)$, and the difference quotients $u'/h\sqrt{2}$ there converge uniformly to a continuously differentiable function $g(x)$. Evidently the right-hand side of (3) goes over uniformly to the expression

$$f(x - t) + \frac{1}{\sqrt{2}} \int_{x-t}^{x+t} g(\xi)\, d\xi \tag{4}$$

if S converges to the point (x, t). This is the well-known expression for the solution of the wave equation (1) with initial values $u(x, 0) = f(x)$ and $\partial u(x, 0)/\partial t = f'(x) + \sqrt{2}g(x)$. Thus it is shown that as $h \to 0$ the solution of the difference equation converges to the solution of the differential equation provided the initial values converge appropriately (as above).

Section 2. On the influence of the choice of mesh. The domains of dependence of the difference and differential equations

The relationships developed in Section 1 suggest the following considerations.

In the same way that the solution of a linear hyperbolic equation at a point S depends only on a certain part of the initial line—namely the "domain of dependence" lying between the two characteristics drawn through S, the solution of the difference equation has also at the point S a certain domain of dependence defined by the lines of determination drawn through S. In Section 1 the directions for the lines of determination of the difference equation coincided with the characteristic directions for the differential equation so that the domains of dependence coincided in the limit. This result, however, was based essentially on the orientation of the mesh in the (x,t)-plane, and depended furthermore on the fact that a square mesh had been chosen. We shall now consider a more general rectangular mesh with mesh size equal to h (time interval) in the t-direction and equal to kh (space interval) in the x-direction, where k is a constant. The domain of dependence for the difference equation, $u_{\bar{t}t} - u_{\bar{x}x} = 0$ for this mesh will now either lie entirely within the domain of dependence of the differential equation, $\partial^2 u/\partial t^2 - \partial^2 u/\partial x^2 = 0$, or on the other hand will contain this latter region inside its own domain according as $k < 1$ or $k > 1$ respectively.

From this follows a remarkable fact: if for the case $k < 1$ one lets $h \to 0$, then the solution to the difference equation in general *cannot* converge to the solution of the differential equation. In this case a change in the initial values of the solution of the differential equation in the neighborhood of the endpoints α and β of the domain of dependence (see Fig. 7) causes, according to formula (4), a change in the solution itself at the point (x, t). For the solution of the difference equation at the point S, however, the changes at the points α and β are not relevant since these points lie outside of the domain of dependence of the difference equations. That convergence does occur for the case $k > 1$ will be proved in Section 3. See for example Fig. 9.

If we consider the differential equation

$$2\frac{\partial^2 u}{\partial t^2} - \frac{\partial^2 u}{\partial x^2} - \frac{\partial^2 u}{\partial y^2} = 0 \tag{5}$$

in two space variables, x and y, and time, t, and if we replace it by the corresponding difference equation on a rectilinear grid, then in contrast to the case of only two independent variables it is impossible to choose the mesh division so that the domain of dependence of the difference and differential equations coincide, since the domain of dependence of the difference equation is a quadrilateral while that of the differential equation is a circle. Later (cf. Section 4) we shall choose the mesh division so that the domain of determination of the difference equation contains that of the differential equation in its interior, and shall show that once again convergence occurs.

On the whole an essential result of this section will be that in the case of each linear homogeneous hyperbolic equation of second order one can choose the mesh so that the solution of the difference equation converges to the solution of the differential equation as $h \to 0$, (see for instance Sections 3, 4, 7, 8).

Section 3. Limiting values for arbitrary rectangular grids

Now we consider further the wave equation

$$\frac{\partial^2 u}{\partial t^2} - \frac{\partial^2 u}{\partial x^2} = 0, \tag{1}$$

but impose it now on a rectangular grid with time interval h and space interval kh. The corresponding difference equation is

$$L(u) = \frac{1}{h^2}(u_1 - 2u_0 + u_3) - \frac{1}{k^2 h^2}(u_2 - 2u_0 + u_4) = 0, \tag{6}$$

where the indices represent a "fundamental rhombus" with midpoint P_0 and corners P_1, P_2, P_3, P_4 (see Fig. 8). According to the equation $L(u) = 0$ the value of the function u at a point S can be represented by its values on that

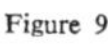
Figure 9

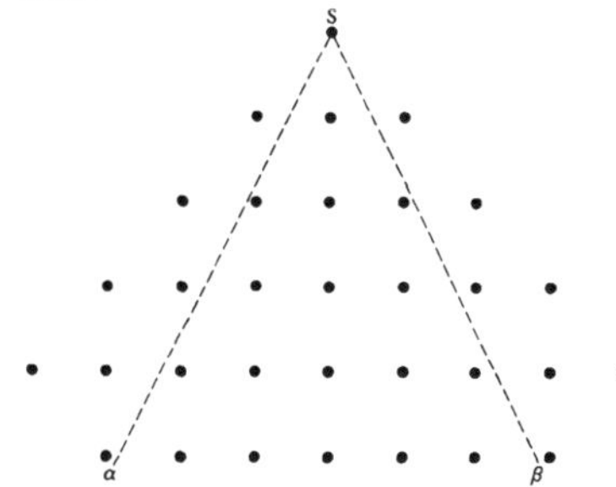

section of the initial rows $t = 0$ and $t = h$ cut out by lines of determination through S parallel to the sides of an elementary rhombus. We assume that the initial values are prescribed in such a way that as $t \to 0$ for fixed k the first difference quotients formed from them converge uniformly to given continuous functions on the line $t = 0$. The initial values can be used to form an explicit representation of the solution of the difference equation (corresponding to (3) in Section 1); however it is too complicated to yield a limiting value easily as $h \to 0$. Thus we will try another approach which will also make it possible for us to treat the general problem.[24]

We multiply the difference expression $L(u)$ by $(u_1 - u_3)$ and form the product using the following identities:

$$(u_1 - u_3)(u_1 - 2u_0 + u_3) = (u_1 - u_0)^2 - (u_0 - u_3)^2, \tag{7}$$

$$\begin{aligned}(u_1 - u_3)(u_2 - 2u_0 + u_4) &= (u_1 - u_0)^2 - (u_0 - u_3)^2 - \tfrac{1}{2}[(u_1 - u_2)^2 \\ &\quad + (u_1 - u_4)^2 - (u_2 - u_3)^2 - (u_4 - u_3)^2].\end{aligned} \tag{8}$$

Then we obtain

$$\begin{aligned}2(u_1 - u_3)L(u) &= \frac{2}{h^2}\left(1 - \frac{1}{k^2}\right)[(u_1 - u_0)^2 \\ &\quad - (u_0 - u_3)^2] + \frac{1}{h^2k^2}[(u_1 - u_2)^2 \\ &\quad + (u_1 - u_4)^2 - (u_2 - u_3)^2 - (u_4 - u_3)^2].\end{aligned} \tag{9}$$

The product (9) is now summed over all elementary rhombuses of the domain of determination, $S\alpha\beta$. The quadratic difference terms on the right-hand side always appear with alternate signs in two neighboring rhombuses so that they cancel out for any two rhombuses belonging to the triangle $S\alpha\beta$. Only the sums of squared differences over the "boundary" of the triangle remain, and we obtain the relation:

$$\begin{aligned}h^2 \sum_{S\alpha\beta}\sum 2\,\frac{u_1 - u_3}{h}\,L(u) &= h\sum_{S\alpha}\left[2\left(1 - \frac{1}{k^2}\right)\left(\frac{\dot u}{h}\right)^2 + \frac{1}{k^2}\left(\frac{u'}{h}\right)^2\right] \\ &\quad + h\sum_{S\beta}\left[2\left(1 - \frac{1}{k^2}\right)\left(\frac{\dot u}{h}\right)^2 + \frac{1}{k^2}\left(\frac{u^{\backprime}}{h}\right)^2\right] \\ &\quad - h\sum_{\mathrm{I}\ \mathrm{II}}\left[2\left(1 - \frac{1}{k^2}\right)\left(\frac{\dot u}{h}\right)^2 \right. \\ &\quad \left. + \frac{1}{k^2}\left(\frac{u'}{h}\right)^2 + \frac{1}{k^2}\left(\frac{u^{\backprime}}{h}\right)^2\right].\end{aligned} \tag{10}$$

---

[24] For the following see K. Friedrichs and H. Lewy, "Über die Eindeutigkeit...etc.," *Math. Ann.* 98, 192 ff. (1928), where a similar transformation is used for integrals.

Here u' and $u^{\backprime}$ denote differences in the direction of determination defined in Section 1, while $\dot{u}$ designates the difference of the functional values at two neighboring points on a mesh line parallel to the t axis. The range in $\sum_{S\alpha}$ over which $(u')^2$ is taken is the outermost boundary line $S\alpha$ and its nearest parallel neighbor found by shifting the points of $S\alpha$ downward by the amount h. There is a similar range for $(u^{\backprime})^2$ in $\sum_{S\beta}$, and so all of the differences, u', $u^{\backprime}$, and $\dot{u}$ appear once and only once.

For the solution to the problem $L(u) = 0$ the right-hand side of (10) vanishes. The sum over the initial rows I and II which occurs there remains bounded as $h \to 0$ (for fixed k); specifically it goes over into an integral of the prescribed function along the initial line. Consequently the sums over $S\alpha$ and $S\beta$ in (10) also remain bounded. If now $k \geq 1$ as we must require (see previous discussion), then $1 - 1/k^2$ is non-negative, and we find that the individual sums

$$h \sum_{S\alpha} \left(\frac{u'}{h}\right)^2, \qquad h \sum_{S\beta} \left(\frac{u^{\backprime}}{h}\right)^2,$$

extended over any line of determination whatever, remain bounded.

From this we can derive the "uniform continuity" (equicontinuity) (cf. Section 4 of the first part of the paper) of the sequence of grid functions in all directions in the plane;[25] since the values of u on the initial line are bounded, there must exist a subset which converges uniformly to a limit function $u(x, t)$.

Both the first and second difference quotients of the function u also satisfy the difference equation $L(u) = 0$. Their initial values can be expressed through the equation $L(u) = 0$ in terms of the first, second and third difference quotients of u involving initial values at points on the two initial lines I and II only. We require that they tend to continuous limit functions, that is, that the given initial values $u(x, 0)$, $u_t(x, 0)$ be three times or respectively twice continuously differentiable with respect to x.

From this we can apply the convergence considerations set forth above to the first and second difference quotients of u, as well as to u itself, and we can choose a subsequence such that these difference quotients converge uniformly to certain functions, which must be the first or respectively second derivatives of the limit function $u(x, t)$. Hence the limit function satisfies the differential equation $\partial^2 u/\partial t^2 - \partial^2 u/\partial x^2 = 0$ which results as the limit of the difference equation $L(u) = 0$; it represents indeed the solution of the initial value problem. Since such a solution is uniquely determined, every subsequence of mesh functions converges, and therefore the sequence itself converges to the limit function.

[25] If S_1 and S_2 are two points at a distance δ, then one connects them by a path of two segments, S_1S and SS_2, where the former is parallel to one line of determination and the latter to the other. Then one finds the appraisal,

$$|u_{S_1} - u_S| \leq |u_{S_1} - u_S| + |u_S - u_{S_2}|$$
$$\leq \sqrt{\delta}\sqrt{h \sum_{S_1S} \left(\frac{u'}{h}\right)^2} + \sqrt{\delta}\sqrt{h \sum_{SS_2} \left(\frac{u^{\backprime}}{\delta}\right)^2}.$$

Section 4. The wave equation in three variables

We treat next the wave equation,

$$2\frac{\partial^2 u}{\partial t^2} - \frac{\partial^2 u}{\partial x^2} - \frac{\partial^2 u}{\partial y^2} = 0 \tag{11}$$

and consider its relation to the observations on the domain of dependence discussed in Section 2. The domain of dependence of the differential equation (11) is a circular cone with axis parallel to the t-direction and with apex angle α, where $\tan \alpha = 1/\sqrt{2}$. In any rectilinear grid parallel to the axes we introduce the corresponding difference equation

$$2u_{t\bar{t}} - u_{x\bar{x}} - u_{y\bar{y}} = 0. \tag{12}$$

This equation relates the functional values of u at points of an elementary tetrahedron. In fact it allows the value of the function u at a point S to be expressed uniquely in terms of the values of the function at certain points of the two initial planes $t = 0$ and $t = h$. At each point S we obtain a pyramid of determination which cuts out from the two base planes two rhombuses as domains of dependence.

If we let the mesh widths tend to zero, keeping their ratios fixed, then we can expect convergence of the sequence of mesh functions to the solution of the differential equation only provided the pyramid of determination contains the cone of determination of the differential equation in its interior. The simplest grid with this property will be one constructed in such a way that the pyramid of determination is tangent to the exterior of the cone of determination. Our differential equation is chosen so that this occurs for a grid of cubes parallel to the axes.

The difference equation (12), in the notation of Fig. 10, assumes for such a grid the form:

$$L(u) = \frac{2}{h^2}(u'_\alpha - 2u_0 + u_\alpha) - \frac{1}{h^2}(u_1 - 2u_0 + u_3)$$
$$- \frac{1}{h^2}(u_2 - 2u_0 + u_4), \tag{13}$$

in which the functional value, u_0, at the midpoint actually cancels out. The values of the solution on the two initial planes must be the values of a function having four continuous derivatives with respect to $x_1 y_2 t$.

For the convergence proof we again use the method developed in Section 3. We construct the triple sum

$$h^2 \sum \sum \sum 2\frac{u'_\alpha - u_\alpha}{h} L(u) = 0$$

for the solution to the difference equation, where the summation is to be extended over all elementary octahedrons of the pyramid of determination emanating from

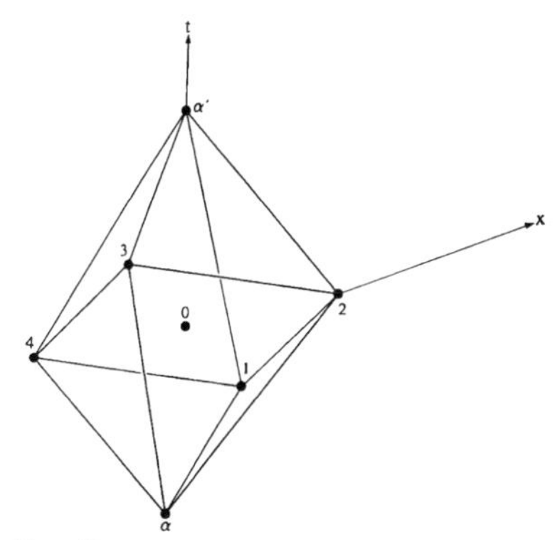

Figure 10

the point S. Then almost exactly as before we find that the values of the function u at the interior points of the pyramid of determination cancel out in the summation and that only the values on the two pyramids called F, and on the two base surfaces I and II remain.

If we denote by u' the difference of the values of the function at two points connected by a line of an elementary octahedron, then we can write the result as

$$\sum_F \sum (u')^2 - \sum_{\mathrm{I}} \sum_{\mathrm{II}} (u')^2 = 0, \tag{14}$$

where the sum is extended over all surfaces containing differences u'; each such difference is to appear only once.[26] The double sum over the two initial surfaces stays bounded since it goes over into an integral of the initial values. Therefore the sum over the "surface of determination" F remains bounded.

We now apply these results not to u itself, but to its first, second and third difference quotients, which themselves satisfy the difference equation (13). Their initial values can be expressed using only values on the first two initial planes by means of (13) using first through fourth difference quotients. If $w = w_h$ is one of the difference quotients of any order up to third order, then we know that the sum $h^2 \sum \sum (w'/h)^2$ extended over a surface of determination remains bounded. From this it follows, through exactly the methods used in Section 4 of the first part of the paper, that the function u and its first and second difference quotients are uniformly continuous (equicontinuous). Thus there exists a sequence of mesh widths decreasing to zero such that these quantities, which are bounded initially, converge to continuous limit functions and, in fact, converge to the solution of the differential equation and to the first and second derivatives of this solution, all exactly as we found earlier (Section 3).

[26] The grid ratio has been chosen in such a way that the differences between values of u appearing on the two neighboring surfaces in F do not occur, (as they did in the general case in one dimension treated in Section 3).

Appendix. *Supplements and generalizations*

Section 5. Example of a differential equation of first order

We have seen in Section 2 that in the case when the region of dependence of the differential equation covers only a part of the region of dependence of the difference equation, the influence of the rest of the region is not included in the limit. We can demonstrate this phenomenon explicitly by the example of the differential equation of first order, $\partial u/\partial t = 0$ if we replace it by the difference equation

$$2u_t - u_x + u_{\bar{x}} = 0. \tag{15}$$

In the notation of Fig. 5 this becomes

$$u_1 = \frac{u_2 + u_4}{2}. \tag{16}$$

As before, the difference equation connects only the points of a submesh with one another. The initial value problem consists of assigning as initial values for u at points $x = 2ih$ on the row $t = 0$ the values, f_{2i}, assumed there by a continuous function $f(x)$.

We consider the point S at a distance $2nh$ up along the t-axis. It is easy to verify that the solution u at S is represented as

$$u_S = \sum_{i=-n}^{n} \frac{1}{2^{2n}} \begin{pmatrix} 2n \\ n+i \end{pmatrix} f_{2i}. \tag{17}$$

As the mesh size decreases, that is as $n \to \infty$, the sum on the right-hand side tends simply to the value f_0. This can be demonstrated from the continuity of $f(x)$ and from the behavior of the binomial coefficients as n increases (see the following paragraph).

Section 6. The equation of heat conduction

The difference equation (16) of Section 5 can also be interpreted as the analogue of an entirely different differential equation, namely the equation of heat conduction,

$$2\frac{\partial u}{\partial t} - \frac{\partial^2 u}{\partial x^2} = 0. \tag{18}$$

In any rectangular mesh with mesh spacing l and h in the time and space directions, respectively, the corresponding difference equation becomes

$$2\left(\frac{u_1 - u_0}{l}\right) = \left(\frac{u_2 + u_4 - 2u_0}{h^2}\right). \tag{19}$$

In the limit as the mesh size goes to zero the difference equation preserves its form only if l and h^2 are decreased proportionately. In particular if we set $l = h^2$, then the value u_0 drops out of the equation and the difference equation becomes

$$u_1 = \frac{u_2 + u_4}{2}. \tag{16}$$

The solution to (16) is given by formula

$$u(0, t) = \sum_{i=-n}^{n} \frac{1}{2^{2n}} \binom{2n}{n+i} f_{2i}.$$

As the mesh width decreases, a point ξ on the x-axis is always characterized by the index

$$2i = \xi/h. \tag{20}$$

The mesh width h is related to the ordinate t of a particular point by

$$2nh^2 = t. \tag{21}$$

We shall investigate what happens to formula (17) as $h \to 0$, that is $n \to \infty$. Using (21) we write (17) in the form

$$u(0, t) = \sum_{i=-n}^{n} \frac{\sqrt{2n}}{2 \cdot 2^{2n} \sqrt{t}} \binom{2n}{n+i} f_{2i} \cdot 2h. \tag{22}$$

For the coefficient of $2hf_{2i} = 2hf(\xi)$ we use the abbreviation

$$\frac{1}{2\sqrt{t}} g_{2n}(\xi) = \frac{\sqrt{2n}}{2 \cdot 2^{2n} \sqrt{t}} \binom{2n}{n + \frac{\xi}{\sqrt{2t}} \sqrt{n}}.$$

The limiting value of the coefficient, which is usually determined by using Stirling's formula, we will calculate here by considering the function $g_{2n}(\xi)$ as the solution of an ordinary difference equation which approaches a differential equation as $h \to 0$. As the difference equation one finds

$$\frac{1}{2h}[g_h(\xi + 2h) - g_h(\xi)] = -\frac{1}{2h} g_h(\xi) \frac{2i+1}{n+i+1}$$

(in which we have written $g_h(\xi)$ instead of $g_{2n}(\xi)$). Or

$$\frac{1}{2h}[g_h(\xi + 2h) - g_h(\xi)] = -g_h(\xi) \frac{\xi + h}{t + h\xi + 2h^2}.$$

$g_h(\xi)$ satisfies the normalization condition

$$\sum_{i=-n}^{n} g_h(\xi) \cdot 2h = 2\sqrt{t}.$$

This sum is over the region of dependence of the difference equation, and as $h \to 0$ this covers the entire x-axis.

It can be shown easily that $g_h(\xi)$ converges uniformly to the solution $g(x)$ of the differential equation

$$g'(x) = -g(x)x/t$$

with the auxiliary condition

$$\int_{-\infty}^{\infty} g(x)\, dx = 2\sqrt{t}.$$

From formula (22) after passing to the limit we find

$$u(0, t) = \int_{-\infty}^{\infty} \frac{1}{\sqrt{2\pi t}} e^{-\xi^2/2t} f(\xi)\, d\xi,$$

which is the known solution of the heat conduction equation.

The results of this section can be carried over directly to the case of the differential equation,

$$4\frac{\partial u}{\partial t} - \frac{\partial^2 u}{\partial x^2} - \frac{\partial^2 u}{\partial y^2} = 0$$

and so on for even more independent variables.

Section 7. The general homogeneous linear equation of second order in the plane

We consider the differential equation

$$\frac{\partial^2 u}{\partial t^2} - k^2 \frac{\partial^2 u}{\partial x^2} + \alpha \frac{\partial u}{\partial t} + \beta \frac{\partial u}{\partial x} + \gamma u = 0. \tag{23}$$

The coefficients are assumed to be twice continuously differentiable with respect to x and t, while the initial values on the line $t = 0$ are assumed three times continuously differentiable with respect to x. We replace the differential equation by the difference equation

$$L(u) = u_{\bar{t}t}(x, t) - k^2 u_{\bar{x}x}(x, t)$$
$$+ \alpha u_t + \beta u_x + \gamma u = 0 \tag{24}$$

in a grid with time mesh width h and space mesh width ch so that in a neighborhood of the appropriate part of the initial value line the inequality $1 - k^2/c^2 > \varepsilon > 0$ holds for the constant, c. The initial values are to be chosen as in Section 3.

For the proof of convergence we again transform the sum,

$$h^2 \sum_{S\alpha\beta} \sum 2 \frac{u_1 - u_3}{h} L(u)$$

by using identities (7) and (8). In addition to a sum (see for example (10)) over the doubled boundary of the triangle $S\alpha\beta$ (Fig. 6) one obtains a sum over the entire triangle $S\alpha\beta$ because of the variability of the coefficient k and the presence of lower order derivatives in the differential equation. By using the differentiability of k and the Schwarz inequality one can show that this latter sum is bounded from above by

$$Ch^2 \sum_{S\alpha\beta} \sum \left[\left(\frac{u^{\backprime}}{h}\right)^2 + \left(\frac{u'}{h}\right)^2 + \left(\frac{\dot{u}}{h}\right)^2 + u^2\right]$$

where the constant C is independent of the function u,

the mesh width h, and, in a certain neighborhood of the initial line, also independent of the point S.

Again we can estimate an upper bound for $h^2 \sum_{S\alpha\beta} \sum u^2$ by[27]

$$C_1 h^2 \sum_{S} \sum_{\alpha\beta} \left(\frac{\dot{u}}{h}\right)^2 + C_2 h \sum_{\mathrm{I\,II}} u^2,$$

where the same properties hold for C_1 and C_2 as are stated above for C.

Thus we obtain an inequality of the form

$$\begin{aligned} h \sum_{S\alpha} &\left[2\left(1 - \frac{k^2}{c^2}\right)\left(\frac{\dot{u}}{h}\right)^2 + \frac{k^2}{c^2}\left(\frac{u'}{h}\right)^2\right] \\ &+ h \sum_{S\beta}\left[2\left(1 - \frac{k^2}{c^2}\right)\left(\frac{\dot{u}}{h}\right)^2 + \frac{k^2}{c^2}\left(\frac{u^{\backprime}}{h}\right)^2\right] \\ &\leq C_3 h^2 \sum_{S\alpha\beta} \sum \left[\left(\frac{\dot{u}}{h}\right)^2 + \left(\frac{u'}{h}\right)^2 + \left(\frac{u^{\backprime}}{h}\right)^2\right] + D, \qquad (25) \end{aligned}$$

where D, for all points S and mesh widths h, is a fixed bound for the sums over the initial line.

As vertices of our triangles we choose a sequence of points $S_0, S_1, \cdots, S_n = S$ lying on a line parallel to the t-axis. By summing the corresponding sequence of inequalities (25) after multiplying by h we obtain the following inequality

$$\begin{aligned} h^2 \sum_{SS_0\alpha} \sum &\left[2\left(1 - \frac{k^2}{c^2}\right)\left(\frac{\dot{u}}{h}\right)^2 + \frac{k^2}{c^2}\left(\frac{u'}{h}\right)^2\right] \\ &+ h^2 \sum_{SS_0\beta} \sum \left[2\left(1 - \frac{k^2}{c^2}\right)\left(\frac{\dot{u}}{h}\right)^2 + \frac{k^2}{c^2}\left(\frac{u^{\backprime}}{h}\right)^2\right] \\ &\leq nh^3 C_3 \sum_{S\alpha\beta} \sum \left[\left(\frac{\dot{u}}{h}\right)^2 + \left(\frac{u'}{h}\right)^2 + \left(\frac{u^{\backprime}}{h}\right)^2\right] + nhD. \qquad (26) \end{aligned}$$

Now if we notice that one can express a difference u' or $u^{\backprime}$ in terms of two differences $\dot{u}$ and a difference $u^{\backprime}$ or respectively u', then we see that the left-hand side of (26) is larger than the simpler form

$$C_4 h^2 \sum_{S\alpha\beta} \sum \left[\left(\frac{\dot{u}}{h}\right)^2 + \left(\frac{u'}{h}\right)^2 + \left(\frac{u^{\backprime}}{h}\right)^2\right],$$

with a suitable constant C_4.

Then by confining the discussion to a sufficiently small neighborhood, $0 \leq t \leq nh = \delta$ of the initial line where δ is small enough so that

$$C_4 - nhC_3 = C_5 > 0,$$

we find from (26),

$$C_5 h^2 \sum_{S\alpha\beta} \sum \left[\left(\frac{\dot{u}}{h}\right)^2 + \left(\frac{u'}{h}\right)^2 + \left(\frac{u^{\backprime}}{h}\right)^2\right] \leq \frac{C_4}{C_3} D. \qquad (27)$$

[27] For proof one makes use of the inequality used in Footnote 25.

The bound given by (27) when combined with (25) gives a bound on

$$h \sum_{S\alpha} \left(\frac{u'}{h}\right)^2 + h \sum_{S\beta} \left(\frac{u^{\backprime}}{h}\right)^2,$$

from which, as in Section 3, one can prove the uniform continuity of u.

We apply the inequality (25) now, instead of to the function u itself, to its first and second difference quotients, w, which also satisfy difference equations whose second-order terms are the same as those of (24). In the rest of the terms there will appear lower order differences of u which cannot be expressed in terms w, but they will appear in the above argument in the form of quadratic double sums multiplied by h^2. This is enough to let us reach the same conclusions for the difference equation for w as we found above for u. So we can conclude from this the uniform continuity (equicontinuity) and boundedness of the function u and its first and second derivatives. Consequently a subsequence exists which converges uniformly to the solution of the initial value problem for the differential equation. Again from the uniqueness of the solution we find that the sequence of functions itself converges.

In all of this the assumption must be made that the difference quotients up to third order involving values on the two initial lines converge to continuous limit functions.[28]

Section 8. The initial value problem for an arbitrary linear hyperbolic differential equation of second order

We shall now show that the methods developed so far are adequate for solving the initial value problem for an arbitrary homogeneous linear hyperbolic differential equation of second order. It suffices to limit the discussion to the case of three variables. The development can be extended immediately to the case of more variables. It is easy to see that a transformation of variables can reduce the most general problem of this type to the following: a function $u(x, y, t)$ is to be found which satisfies the differential equation

$$\begin{aligned} u_{tt} &- (au_{xx} + 2bu_{xy} + cu_{yy}) \\ &+ \alpha u_t + \beta u_x + \gamma u_y + \delta u = 0, \qquad (28) \end{aligned}$$

and which, together with its first derivative, assumes prescribed values on the surface $t = 0$. The coefficients in Eq. (28) are functions of the variables x, y, and t and satisfy the condition

$$a > 0, \quad c > 0, \quad ac - b^2 > 0.$$

We assume that the coefficients are three times differentiable with respect to x, y, and t, and that the initial

[28] This assumption and also the assumptions on the differentiability of the coefficients of the differential equation, and further on the restriction to a sufficiently small region of the initial line can be weakened in special cases.

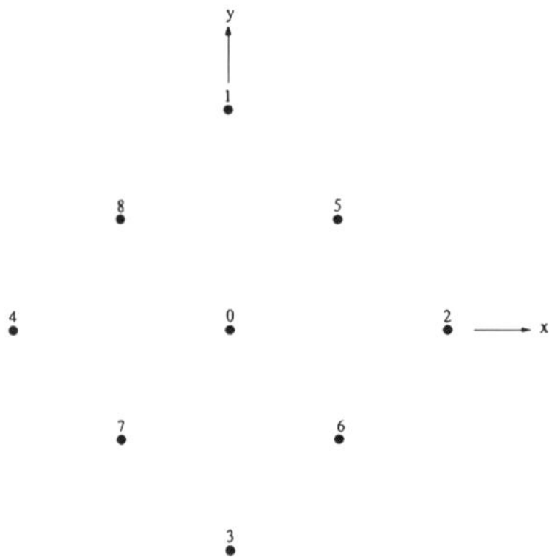

Figure 11

values u and u_t are respectively four and three times continuously differentiable with respect to x and y.

The coordinates x and y can be drawn from a given point on the initial plane in such a way that $b = 0$ there. Then of course in a certain neighborhood of this point the conditions

$$a - |b| > 0, \qquad c - |b| > 0$$

hold. We restrict our investigation to this neighborhood. We can choose a three times continuously differentiable function $d > 0$ so that

$$\left.\begin{matrix} a - d \\ c - d \\ d - |b| \end{matrix}\right\} > \varepsilon > 0 \tag{29}$$

for some constant ε. Then we put the differential equation into the form

$$\begin{aligned} u_{tt} - (a - d)u_{xx} - (c - d)u_{yy} \\ - \tfrac{1}{2}(d + b)(u_{xx} + 2u_{xy} + u_{yy}) \\ - \tfrac{1}{2}(d - b)(u_{xx} - 2u_{xy} + u_{yy}) \\ + \alpha u_t + \beta u_x + \gamma u_y + \delta u = 0. \end{aligned} \tag{30}$$

We now construct in the space a grid of points, $t = lh$, $x + y = mkh$, $x - y = nkl$ $(l, m, n = \cdots -1, 0, 1, 2 \cdots)$ and we replace Eq. (30) by a difference equation $L(u) = 0$ over this mesh. We do this by assigning to each point P_0 the following neighborhood: The point $P_{\alpha'}$ or the point P_α which lies at a distance h or $-h$ respectively along the t-axis from P_0; also the points $P_1, \cdots, P_8$ which lie in the same (x, y)-plane with P_0 (see Fig. 11). These points constitute an "elementary octahedron" with vertices $P_{\alpha'}$, P_α, P_1, P_2, P_3, P_4. For each grid point lying in the interior of G we replace the derivatives appearing in Eq. (30) by difference quotients over the elementary octahedron about P_0.

We replace

$$\begin{aligned} u_{tt} \quad &\text{by} \quad \frac{1}{h^2}(u_{\alpha'} - 2u_0 + u_\alpha) \\ u_{xx} \quad &\text{by} \quad \frac{1}{k^2h^2}(u_2 - 2u_0 + u_4) \\ u_{yy} \quad &\text{by} \quad \frac{1}{k^2h^2}(u_1 - 2u_0 + u_3) \\ u_{xx} + 2u_{xy} + u_{yy} \quad &\text{by} \quad \frac{4}{k^2h^2}(u_6 - 2u_0 + u_8) \\ u_{xx} - 2u_{xy} + u_{yy} \quad &\text{by} \quad \frac{4}{k^2h^2}(u_5 - 2u_0 + u_7). \end{aligned}$$

The first derivatives in (30) are replaced by the corresponding difference quotients in the elementary octahedron. The coefficients of the difference equation are given the values assumed by the coefficients of the differential equation at the point P_0.

On the initial planes $t = 0$ and $t = h$ we assume that the values of the function are assigned in such a way that as the mesh size approaches zero for fixed k, the function approaches the prescribed initial value, and the difference quotients over the two planes up through differences of fourth order converge uniformly to the prescribed derivatives.

The solution of the difference equation $L(u) = 0$ at a point is uniquely determined by the values on the two base surfaces of the pyramid of determination passing through the point.

To prove convergence we construct a sum

$$h^3 \sum \sum \sum 2 \frac{u_{\alpha'} - u_\alpha}{h} L(u)$$

over all the elementary octahedrons of the pyramid of determination, and we transform it by using identities (7) and (8). This gives one space summation multiplied by h^3 and quadratic in the first difference quotients, and also over a double surface a sum which is multiplied by h^2 and contains the squares of all the difference quotients of the type $u_\alpha - u_0$, $u_\alpha - u_1$, $\cdots$, $u_\alpha - u_8$. In this expression according to (29) the coefficients are larger than some fixed positive constant in all those cases where the ratio $1/k$ between the time and space mesh sizes is taken sufficiently small.

From here on one can proceed exactly as before (Sections 7, 4) and prove that the solution of the difference equation converges to the solution of the differential equation.

(Submitted to Math. Ann. September 1, 1927)

Appendix D

Event schedule and list of supporters, organizers, lecturers, and attendees

Technical sessions schedule

Monday, May 3rd

12:30–13:30 **Barbara Lee Keyfitz**
Singular Shocks and Loss of Hyperbolicity in Conservation Laws
Chair: Leon Sinay

15:30–16:30 **Philippe G. LeFloch**
Undercompressive shocks and kinetic relations
Chair: Wladimir Neves

Contributed Papers – Session 1
16:30–17:00 Stefan Berres
30 years MLB model for polydisperse suspensions: a particular non-convex system modeling particle flow
17:00–17:30 José Pontes (speaker), Daniel Walgraef, Christo I. Christov
A second order in time numerical scheme for solving reaction–diffusion equations modeling dislocation dynamics in materials subjected to cyclic loading
Chair: Sandra Malta

C.A. de Moura, C.S. Kubrusly (eds.), *The Courant–Friedrichs–Lewy (CFL) Condition*, DOI 10.1007/978-0-8176-8394-8, © Springer Science+Business Media New York 2013

Tuesday, May 4th

10:00–11:00 **Carlos Tomei**
The weird topology of the discretization of simple nonlinear DEs
Chair: Carlos A. de Moura

11:30–12:30 **Poster session**
Válter A. Nascimento
Study of solitons in the Gross–Pitaevskii equation using variational approximation in quantum gases
Patricia N. da Silva and Carlos F. Vasconcellos
Remarks on the General Linear Kawahara System
Margarete Domingues, S. M. Gomes, O. Roussel and K. Schneider
Adaptive multiresolution methods for 3D Euler equations: a case study
Chair: Augusto Cesar de C. Barbosa

12:30–13:30 **Kai Schneider**
Local CFL condition for space-time adaptive multiresolution methods
Chair: Sônia Gomes

15:30–16:30 **Reuben Hersh**
Computational results leading to mathematical knowledge
Chair: Djairo Guedes de Figueiredo

16:30–17:30 **Round Table**
Parallel, non-deterministic, multiscale algorithms – whatś ahead?
Antônio J. Silva Neto and Haroldo F. Campos Velho
Chair: Carlos A. de Moura

Contributed Papers – Session 2
18:00–18:30 Peter Frolkovic
Relaxing the CFL condition for flux-based finite volume methods
18:30–19:00 Eduardo Abreu
Computation of nonclassical wave for a 2×2 system modeling immiscible three-phase flow in porous media
Chair: Margarete Domingues

Wednesday, May 5th

10:00–11:00 **Maria Assunção F. da Silva Dias**
Numerical weather prediction challenges
Chair: Haroldo F. Campos Velho

11:30–12:30 **Djairo Guedes de Figueiredo**
On semilinear elliptic systems
Chair: Pedro Leite da Silva Dias

12:30–13:30 **Rolf Jeltsch**
Numerical Simulation of Compressible Magnetohydrodynamic Plasma Flow in a Circuit Breaker
Chair: Antônio J. Silva Neto

15:30–17:30 **João Candido Portinari**
The Portinari Project: From the Coffee Plantation to the United Nations
Chair: Mauricio V. Kritz

18:00–19:00 **Peter D. Lax**
Stability of difference schemes for hyperbolic equations
Chair: Pedro Nowosad

19:00–20:00 **Special session honoring Peter D. Lax**

Thursday, May 6th

10:00–11:00 **Enrique Zuazua**
Convergence rates for dispersive approximation schemes to nonlinear Schrödinger Equations
Chair: Mauricio Kischinhevsky

Contributed Papers – Session 3
11:30–12:00 Dmitry Vishnevskiy (speaker) and Vadim Lisitsa
Explicit finite difference scheme for waves propagation in arbitrary anisotropic elastic media
12:00–12:30 José Cal Neto
Obtaining numerically all the solutions of Ambrosetti–Prodi type PDEs
Chair: Sanderson Gonzaga

12:30–13:30 **Uri Ascher**
The chaotic nature of faster gradient descent
Chair: Jorge Zubelli

15:30–16:30 **Ken McLaughlin**
Asymptotic analysis of Riemann–Hilbert problems: singular limits and a few tricks
Chair: Carlos Tomei

Contributed Papers – Session 4
16:30–17:00 Nguyen H. Hoang
L^p-weighted theory for Navier–Stokes equations in exterior domains
17:00–17:30 V. Ruas (speaker), J. Henrique Carneiro de Araujo, P. Dias Gomes
Explicit finite element solution of time-dependent viscoelastic flow problems
Chair: Helio Pedro Amaral Souto

18:00 Closing session

List of supporting institutions

UERJ Rio de Janeiro State University
- Graduate Program of Computational Sciences
- Graduate Program of Mechanical Engineering
- IME—Mathematics and Statistics Institute
- IPRJ—Rio de Janeiro Polytechnic Institute

LNCC Brazilian National Laboratory for Scientific Computing

INCTMat Brazilian National Institute for Mathematics in Science and Technology

UFRJ Rio de Janeiro Federal University
- COPPE—Engineering Graduate Program
- IM—Mathematics Institute

PUC-Rio Rio de Janeiro Pontifical Catholics University
- Electrical Engineering Graduate Program
- Mathematics Department

IMPA Brazilian National Pure and Applied Mathematics Institute

UFF Rio de Janeiro State Federal University
- IC—Computing Science Institute

SBMAC Brazilian Society for Computational and Applied Mathematics

SBM Brazilian Mathematics Society

ABC Brazilian Sciences Academy

SBPC Brazilian Society for the Advancement of Sciences

Portinari Project

CNPq Brazilian Council for Scientific and Technological Development

CAPES Brazilian Coordination for Graduate Studies Support

FAPERJ Rio de Janeiro Foundation for Scientific Research Support

List of lecturers, attendees, and organizers

1	Alexandre Madureira	LNCC	Brazil
2	Alvaro L. G. A. Coutinho Program & Organizing Committee	Coppe-UFRJ	Brazil
3	Américo Barbosa da Cunha Jr.	PUC-Rio	Brazil
4	Anna Karina Fontes Gomes	INPE	Brazil
5	Antônio J. Silva Neto Invited Panelist	IPRJ-UERJ	Brazil
6	Antonio Roberto Mury	LNCC	Brazil
7	Aylton Cordeiro Neto	IME-UERJ	Brazil
8	Barbara Keyfitz Invited Lecturer	Ohio St. University	USA
9	Bruno Osório Rodrigues	UERJ	Brazil
10	Carlos A. de Moura Program & Organizing Committee	IME-UERJ Chair	Brazil
11	Carlos Eduardo Hirakawa	IME-UERJ	Brazil
12	Carlos F. Vasconcellos Contributed paper	IME-UERJ	Brazil
13	Carlos E. Mathias Motta	IM-UFF	Brazil
14	Carlos S. Kubrusly Program & Organizing Committee	PUC-Rio Editor	Brazil
15	Carlos Saver	IMPA	Brazil
16	Carlos Tomei Program & Organizing Committee Invited Lecturer	PUC-Rio	Brazil
17	Caroline dos Santos de Moraes	IME-UERJ	Brazil
18	Christo I. Christov Contributed paper	Univ. Louisiana at Lafayette	USA
19	Daniel Walgraef Contributed paper	Université Libre de Bruxelles	Belgium
20	Danielle Gonçalves Teixeira	IME-UERJ	Brazil
21	Deise Lilian de Oliveira	IME-UERJ	Brazil
22	Diego Brandão	UFF	Brazil
23	Djairo Guedes de Figueiredo Invited Lecturer	IMECC-UNICAMP	Brazil
98	Dmitriy Vishnevskiy Contributed paper	Inst. Petroleum Geology and Geophysics, Novosibirsk	Russia
24	Douglas Milanez Marques	IF-UERJ	Brazil
25	Eduardo Cardoso de Abreu Contributed paper	IMPA	Brazil
26	Eduardo Teles da Silva	PUC-Rio	Brazil
27	Emílio Jorge Lydia	IF-UERJ	Brazil

28	Enrique Zuazua Invited Lecturer	Basque Center Applied Math.	Spain
29	Erick Edgar Auiaga Sanz	IME-UERJ	Brazil
30	Fabiana Fortes Rodrigues	IME-UERJ	Brazil
31	Felipe Bottega Diniz	IME-UERJ	Brazil
32	Haroldo Fraga de Campos Velho Invited Panelist	INPE	Brazil
33	Helio Pedro Amaral Souto Program & Organizing Committee	IPRJ-UERJ	Brazil
34	Hilário Alencar Program & Organizing Committee	SBM	Brazil
35	Huy Hoang Nguyen Contributed paper	Unicamp	Brazil
36	Italo Marinho Sá Barreto	UERJ	Brazil
37	Jacob Palis Program & Organizing Committee	ABC	Brazil
38	Jacqueline Telles de S. Raposo Secretary chair	UERJ	Brazil
39	Jeanne Denise Bezerra de Barros	IME-UERJ	Brazil
40	Jhoab Pessoa de Negreiros	IME-UERJ & UniGranrio	Brazil
41	João Cândido Portinari Invited Lecturer	PUC-Rio	Brazil
42	Jorge Passamani Zubelli Program & Organizing Committcc	IMPA	Brazil
43	José Alberto Cuminato Program & Organizing Committee	ICMC-USP	Brazil
44	José Cal Neto Contributed paper	PUC-Rio	Brazil
45	José da Rocha Miranda Pontes Contributed paper	UFRJ	Brazil
46	Júlio Daniel Machado Silva	IMPA	Brazil
47	Kai Schneider Invited Lecturer	Université de Provence	France
48	Ken McLaughlin Invited Lecturer	University of Arizona	USA
49	Lená Medeiros de Menezes	SR1-UERJ	Brazil
50	Leon Sinay Session chair	LNCC	Brazil
51	Leonardo Espin	IMPA	Brazil
52	Lori Courant Lax Invited Musician		USA
53	Lucianna Helene S dos Santos	IME-UERJ	Brazil
54	Luiz Felipe N. Soares	IME-UERJ	Brazil
55	Márcio Coutinho B. Cortes Filho	IME-UERJ	Brazil

56	Marco Antonio Raupp Program & Organizing Committee	SBPC	Brazil
57	Margarete Oliveira Domingues Session chair	INPE	Brazil
58	Maria Assunção F. Silva Dias Invited Lecturer	IAG-USP	Brazil
59	Maria Georgina M. Washington	CTC-UERJ	Brazil
60	Maria Herminia de P. R. Mello	IME-UERJ	Brazil
61	Maurício Kischinhevsky Program & Organizing Committee	UFF	Brazil
62	Maurício Kritz Session chair	LNCC	Brazil
63	Moisés Ceni de Almeida	IME-UERJ	Brazil
64	Monica da Costa P. L. Heilbron	SR2-UERJ	Brazil
65	Nirzi Gonçalves de Andrade	UFF-EEIMVR	Brazil
66	Norberto Mangiavacchi Program & Organizing Committee	FEN-UERJ	Brazil
67	Patrícia Nunes da Silva Contributed paper	IME-UERJ	Brazil
68	Pedro Henrique Gonzalez Silva	IME-UERJ	Brazil
69	Pedro Antonio Santoro Salomão	IME-USP	Brazil
70	Pedro Leite da Silva Dias Program & Organizing Committee	LNCC	Brazil
71	Pedro Nowosad Session chair	IMPA & USP	Brazil
72	Peter Frolkovic Contributed paper	Slovakia Univ. of Technology	Slovakia
73	Peter Lax Main Lecturer	NYU	USA
74	Philippe G. LeFloch Invited Lecturer	Université de Paris	France
75	Priscilla Fonseca de Abreu	IME-UERJ	Brazil
76	Ralph dos Santos Mansur	IME-UERJ	Brazil
77	Raquel Cupolillo S. de Sousa	IME-UERJ	Brazil
78	Regia Gonçalves	UFF	Brazil
79	Reuben Hersh Invited Lecturer	New Mexico University	USA
80	Roger Cristiano Brock	UERJ	Brazil
81	Rolf Jeltsch Invited Lecturer	ETH, Zürich	Germany
82	Rose Pereira Maria	IME-UERJ	Brazil
83	Roseli S. Wedemann	IME-UERJ	Brazil

84	Sanderson Gonzaga Session chair	DCC-UFLA	Brazil
85	Sandra Malta Session chair	LNCC	Brazil
86	Sandra Moura Web design		Brazil
87	Sergio Luiz Silva	IME-UERJ	Brazil
88	Simone Santana Franco Secretary co-chair	LNCC	Brazil
89	Sonia Maria Gomes Session chair	UNICAMP	Brazil
90	Stephan Berres Contributed paper	Univ. Católica de Temuco	Chile
91	Tânia Maria R. de Souza Art design		Brazil
92	Uri Ascher Invited Lecturer	Univ. British Columbia	Canada
93	Vadim Lisitsa Contributed paper	Inst. Petroleum Geology and Geophysics, Novosibirsk	Russia
94	Válter A. Nascimento Contributed paper	UFMS	Brazil
95	Vanessa de Freitas Rodrigues	IME-UERJ	Brazil
96	Vanessa Simões	IMPA	Brazil
97	Vinicius Buçard de Castro	IME-UERJ	Brazil
99	Vitoriano Ruas Contributed paper	UFF	Brazil
100	Wladimir Neves Program & Organizing Committee	IM-UFRJ	Brazil
101	Walcy Santos	IM-UFRJ	Brazil

Printed by Printforce, the Netherlands